A NOTE FROM THE AUTHORS

Congratulations on your decision to take the AP Physics exams! Whether or not you're completing an AP Physics course, this book can help you prepare for both the Physics B and C exams. Inside, you'll find information about the exams as well as Kaplan's test-taking strategies, targeted reviews of both Physics B and Physics C that highlight the most frequently tested concepts, and a diagnostic and a practice test for both exams. Don't miss the strategies for answering the free-response questions: You'll learn how to cover the key points AP graders will want to see.

By studying college-level physics in high school, you've placed yourself a step ahead of other students. You've developed your critical-thinking and time-management skills, as well as your understanding of the practice of physics research. Now it's time for you to show off what you've learned by acing the exams.

Best of luck,

Bruce Brazell

Paul Heckert

Joscelyn Nittler

Matthew Vannette

Michael Willis

RELATED TITLES

AP® PHYSICS B & C

2014

Kaplan offers resources and options to help you prepare for the PSAT, SAT, ACT, AP exams, and other high-stakes exams. Go to www.kaptest.com or scan this code below with your phone (you will need to download a QR code reader) for free events and promotions.

snap.vu/m87n

AP® PHYSICS B & C

2014

Bruce Brazell

Paul Heckert

Joscelyn Nittler

Matthew Vannette

Michael Willis

KAPLAN

PUBLISHING

New York

AP® is a registered trademark of the College Board, which neither sponsors nor endorses this product.

© 2013 by Anaxos Inc.
Published by Kaplan Publishing, a division of Kaplan, Inc.
395 Hudson Street
New York, NY 10014

Printed in the United States of America

10 9 8 7 6 5 4 3 2 1

ISBN: 978-1-61865-257-7

TABLE OF CONTENTS

PART FOUR: PRACTICE TESTS

ABOUT THE AUTHORS

Anaxos Inc. was founded in 1999 by Drew and Cynthia Johnson. Anaxos is a leading provider of educational content for print and electronic media.

Bruce Brazell is the Director of the Cook Planetarium at Navarro College, where he also teaches astronomy and physics.

Paul Heckert is a Professor of physics and astronomy at Western Carolina University with over 25 years of college-level teaching experience. He holds a PhD in physics from the University of New Mexico.

Joscelyn Nittler has taught physics at Queensborough Community College and the University of Oklahoma; she currently teaches high school sciences in Kansas.

Matthew Vannette taught high school and undergraduate science courses, including AP Physics, after earning his master's degree in physics from Boston College in 2001. He is now working toward his PhD in physics at the University of South Carolina.

Michael Willis holds an MS in Nuclear Engineering from Pennsylvania State University. He spent 10 years as a nuclear engineer and has been teaching AP Physics for the past 4 years at Glen Burnie High School, Maryland.

KAPLAN PANEL OF AP EXPERTS

Congratulations—you have chosen Kaplan to help you get a top score on your AP exam.

Kaplan understands your goals, and what you're up against—achieving college credit and conquering a tough test—while participating in everything else that high school has to offer.

You expect realistic practice, authoritative advice, and accurate, up-to-the-minute information on the test. And that's exactly what you'll find in this book, as well as every other book in Kaplan's AP series. To help you (and us!) reach these goals, we have sought out leaders in the AP community. Allow us to introduce our experts.

AP PHYSICS B & C EXPERTS

Brad Allen has been teaching AP Physics B for 10 years. He teaches in the Rochester, New York, area at Brighton High School, which earned the 2007–08 New York State Siemens Award for AP Excellence in Teaching based on its AP Physics B performance. He was awarded a Churchill Scholarship and earned his master's degree in physics from Cambridge University in the United Kingdom.

Jeff Funkhouser has taught high school physics since 1988. Since 2001, he has taught AP Physics B and C at Northwest High School in Justin, Texas. He has been a reader and a table leader since 2001.

Dolores Gende has an undergraduate degree in chemical engineering from Iberoamericana University in Mexico City. She has 13 years of experience teaching college-level introductory physics courses and presently teaches at the Parish Episcopal School in Dallas, Texas. Dolores serves as an AP Physics table leader, an AP Physics workshop consultant, and the College Board Advisor for the AP Physics Development Committee. She received the Excellence in Physics Teaching Award by the Texas section of the American Association of Physics Teachers in March 2006.

Martin Kirby has taught AP Physics B and tutored students for Physics C at Hart High School in Newhall, California, for the past 18 years. He has been a reader for the AP Physics B and C exams and a workshop presenter for the College Board for the past 10 years.

Part One

THE BASICS

CHAPTER 1:   INSIDE THE AP PHYSICS EXAMS

INTRODUCTION

If you are reading this, chances are that you're already in (or thinking about taking) an Advanced Placement (AP) Physics course. If you score high enough on the exams, many universities will give you six to eight hours of credit (two classes) in a related introductory physics course.

Depending on the college, a score of 4 or 5 on the AP Physics exams will allow you to leap over the freshman intro course and jump right into more advanced classes. These classes are usually smaller in size, better focused, more intellectually stimulating, and, simply put, more interesting than a basic course. If you are concerned about fulfilling your science requirement solely so you can get on with your study of Elizabethan music or French literature, AP exams can help you there, too. Ace one or both of the AP Physics exams, and depending on the requirements of the college you choose, you may never have to take a science class again.

If you're currently taking an AP physics course, chances are that your teacher has spent the year cramming your head full of the physics know-how you will need for the exams. But there is more to the AP Physics exams than know-how. If you want your score to reflect your abilities, you will have to be able to work around the challenges and pitfalls of the tests. Studying physics and preparing for the AP Physics exams are not the same thing. Rereading your textbook is helpful, but it's not enough.

That's where this book comes in. We'll show you how to marshal your knowledge of physics and put it to brilliant use on test day. We'll explain the ins and outs of the test structure and question format so you won't experience any nasty surprises. We'll even give you answering strategies designed specifically for the AP Physics exams.

AP EXPERT TIP

There is no substitute for trying new problems. Make sure to read the solutions to this book's problems after you have tried them first on your own.

Preparing yourself effectively for the AP Physics exams means doing some extra work. You need to review your text *and* master the material in this book. Is the extra push worth it? If you have any doubts, think of all the interesting things you could be doing in college instead of taking an intro course filled with facts you already know.

Advanced Placement exams have been around for a half century. While the format and content have changed over the years, the basic goal of the AP Program remains the same: to give high school students a chance to earn college credit or placement in more advanced college courses. To do this, a student needs to do two things:

1. Find a college that accepts AP scores.

2. Do well enough on the exams to place out of required subjects.

The first part is easy, since a majority of colleges accept AP scores in some form or another. The second part requires a little more effort. If you have worked diligently all year in your coursework, you've laid the groundwork. The next step is familiarizing yourself with the test itself.

As you may already know, there are actually two AP Physics exams: AP Physics B and AP Physics C. Physics B is the more general and comprehensive of the two exams and is, therefore, the one most students prefer and prepare to take. The Physics B exam is three hours long and covers a broad range of topics, and a successful score on this exam will give you three hours of college credit toward an introductory physics course.

Physics C is a more specific exam and is actually split into two parts: One focuses on Newtonian mechanics, and the other focuses on electricity and magnetism. Students are allowed to take either or both parts. The Physics C exam not only requires a more in-depth knowledge of the subjects covered but also requires a working knowledge of calculus as it applies to physics calculations. Each part of the exam is 90 minutes long, and each part is graded separately. Students who choose to take the entire three-hour exam can earn up to two classes' worth of credit, depending on how they score on each part.

OVERVIEW OF THE EXAM STRUCTURE

The College Board—the company that creates the AP exams—releases a list of the topics covered on each exam. It even provides the percentage of questions that deal with each topic on a given exam. Check out the following breakdown of topics on the AP Physics exams.

NOTE: The material on the C exam is half Newtonian mechanics and half electricity and magnetism. You may choose to take either or both halves. The material on the B exam represents all the topics outlined below in the given proportions.

Topic	B Exam	C Exam
I. Newtonian Mechanics	**35%**	**50%**
A. Kinematics	7%	9%
B. Newton's laws of motion	9%	10%
C. Work, energy, and power	5%	7%
D. Systems of particles, linear momentum	4%	6%
E. Circular motion and rotation	4%	9%
F. Oscillations and gravitation	6%	9%
II. Fluid Mechanics and Thermal Physics	**15%**	**N/A**
A. Fluid mechanics	6%	N/A
B. Temperature and heat	2%	N/A
III. Electricity and Magnetism	**25%**	**50%**
A. Electrostatics	5%	15%
B. Conductors, capacitors, and dielectrics	4%	7%
C. Electric circuits	7%	10%
D. Magnetostatics	4%	10%
E. Electromagnetism	5%	8%
IV. Waves and Optics	**15%**	**N/A**
A. Wave motion	5%	N/A
B. Physical optics	5%	N/A
C. Geometric optics	5%	N/A
V. Atomic and Nuclear Physics	**10%**	**N/A**
A. Atomic physics and quantum effects	7%	N/A
B. Nuclear physics	3%	N/A

HOW THE EXAMS ARE SCORED

Scores are based on the number of questions answered correctly. **No points are deducted for wrong answers.** No points are awarded for unanswered questions. Therefore, you should answer every question, even if you have to guess.

The AP Physics B exam is three hours long and consists of two parts: a multiple-choice section and a free-response section. In Section I, you have 90 minutes to answer 70 multiple-choice questions with five answers each. This section is worth 50 percent of your total score.

Section II of the AP Physics B exam consists of six to eight free-response questions that are worth the other 50 percent of your total score. The term *free-response* means roughly the same thing as "large, multistep, and involved," since you will spend 90 minutes answering only six to eight problems. Although these free-response problems are long and often broken down into multiple parts, they usually do not cover an obscure topic. Instead, they take a fairly basic physics concept and ask you a set of related questions about it.

The AP Physics C exam is three hours long and broken up into two parts. If you are taking only one part of the exam (either Newtonian mechanics or electricity and magnetism), the exam is 90 minutes long and consists of two sections: multiple-choice and free-response. In Section I, you have 45 minutes to answer 35 multiple-choice questions with five answer choices each. This section is worth 50 percent of your score. In Section II, you have 45 minutes to answer three free-response questions that are worth the other 50 percent of your total score.

If you choose to take both parts of the AP Physics C exam, the entire test will be broken up into four 45-minute parts. The first segment will consist of the 35 Newtonian mechanics multiple-choice questions, the second segment will consist of the 35 electricity and magnetism multiple-choice questions, the third segment will consist of the three free-response Newtonian mechanics questions, and the fourth segment will consist of the three free-response electricity and magnetism questions.

When your three hours of testing are up (or $1\frac{1}{2}$ hours, depending on the test you took), your exam is sent away for grading. The multiple-choice section is handled by a machine, while qualified graders—called "readers"—grade your responses to Section II. The group of readers is made up of current AP Physics teachers and college professors of principles-level physics. After a seemingly interminable wait, your composite score will arrive. Your results will be placed into one of the following categories, reported on a 5-point scale.

AP EXPERT TIP

Making it as easy as possible on the readers by submitting legible work will likely help your score. They may not take the time to decipher messy handwriting.

5 = Extremely well qualified (to receive college credit or advanced placement)

4 = Well qualified

3 = Qualified

2 = Possibly qualified

1 = No recommendation

Some colleges will give you college credit for a score of 3 or higher, but it's much safer to score a 4 or a 5. If you have an idea of which colleges you might attend, check out their websites or call their admissions offices to find out their particular criteria regarding AP scores.

CALCULATORS

Calculators are *not* permitted on the multiple-choice sections of the Physics B and Physics C exams. However, they are allowed on the free-response sections of both exams. Minicomputers, pocket organizers, electronic writing pads, and calculators with QWERTY keyboards are not allowed. Essentially, your calculator memory can be used only to store and access programs that are relevant to the problem. Accessing any notes on the exam or copying and storing any part of the exam in your calculator will be considered cheating. If you forget your calculator, you will be up the proverbial creek without a paddle, since calculators cannot be shared.

TABLES AND FORMULAS

Formulas and tables are an intricate part of understanding and studying physics. Therefore, tables containing commonly used physics equations will be printed on the green insert provided with the free-response portions of both exams. The equation table *cannot* be used in the multiple-choice sections. You are expected to have a working familiarity with these formulas and graphs, so this information is to be used only as a reference.

The table of formulas changes each test year. However, ETS releases copies of the formulas each year in the *AP Physics Course Description Booklet*. Get a copy of these formulas at the beginning of your school year so you can become accustomed to using them. Your physics teacher will also be using these equations throughout the school year.

REGISTRATION AND FEES

You can register for the exams by contacting your guidance counselor or AP coordinator. If your school doesn't administer AP exams, contact AP Services for a listing of schools in your area that do. As of this printing, the fee for each AP exam is $89 within the United States, and $117 at schools and testing centers outside of the United States. For students with acute financial need, the College Board offers a $28 credit. In addition, most states offer exam subsidies to cover all or part of the remaining cost for eligible students. To learn about other sources of financial aid, contact your AP coordinator.

AP EXPERT TIP

Even though you will not have an equation table for the multiple-choice section, the questions are written assuming that you know the equations. It's best to memorize as many as you can.

For more information on all things AP, visit **collegeboard.com** or contact AP Services:

AP Services
P.O. Box 6671
Princeton, NJ 08541-6671
Phone: (609) 771-7300 or 888-225-5427 (toll-free in the United States and Canada)
Email: apexams@info.collegeboard.org

ADDITIONAL RESOURCES

AP PHYSICS B RESOURCES

The following textbooks are commonly used in colleges and typify the level of the B course:

- Cutnell, John D., and Kenneth W. Johnson. 2004. *Physics*, 6th ed. Hoboken, N.J.: John Wiley & Sons.

- Giancoli, Douglas C. 2005. *Physics: Principles with Applications*, 6th ed. Upper Saddle River, N.J.: Prentice Hall.

- Hecht, Eugene. 2003. *Physics: Algebra/Trigonometry*, 3rd ed. Pacific Grove, Calif.: Brooks/Cole.

- Serway, Raymond A., and Jerry S. Faughn. 2003. *College Physics*, 6th ed. Pacific Grove, Calif.: Brooks/Cole.

- Wilson, Jerry D., and Anthony J. Buffa. 2003. *College Physics*, 5th ed. Upper Saddle River, N.J.: Prentice Hall.

AP PHYSICS C RESOURCES

The following textbooks are commonly used in colleges and typify the level of the C course:

- Chabay, Ruth W., and Bruce A. Sherwood. 2003. *Matter & Interaction II: Electric & Magnetic Interactions*, Version 1.2. Hoboken, N.J.: John Wiley & Sons.

- Fishbane, Paul M., Stephen Gasiorowicz, and Stephen T. Thornton. 2005. *Physics for Scientists and Engineers*, 3rd ed. Upper Saddle River, N.J.: Prentice Hall.

- Giancoli, Douglas C. 2000. *Physics for Scientists and Engineers*, 3rd ed. Upper Saddle River, N.J.: Prentice Hall.

- Halliday, David, Robert Resnick, and Kenneth Krane. 2001. *Physics, Parts I and II*, 5th ed. Hoboken, N.J.: John Wiley & Sons.

- Halliday, David, Robert Resnick, and Jearl Walke. 2005. *Fundamentals of Physics*, 7th ed. Hoboken, N.J.: John Wiley & Sons.

- Knight, Randall D. 2004. *Physics for Scientists and Engineers: A Strategic Approach with Modern Physics*. Boston: Addison-Wesley.

- Serway, Raymond A., Robert J. Beichner, and John J. Jewett. 2000. *Physics for Scientists and Engineers*, 5th ed. Pacific Grove, Calif.: Brooks/Cole.

- Serway, Raymond A., and John W. Jewett. 2002. *Principles of Physics*, 3rd ed. Pacific Grove, Calif.: Brooks/Cole.

- Tipler, Paul A., and Gene P. Mosca. 2004. *Physics for Scientists and Engineers*, 5th ed. New York: W. H. Freeman.

- Wolfson, Richard, and Jay M. Pasachoff. 1999. *Physics for Scientists and Engineers*, 3rd ed. Boston: Addison-Wesley.

- Young, Hugh D., and Roger A. Freedman. 2004. *University Physics*, 11th ed. Boston: Addison-Wesley.

CHAPTER 2:   STRATEGIES FOR SUCCESS: IT'S NOT ALWAYS HOW MUCH YOU KNOW

INTRODUCTION

When deciding on which AP Physics exam to sign up for, consult your AP Physics teacher. **Choose only the exam for which you have prepared and studied**. Although the Physics C exam might be tempting, keep in mind that it requires you to know additional material. Only students who prepare specifically for the Physics C exam should take it.

The topics presented in this book will cover **both** exams. Topics that are specific to only one of the exams will be noted (exam B only or exam C only). Feel free to focus on only those topics that will be covered in your specific exam. However, the test structure and question format are the same for both exams, so the testing strategies outlined in this chapter apply to both exams. You should read over and study the strategies carefully, regardless of the exam you plan to take.

We have scrutinized and analyzed the AP Physics exams more times than Newton's critics analyzed his theory of gravity to learn everything we could about them, so we could then pass on this information to you. This book contains precisely the information you will need to ace the tests. There's no extra material in here to waste your time—no pointless review material that won't be tested, no rah-rah speeches. Just the most potent test preparation tools available: **strategies, review, and practice**.

1. STRATEGIES

Process of elimination is a time-tested test-taking strategy. We're going to talk about process of elimination as it applies to the AP Physics exams and only as it applies to those exams. This chapter covers the several skills and general strategies that work for these particular tests.

2. REVIEW

The best test-taking strategies in the world won't get you a good score if you can't tell the difference between static equilibrium and parallel plate capacitors. At their core, the AP Physics exams cover a wide range of physics topics, and learning these topics is **absolutely** necessary. However, chances are you're already familiar with these subjects, so we don't need to start from scratch. This is not a physics textbook; we've tailored our review section to focus on the most relevant topics and how they typically appear on the exams. We also cover the things you should know to answer the questions correctly and the connections between the concepts that will help you think through the questions.

3. PRACTICE

Few things are better than experience when it comes to standardized testing. Taking a practice test gives you an idea of what it is like to answer physics questions for three hours. It's definitely not a fun experience, but it is a helpful one. Practice exams give you the opportunity to find out what areas are your strongest and what topics you should spend some additional time studying. And the best part is that it doesn't count—the mistakes you make on our practice tests are mistakes you won't make on the real tests.

The preceding three points describe the general outline of this book. Now let's look at some specific strategies you can use on the AP Physics exams.

STRATEGIES FOR THE AP PHYSICS EXAMS

Sixty years ago, there was only one standardized test; it was administered by the U.S. Army to determine which enlistees were qualified for officer training. The idea behind the U.S. Army officers' exam has been adapted and has flourished in both the public and private sectors. Nowadays, you can't get through a semester of school without taking some letter-jumble exam like the PSAT, SAT, ACT, ASVAB, BLAM, ZORK, or FWOOSH (some of those tests are fake!). As you may already know, developing certain test-taking strategies is the best way to help prepare yourself for these exams. Since everyone reading this has already taken a standardized test of some sort, you are probably familiar with some of the general strategies that will help you increase your score on this kind of test. The following are some of these strategies.

1. PACING

Since many tests are timed, proper pacing allows you to attempt every question in the time allotted. Poor pacing causes you to spend too much time on some questions to the point where you may run out of time before attempting every problem.

2. PROCESS OF ELIMINATION

On every multiple-choice test you ever take, the answer is given to you. The only difficulty resides in the fact that the correct answer is hidden among incorrect choices. Even so, the multiple-choice format means you don't have to pluck the answer out of thin air. Instead, if you can eliminate answer choices you know are incorrect and only one choice remains, then it must be the correct answer.

3. PATTERNS AND TRENDS

Standardized is the key word in "standardized testing." Standardized tests don't change greatly from year to year. Sure, each question won't be the same, and different topics will be covered from one administration to the next, but there will also be a lot of overlap from one year to the next. That's the nature of *standardized* testing: If the test changed wildly each time it came out, it would be useless as a tool for comparison. Because of this, certain patterns can be uncovered about any standardized test. Learning about a test's patterns and trends will help you feel confident, even though you haven't seen particular questions before.

4. THE RIGHT APPROACH

Having the right mindset plays a large part in how well you will do on a test. If you're nervous about the exam and hesitant to make guesses, you'll fare much worse than students with an aggressive, confident attitude. Students who start with question 1 and plod on from there don't score as well as students who pick and choose the easy questions first before tackling the harder ones. People who take a test cold have more problems than those who take the time to learn about the test beforehand. In the end, factors like these create people who are good test takers and those who struggle even when they know the material.

These points are valid for every standardized test, but they are quite broad in scope. The rest of this chapter will discuss how these general ideas can be modified to apply specifically to the AP Physics exams. These test-specific strategies—combined with the factual information covered in your course and in this book's review section and practice tests—are the one-two punch that will help you succeed on test day.

SECTION I: MULTIPLE-CHOICE QUESTIONS

The worst thing that can be said about the AP Physics multiple-choice questions is that they count for 50 percent of your total score. Although you might not like this question type, there's no denying the fact that it's easier to guess on a multiple-choice question than on an open-ended question. The answer is always there in front of you; the trick is to find it among the incorrect answers.

There are 70 multiple-choice questions on the AP Physics B exam and 35 multiple-choice questions on each part of the AP Physics C exam. They come in two distinct question types.

1. CONCEPT QUESTIONS

These are the questions that do *not* deal with math. As their name implies, these questions are concerned with physics concepts expressed through words, not numbers. You can easily identify these questions by looking at the answer choices, because all the answers will be expressed verbally, not numerically. Since very little to no calculation is required to answer these questions, they tend to take less time to answer and should be noted for this expediency.

A typical concept question looks like this:

Two people of different mass are standing still wearing roller skates on a waxed, level surface made of wood. They then simultaneously push each other horizontally. Which of the following would be true after the push?

(A) The person with less mass has a smaller initial acceleration than the person with more mass.

(B) The center of mass between the two people moves in the direction of the less massive person.

(C) The momenta magnitudes of the two people are equal.

(D) The speeds of the two people are equal.

(E) The kinetic energies of the two people are equal.

Where this question occurs on the test makes no difference because there's no obvious order of difficulty on the AP Physics exams. Tough questions are scattered among easy and moderately difficult questions.

2. CALCULATION QUESTIONS

As their name implies, calculation questions involve math. Since these questions require both conceptual knowledge and a familiarity with formulas, they tend to be more time consuming than the content questions. You can easily identify these questions by the equations or numerical values in the answer choices.

A typical calculation question looks like this:

> A student places an object 3.0 cm from a converging lens of focal length 6.0 cm. What is the magnitude of the magnification of the image produced?

(A) 0.5

(B) 1.0

(C) 2.0

(D) 3.0

(E) 3.5

$E = MC^2$ (OR "AP EFFICIENCY = MULTIPLE-CHOICE^2")

There are two factors that you can use on the multiple-choice section of the AP Physics exams to help increase your score: **time** and **knowledge**.

TIME

Most standardized test takers start with question 1 and work consecutively through the exam until they get to the end or run out of time, whichever comes first. Students run out of time because they get stuck on some problems that are more difficult and, thus, time consuming. Then students not only miss the difficult questions but also, by running out of time, don't get a chance to answer the easier questions that may follow the question that stumped them. The result is a lower score.

You know that you have just 90 minutes to answer 70 multiple-choice questions, so make sure you don't spend too much time on any one question. **Answer the questions that take the least amount of time first**; this way, you won't get bogged down on one question and miss the opportunity to answer three or four other questions.

KNOWLEDGE

What many students fail to realize is that they don't need to get every multiple-choice question right to score well on the AP Physics exams. You should strive to do as well as you can, but very few people will get perfect scores. All you need is a 4 or 5, and that means you need to get a large portion, but not all, of the multiple-choice problems right. Struggling for perfection may lead you to waste time on less valuable questions or even to panic and freeze up if you feel that the exam is not going well.

If you feel as though you are not going to have enough time to get to all the questions, make sure that you get to all the questions that you *know*. This means you should not answer the questions in consecutive order but rather in order of your ability. Take a minute and look over the list of topics covered on the AP Physics exams on page 5 of Chapter 1. Divide these topics into two lists: "Rock-solid physics" and "Physics I have trouble with." Keep these lists in mind when you begin a multiple-choice section. On your first pass through the questions, answer all the questions that deal with the concepts an the "Rock-solid physics" list. Save all the questions dealing with the "Physics

I have trouble with" list for the second pass. The idea here is *not* that you will classify each question on the test before beginning to work on the exam but for you to develop a clear sense of topics and keywords that help you determine whether you should deal with a particular question first or save it for later in the section. **Remember: you want to answer correctly as many questions as possible, so be sure to answer all the ones you *know* in the given amount of time.**

You can tailor your approach to the exams so that it addresses your own strengths and weaknesses. No matter what your approach, you should follow these three basic steps to answering AP Physics questions:

1. **First pass:** Find and answer all the concept questions first. Because these questions require little or no math, they will take less time to answer. Remember, you can identify concept questions by looking at the answer choices. If a question falls into your "Rock-solid physics" list, answer it and move on to the next concept question.

2. **Second pass:** Read through the calculation questions. If a calculation question falls into your "Rock-solid physics" list, answer it and move on to the next calculation question.

3. **Third pass:** Attempt to answer the questions that fall into your "Physics I have trouble with" list. Save these for last so you can maximize your score by answering all the questions you know first. For these remaining questions, it is up to you to determine the topics with which you feel most comfortable and attack those questions first. Be aware of the time limit and focus on eliminating incorrect answers rather than finding the one correct answer.

For the difficult questions, you should always take a stab at eliminating some answer choices and then make an educated guess. Admittedly, the AP Physics exams are tests of specific knowledge, so picking the right answer from the bad answer choices is harder to do than it is on other standardized tests. Still, it can be done, so it's worth a try.

COMPREHENSIVE, NOT SNEAKY

Some tests are sneakier than others. Sneaky tests feature convoluted writing, questions that are designed to trip you up mentally, and a host of other little tricks. You can take a sneaky test armed with the proper facts but still get questions wrong because of sneaky traps in the questions themselves.

The AP Physics exams are *not* sneaky tests. Their objective is to see how much physics knowledge you have stored in that brain of yours. To do this, you are presented with a wide range of questions from an even wider range of physics topics. The exams try to cover as many different physics facts as they can, which is why the problems jump around from electric potential to frictionless momentum. The tests work hard to be as comprehensive as they can be so that students who only know one or two physics topics will soon find themselves struggling.

Understanding how the tests are designed can help you answer questions correctly. The AP Physics exams are comprehensive; the hard questions are difficult because they ask about hard subjects, not because they are trick questions.

That said, trust your instincts when guessing. If you think you know the right answer, chances are you dimly remember the topic being discussed from your AP course. The test is about knowledge, not traps, so trusting your instincts will help more often than not. On the multiple-choice section of the AP Physics exams, points are no longer deducted for an incorrect response. In other words, completely random guessing is fine.

You don't have time to ponder every tough question, so trusting your instincts can prevent you from getting bogged down and wasting time on a problem. You might not get every educated guess correct, but again, the point is not to get a perfect score. It's to get a good score, and surviving hard questions by trusting your gut feelings is a good way to achieve this.

On some problems, though, you might have no inkling of what the correct answer should be. In that case, turn to the following key idea: "good physics."

THINK "GOOD PHYSICS"

The AP Physics exams rewards good physicists. The tests want to foster future physicists by covering fundamental topics. What they do not want is bad physics. They do *not* want answers that are **factually incorrect, too extreme to be true**, or **irrelevant to the topic at hand**.

Yet these bad physics answers invariably appear on the exams because they're multiple-choice tests and you have to have four incorrect answer choices around the one right answer. So if you don't know how to answer a problem, look at the answer choices and think "good physics." This will lead you to find some poor answer choices that can be eliminated. Look at the multiple-choice question from earlier in the chapter to see if you can spot bad physics in the answer choices.

> Two people of different mass are standing still wearing roller skates on a waxed, level surface made of wood. They then simultaneously push each other horizontally. Which of the following would be true after the push?
>
> (A) The person with less mass has a smaller initial acceleration than the person with more mass.
>
> (B) The center of mass between the two people moves in the direction of the less massive person.
>
> (C) The momenta magnitudes of the two people are equal.
>
> (D) The speeds of the two people are equal.
>
> (E) The kinetic energies of the two people are equal.

AP EXPERT TIP

No points are deducted for wrong answers, but no points are awarded for unanswered questions either. Therefore, you should answer every question, even if you have to guess.

How did you do? Even if you forgot exactly what happens when two objects with different mass repel each other, you have still taken a year of physics and have a good idea of basic physical concepts. This sounds like a classical conservation of linear momentum problem, which is choice (C). If you're not sure about the flaw in the other answers, think of one person as a giant and the other as a small child. If they push against each other, the child will go flying and the giant will barely move. Since almost all of the mass is with the giant, the center of mass will stay closer to him, so (B) is incorrect. Choices (A) and (D) are clearly wrong—the child is going to have the greater speed and acceleration. Choice (E) may be tempting, but remember that total energy (kinetic plus potential) is conserved. Not all of the energy in the giant-child system is kinetic.

You would be surprised how many times the correct answer on a multiple-choice question is a simple, blandly worded fact like "For every action, there is an equal and opposite reaction." No breaking news there, but it is "good physics": a carefully worded statement that is factually accurate.

Thinking "good physics" can help you in two ways:

1. It helps you eliminate extreme answer choices or choices that are untrue or out of place.

2. It can occasionally point you toward the correct answer, because the correct answer will be a factual piece of information sensibly worded.

Part of thinking "good physics" is looking at the units in the answer choices for quantitative questions. The problem setup generally provides you with clues about the units of the correct answer, information that can sometimes be used to eliminate incorrect answer choices.

> **AP EXPERT TIP**
>
> Cross out any choices that don't make physical sense, like a car traveling 500 m/s or an index of refraction less than 1.

That's all that can be said for the multiple-choice section of the AP Physics exams. Make sure to practice these strategies on the practice tests in this book so that you'll actually use them on the real tests. Once you implement these techniques, your mindset, approach, and score should benefit.

Remember:

- There is no obvious order of difficulty. Easy questions are the ones you know, medium questions are the ones you sort of know, and harder questions are ones you do not know.

- No two questions are connected to each other in any way.

- There's no system to when certain physics concepts will appear on the test or in which question they will appear.

Of course, the multiple-choice questions only account for 50 percent of your total score. To get the other 50 percent, you have to tackle Section II.

SECTION II: FREE-RESPONSE QUESTIONS

In Section II, you have 90 minutes to write down answers to six to eight free-response questions. Some questions may count more heavily toward your total score; if so, this will be noted.

The free-response questions focus on solving in-depth problems. Knowledge of physics principles and how to apply those principles are the keys to earning points on this section. On *both* AP Physics exams, the free-response section allows the use of a programmable or graphing calculator (typewriter-style, or QWERTY, keyboards will not be permitted). The test writers feel that the calculator should be used as an aid and not as the sole resource in helping you solve the free-response questions.

You will also be provided with an equation sheet containing commonly used physics equations for use *only* on the free-response section. You should be familiar with these equations after working with them throughout your AP Physics course. The test writers expect you to arrive at the test center with a working knowledge of the equations; the sheet is to be used only as a reference. Because the equations are provided, you will not receive credit for writing them in your answers unless they are followed by explanations relating them to the question at hand. The equation sheet can change from year to year, but ETS releases copies of each year's formulas in the *AP Physics Course Description Booklet.* You should get a copy of these formulas (by downloading it from **www.collegeboard.com/prod_downloads/ap/students/ physics/physics_equation_tables.pdf** or asking your guidance office) at the beginning of the school year so you can become accustomed to using them.

Numerical calculations are an important factor in answering free-response questions correctly, but are not the only factor. The AP Physics exams try to minimize numerical calculations and provide equations with the free-response section because the test makers are more interested in the understanding and application of fundamental physical principles and concepts. Therefore, a good score on the free-response section will require you to express your thorough grasp of the physics involved in the question rather than just perform the calculations.

Free-response questions can come in many forms, but some important points apply to all of them.

AP EXPERT TIP

The point value of a question is given at the beginning, and you should spend about one minute per available point, which will leave you a little time at the end to fill in any gaps.

1. MOST QUESTIONS CONTAIN SMALLER QUESTIONS.

You usually won't get one broad question like "Discuss the impact of Newton's law of gravity in our world." Instead, you'll get an initial setup followed by questions labeled (a), (b), (c), and so on. Expect to spend about one-third to one-half of a page writing about each sub-question. You should not write outside of the space provided for each part.

2. WRITING SMART THINGS EARNS YOU POINTS.

For each sub-question on a free-response question, you receive points for writing a correct response to the prompt. The more points you score, the better off you are on that question. The complete details about how points are scored would make your head spin, but in general, the free-response readers use a rubric that acts as a blueprint for a good answer. Every subsection of a question has two to five key ideas attached to it. If you write about one of those ideas or provide part of the solution, you earn yourself a point. Readers always use the same rubric for a question, and all questions must be evaluated using this rubric. Every reader who reads your exam should, in theory, award you the same number of points. Readers check and cross-check each other to ensure that each answer is evaluated in the same way. Two readers rarely differ on a score.

There's an array of other complex rules regarding free-response scoring (e.g., there is a limit to how many points you can earn on a single sub-question) but it basically boils down to this: Writing smart things about each question will earn you points toward that question. Even if you don't know the exact answer but know certain principles that are involved in solving for the answer, write them out—you may get a few points that can help you out in your overall Section II score.

So don't be terse and don't rush. You have a little more than 10 minutes for each question. Use that time to be as precise as you can on each sub-question. Sometimes doing well on one sub-question earns you enough points to make up for another sub-question you have trouble with. When all the points are tallied for that free-response question, you come out strong on total points, even though you didn't ace every sub-question.

Beyond these points, the fact that there are only between six and eight questions on Section II poses a bit of a risk. If you get a question on a subject you're weak in, things might look pretty grim for that problem. Still, take heart. Quite often, you'll earn some points on every question since you'll be familiar with some sub-questions or segments. Remember, the goal is not perfection. If you can ace the longer question and slug your way to partial credit on the shorter ones, or vice versa, you will put yourself in position to get a good score on the entire test. The total test score is the big picture, so don't lose sight of it just because you don't know the answer to one sub-question on Section II.

AP EXPERT TIP

If you know you have a wrong answer on part (a) of a problem, don't panic. Use your answer in subsequent parts or even make up a reasonable answer. As long as you use the wrong result correctly, you can still earn a lot of points later on.

Be sure to use all the strategies discussed in this chapter when taking the practice tests. Practicing these strategies will help you become comfortable with them so you are able to put them to good use on test day.

Lab-based questions may be a part of the free-response sections of both the B and C tests. This is because laboratory experience is a required part of the education of AP Physics students, just as it is for college physics students. AP students should be able to do the following:

- Design experiments.
- Observe and measure real phenomena.
- Organize, display, and critically analyze data.
- Analyze sources of error and determine uncertainties in measurement.
- Draw inferences from observations and data.
- Communicate results, including suggestions for improving experiments and proposed questions for further study.

Whereas re-creating lab experiences is out of the scope of this book, we have provided a few examples of possible lab-based questions. One example is the optics free-response question at the end of Chapter 11 (an optics question appropriate for B-level students). Another example is #3 in the free-response section in the AP Physics B Practice Test at the end of this book (a magnetism question appropriate for both B-level and C-level students).

STRESS MANAGEMENT

You can beat anxiety the same way you can beat the AP Physics exams—by knowing what to expect beforehand and developing strategies to deal with it.

SOURCES OF STRESS

In the space provided, write down your sources of test-related stress. The idea is to pin down any sources of anxiety so you can deal with them one by one. We have provided common examples—feel free to use them and any others you think of.

- I always freeze up on tests.
- I'm nervous about the heat transfer questions (or the electrostatics questions, the photoelectric effect questions, etc.).
- I need a good/great score to get into my first-choice college.
- My older brother/sister/best friend/girlfriend/boyfriend did really well. I must match their scores or do better.

- My parents, who are paying for school, will be quite disappointed if I don't do well.

- I'm afraid of losing my focus and concentration.

- I'm afraid I'm not spending enough time preparing.

- I study like crazy, but nothing seems to stick in my mind.

- I always run out of time and get panicky.

- The simple act of thinking, for me, is like wading through refrigerated honey.

MY SOURCES OF STRESS

_____

_____

_____

_____

_____

Read through the list. Cross out things or add things. Now rewrite the list in order of most disturbing to least disturbing.

MY SOURCES OF STRESS, IN ORDER

_____

_____

_____

_____

_____

Chances are, the top of the list is a fairly accurate description of exactly how you react to test anxiety, both physically and mentally. The later items usually describe your fears (disappointing mom and dad, looking bad, etc.). Taking care of the major items from the top of the list should go a long way towards relieving overall test anxiety. That's what we'll do next.

HOW TO DEAL

VISUALIZE

Sit in a comfortable chair in a quiet setting. If you wear glasses, take them off. Close your eyes and breathe in a deep, satisfying breath of air. Really fill your lungs until your rib cage is fully expanded and you can't take in any more. Then, exhale the air completely. Imagine you're blowing out a candle with your last little puff of air. Do this two or three more times, filling your lungs to their maximum and emptying them totally. Keep your eyes closed, comfortably but not tightly. Let your body sink deeper into the chair as you become even more comfortable.

With your eyes shut, you can notice something very interesting. You're no longer dealing with the worrisome stuff going on in the world outside of you. Now you can concentrate on what happens inside you. The more you recognize your own physical reactions to stress and anxiety, the more you can do about them. You may not realize it, but you've begun to regain a sense of being in control.

Let images begin to form on TV screens on the back of your eyelids. Allow the images to come easily and naturally; don't force them. Visualize a relaxing situation. It might be in a special place you've visited before or one you've read about. It can be a fictional location that you create in your imagination, but a real-life memory of a place or situation you know is usually better. Make it as detailed as possible and notice as much as you can.

Stay focused on the images as you sink farther into your chair. Breathe easily and naturally. You might have the sensation of any stress or tension draining from your muscles and flowing downward, out your feet, and away from you.

Take a moment to check how you're feeling. Notice how comfortable you've become. Imagine how much easier it would be if you could take the test feeling this relaxed and in this state of ease. You've coupled the images of your special place with sensations of comfort and relaxation. You've also found a way to become relaxed simply by visualizing your own safe, special place.

Close your eyes and start remembering a real-life situation in which you did well on a test. If you can't come up with one, remember a situation in which you did something that you were really proud of—a genuine accomplishment. Make the memory as detailed as possible. Think about the sights, the sounds, the smells, even the tastes associated with this remembered experience. Remember how confident you felt as you accomplished your goal. Now start thinking about the AP Physics exams. Keep your thoughts and feelings in line with that prior, successful experience. Don't make comparisons between them. Just imagine taking the upcoming tests with the same feelings of confidence and relaxed control.

This exercise is a great way to bring the test down to earth. You should practice this exercise often, especially when you feel burned out on test preparation. The more you practice it, the more effective the exercise will be for you.

COUNTDOWN TO THE TESTS

STUDY SCHEDULE

The schedule presented here is the ideal. Compress the schedule to fit your needs. Do keep in mind, though, that research in cognitive psychology has shown that the best way to acquire a great deal of information about a topic is to prepare over a long period of time. Since you have several months to prepare for the AP Physics exams, it makes sense for you to use that time to your advantage. This book, along with your text, should be invaluable in helping you prepare for the tests.

If you have two semesters to prepare, use the following schedule:

September:

Take the Diagnostic Tests in this book and isolate areas in which you need help. The Diagnostics will serve to familiarize you with the type of material you will be asked about on the AP exams.

Begin reading your physics textbooks along with the class outlines.

October–February:

Continue reading this book and use the summaries at the end of each chapter to help guide you to the most salient information for the exams.

March and April:

Take the Practice Tests and get an idea of your score. Identify the areas on which you need to brush up. Then go back and review those topics in both this book and your textbooks.

May:

Do a final review and take the exams.

If you have only one semester to prepare, you'll need a more compact schedule:

January:

Take the Diagnostic Tests in this book.

February–April:

Begin reading this book and identify areas of strengths and weaknesses.

Late April:

Take the Practice Tests and use your performance results to guide you in your preparation.

May:

Do a final review and take the exams.

THREE DAYS BEFORE THE TESTS

It's almost over. Eat an energy bar, drink some soda—do whatever it takes to keep going. Here are Kaplan's strategies for the three days leading up to the test.

Take a full-length Practice Test under timed conditions. Use the techniques and strategies you've learned in this book. Approach the test strategically, actively, and confidently.

WARNING: Do *not* take a full-length practice test if you have fewer than 48 hours left before the test. Doing so will probably exhaust you and hurt your score on the actual test. You wouldn't run a marathon the day before the real thing, right?

TWO DAYS BEFORE THE TESTS

Go over the results of your Practice Tests. Don't worry too much about your score or about whether you got a specific question right or wrong. The Practice Tests don't count. But do examine your performance on specific questions with an eye to how you might get through each one faster and better on the tests to come.

THE NIGHT BEFORE THE TESTS

DO NOT STUDY. Get together an "AP Physics exam kit" containing the following items:

- A watch
- A few No. 2 pencils (pencils with slightly dull points fill the ovals better) and erasers
- Photo ID card
- Your admission ticket from ETS

Know exactly where you're going, exactly how you're getting there, and exactly how long it takes to get there. It's probably a good idea to visit your test center sometime before the day of the test so that you know what to expect—what the rooms are like, how the desks are set up, and so on.

Relax the night before the tests. Do the relaxation and visualization techniques described earlier in this chapter. Read a good book, take a long hot shower, watch something on TV. Get a good night's sleep. Go to bed early and leave yourself extra time in the morning.

THE MORNING OF THE TESTS

First, wake up. After that:

- Eat breakfast. Make it something substantial but not anything too heavy or greasy.

- Don't drink a lot of coffee if you're not used to it. Bathroom breaks cut into your time, and too much caffeine is a bad idea.

- Dress in layers so that you can adjust to the temperature of the test room.

- Read something. Warm up your brain with a newspaper or a magazine. You shouldn't let the exam be the first thing you read that day.

- Be sure to get there early. Allow yourself extra time for traffic, mass transit delays, and/or detours.

DURING THE TESTS

Don't be shaken. If you find your confidence slipping, remind yourself how well you've prepared. You know the structure of the tests; you know the instructions; you've had practice with—and have learned strategies for—every question type.

If something goes really wrong, don't panic. If the test booklet is defective—two pages are stuck together or the ink has run—raise your hand and tell the proctor you need a new book. If you accidentally misgrid your answer page or put the answers in the wrong section, raise your hand and tell the proctor. He or she might be able to arrange for you to regrid your tests after it's over, when it won't cost you any time.

AFTER THE TESTS

You might walk out of the AP Physics exams thinking that you blew it. This is a normal reaction. Lots of people—even the highest scorers—feel that way. You tend to remember the questions that stumped you, not the ones that you knew.

Now, continue your exam prep by taking the Diagnostic Tests that follow this chapter. These short tests will give you an idea of the format of the actual exams and will demonstrate the scope of topics covered. After each Diagnostic, you'll find answers with detailed explanations. Be sure to read these explanations carefully, even when you got the question right, as you can pick up bits of knowledge from them. Use your score to learn which topics you need to review more carefully. Of course, all the strategies in the world can't save you if you don't know anything about physics. The chapters following the Diagnostics will help you review the primary concepts and facts that you can expect to encounter on the AP Physics exams.

AP EXPERT TIP

If you feel yourself starting to panic, close your eyes and take a long, deep breath. You'll be amazed how rejuvenating it can be. Then move on to the next question.

| Part Two |

DIAGNOSTIC TESTS

AP PHYSICS B: DIAGNOSTIC TEST

The chapters that follow in Part Three of this book contain a wealth of information and review questions about all the main topics covered on the Physics B exam. Ideally, you will have the time to go through every chapter and try every review question while working at a steady pace. But one thing often prevents students from doing this—the real world. The fact is, many students have schedules that are already chock-full of activities. Finding large chunks of time to devote to studying for a test—one that isn't even part of your regular schoolwork—isn't just difficult. It's next to impossible.

If this is the case with you, take the following 20-question Diagnostic Test. The questions in this Diagnostic are designed to cover most of the topics you will encounter on the AP Physics B exam. After you take it, you can use the results to give yourself a broad idea of what subjects you are strong in and what topics you need to review more. You can use this information to tailor your approach to the review chapters in Part Three. Ideally you'll still have time to read all the chapters, but if pressed, you can start with the chapters and subjects you know you need to work on.

Give yourself 25 minutes for the 20 multiple-choice questions and 25 minutes for the two free-response questions. Time yourself and take the entire test without interruption.

Be sure to read the explanations for all questions, even those you answered correctly. (This is something you should do on the Physics B Practice Test as well.) Even if you got a problem right, reading another person's answer can give you insights that will prove helpful on the real exam.

Good luck on the Diagnostic!

HOW TO COMPUTE YOUR SCORE

To compute your score for this Diagnostic Test, calculate the number of questions you got right, then divide by 20 to get the percentage of questions that you answered correctly.

The approximate score range is as follows:

5 = 80–100% (extremely well qualified)

4 = 60–79% (well qualified)

3 = 50–59% (qualified)

2 = 40–49% (possibly qualified)

1 = 0–39% (no recommendation)

A score of 49% is a 2, so in this case, you could definitely do better. If your score is low, keep on studying to improve your chances of getting credit for the AP Physics B exam.

AP PHYSICS B DIAGNOSTIC TEST ANSWER GRID

1. Ⓐ Ⓑ Ⓒ Ⓓ Ⓔ
2. Ⓐ Ⓑ Ⓒ Ⓓ Ⓔ
3. Ⓐ Ⓑ Ⓒ Ⓓ Ⓔ
4. Ⓐ Ⓑ Ⓒ Ⓓ Ⓔ
5. Ⓐ Ⓑ Ⓒ Ⓓ Ⓔ
6. Ⓐ Ⓑ Ⓒ Ⓓ Ⓔ
7. Ⓐ Ⓑ Ⓒ Ⓓ Ⓔ
8. Ⓐ Ⓑ Ⓒ Ⓓ Ⓔ
9. Ⓐ Ⓑ Ⓒ Ⓓ Ⓔ
10. Ⓐ Ⓑ Ⓒ Ⓓ Ⓔ

11. Ⓐ Ⓑ Ⓒ Ⓓ Ⓔ
12. Ⓐ Ⓑ Ⓒ Ⓓ Ⓔ
13. Ⓐ Ⓑ Ⓒ Ⓓ Ⓔ
14. Ⓐ Ⓑ Ⓒ Ⓓ Ⓔ
15. Ⓐ Ⓑ Ⓒ Ⓓ Ⓔ
16. Ⓐ Ⓑ Ⓒ Ⓓ Ⓔ
17. Ⓐ Ⓑ Ⓒ Ⓓ Ⓔ
18. Ⓐ Ⓑ Ⓒ Ⓓ Ⓔ
19. Ⓐ Ⓑ Ⓒ Ⓓ Ⓔ
20. Ⓐ Ⓑ Ⓒ Ⓓ Ⓔ

SECTION I

Time—25 Minutes 20 Questions

Directions: Each of the questions or incomplete statements below is followed by five suggested answers or completions. Select the best choice in each case. Remember that you cannot use a calculator or the equation sheet for Section I.

Note: To simplify calculations, you may use $g = 10$ m/s^2 in all problems.

---

1. A car is traveling toward the north at 20 m/s. The driver takes her foot off the gas and allows the car to coast to a stop in a time of 2 minutes. If the stopping acceleration is constant, it is closest to

   (A) 10 m/s^2 to the north.

   (B) 10 m/s^2 to the south.

   (C) 0.17 m/s^2 to the north.

   (D) 0.17 m/s^2 to the south.

   (E) 0.33 m/s^2 to the south.

2. To measure the coefficient of static friction between a small block of plastic and a wooden board, a student lifts one end of the board until the block just begins to slide on the board. He then measures the board's angle, θ, with the horizontal. What is the relationship between the coefficient of static friction, μ_s, and the angle, θ?

   (A) $\mu_s = \sin \theta$

   (B) $\mu_s = \cos \theta$

   (C) $\mu_s = \tan \theta$

   (D) $\mu_s = mg \sin \theta$

   (E) $\mu_s = mg \cos \theta$

3. A ball is thrown at a speed of 10 m/s from the top of a building that is 15 m above the level ground. At what angle should one throw the ball so that it will hit the ground at the maximum possible speed?

   (A) Straight up

   (B) Straight down

   (C) Horizontally

   (D) 45 degrees above the horizontal

   (E) All angles will give the same speed.

4. An object is moving at a constant velocity of 5 m/s along the x-axis. There are a total of five different forces acting on the object. Which of the following statements is true?

   (A) The sum of the forces in the x direction is zero and in the y direction is not zero.

   (B) The sum of the forces in the x direction is positive and in the y direction is zero.

   (C) The sum of the forces in the x direction is negative and in the y direction is zero.

   (D) The sum of the forces in both the x and y directions is positive.

   (E) The sum of the forces in both the x and y directions is zero.

GO ON TO THE NEXT PAGE

5. A 1,500 kg car traveling at 10 m/s has a head-on collision with a 3,000 kg truck traveling at 5 m/s in the opposite direction. They stick together after the impact. What is their velocity after the collision?

   (A) 10 m/s in the original direction of the car

   (B) 5 m/s in the original direction of the car

   (C) 0 m/s

   (D) 5 m/s in the original direction of the truck

   (E) 10 m/s in the original direction of the truck

6. A 5 kg rock is tied to the end of a 2 m long rope and whirled around in a horizontal circular path at a constant speed of 2 m/s. What is the tension in the rope?

   (A) 0 N

   (B) 0.1 N

   (C) 1 N

   (D) 10 N

   (E) 100 N

7. A planet of mass m orbits a star of mass M in a circular orbit with a radius R. The speed of the planet's orbit around the star is given by

   (A) $v = \left(\dfrac{GM}{R} \right)^{\frac{1}{2}}.$

   (B) $v = \left(\dfrac{Gm}{R} \right)^{\frac{1}{2}}.$

   (C) $v = \dfrac{GMm}{R}.$

   (D) $v = \dfrac{GM}{R^2}.$

   (E) $v = \dfrac{Gm}{R^2}.$

8. A block of wood is floating so that 75% of the block is below the water line. What is the density of the wood?

   (A) 750 kg/m^3

   (B) 1,000 kg/m^3

   (C) 1,250 kg/m^3

   (D) 0.75 kg/m^3

   (E) 1.25 kg/m^3

9. When pumping air into a bicycle tire, the pump compresses a volume of air to one-third its initial volume in such a way that the temperature remains constant. What is the final pressure as a fraction or multiple of the original pressure?

   (A) 9

   (B) 3

   (C) 1

   (D) $\dfrac{1}{3}$

   (E) $\dfrac{1}{9}$

10. A positive charge Q is evenly distributed on the surface of a sphere of radius R. The electric field at a point of distance r from the center of the charge distribution, with $r > R$, is

   (A) $E = \dfrac{kQ}{r^2}.$

   (B) $E > \dfrac{kQ}{r^2}.$

   (C) $E < \dfrac{kQ}{r^2}.$

   (D) $E = \dfrac{kQ}{R^2}.$

   (E) $E > \dfrac{kQ}{R^2}.$

11. Three point charges are distributed in a straight line, as shown. The values of the charges are: $q_1 = 2\ \mu C$, $q_2 = 1\ \mu C$, and $q_3 = -4\ \mu C$. Charge q_2 is located in the center, 1 m from both charge q_1 and charge q_3. What is the force on the charge q_1?

$$\vert\longleftarrow 1\ m \longrightarrow\vert\longleftarrow 1\ m \longrightarrow\vert$$
$q_1 = 2\mu C \qquad q_2 = 1\ \mu C \qquad q_3 = -4\ \mu C$

(A) 0.018 N to the right
(B) 0.036 N to the right
(C) 0 N
(D) 0.018 N to the left
(E) 0.036 N to the left

12. Two parallel plate capacitors are nearly identical. One capacitor, C_1, has the plates at twice the separation as the other capacitor, C_2, so that $d_1 = 2d_2$. All other properties of the capacitors are the same. What is the relationship between the capacitance of C_1 and C_2?

(A) $C_1 = \frac{1}{4} C_2$

(B) $C_1 = \frac{1}{2} C_2$

(C) $C_1 = C_2$

(D) $C_1 = 2C_2$

(E) $C_1 = 4C_2$

13. A 1,200 W household appliance operates at 120 V. What is the resistance of the appliance?

(A) 144,000 Ω
(B) 1,200 Ω
(C) 120 Ω
(D) 12 Ω
(E) 10 Ω

14. What is the current through the circuit shown below?

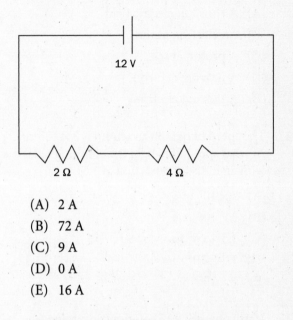

(A) 2 A
(B) 72 A
(C) 9 A
(D) 0 A
(E) 16 A

15. As shown in the diagram, an electron is moving upward in a uniform magnetic field, B, that is directed into the page. What is the direction of the magnetic force on the electron at the instant shown?

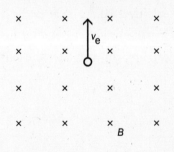

(A) Upward (same direction as the velocity)
(B) Downward
(C) To the right
(D) To the left
(E) Into the page

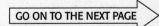 GO ON TO THE NEXT PAGE

16. Which of the following will produce a magnetic field?

(A) A constant electric field

(B) An electron at rest

(C) A proton at rest

(D) A moving neutron

(E) A moving electron

17. The speed of light in a vacuum is 3×10^8 m/s. At what speed does light travel through glass with an index of refraction of 1.5?

(A) 3×10^8 m/s

(B) 2×10^8 m/s

(C) 4.5×10^8 m/s

(D) 1.5×10^8 m/s

(E) 1×10^8 m/s

18. A real object is 1 m before a converging lens with a focal length of 0.5 m. Which best describes the image formed of this object?

(A) Real, inverted, 1 m after the lens

(B) Virtual, inverted, 1 m after the lens

(C) Real, inverted, 0.5 m after the lens

(D) Virtual, upright, 0.5 m after the lens

(E) Real, upright, 1 m after the lens

19. In the fusion reactions that power the sun, each time four H atoms combine to form one He atom, slightly less than 5×10^{-29} kg of mass is lost. Which of the following is closest to the amount of energy generated by each of these individual fusion reactions?

(A) 2×10^{-37} J

(B) 5×10^{-29} J

(C) 5×10^{-46} J

(D) 4×10^{-12} J

(E) 1×10^{-20} J

20. We compare two photons, and one photon has a wavelength twice as long as the other photon. What is the energy of the photon with twice the wavelength?

(A) One-fourth as much

(B) One-half as much

(C) The same

(D) Two times as much

(E) Four times as much

IF YOU FINISH BEFORE TIME IS CALLED, YOU MAY CHECK YOUR WORK ON THIS SECTION ONLY. DO NOT TURN TO ANY OTHER SECTION IN THE TEST.   STOP

SECTION II

Time—25 Minutes 2 Questions

Directions: Write out answers to the following questions. Clearly show the method used and steps involved in arriving at your answers. Partial credit can only be given if your work is clear and demonstrates an understanding of the problem.

Note: Calculators are permitted, except for those with typewriter (QWERTY) keyboards.

1. Beginning on June 30, 1859, Jean Francois Gravelet, the Great Blondin, would frequently amaze crowds by walking across a tightrope stretched high above Niagara Falls. Use $g = 10$ m/s^2 and neglect air resistance for all parts.

   (a) If the rope has enough slack that when Blondin is at the center of the rope both sides of the rope make an angle of θ from the horizontal, draw a free body diagram of the forces acting on the small section of the rope supporting Blondin.

   (b) Calculate the tension on each section of the rope supporting Blondin if he has a mass of 80 kg and $\theta = 2°$. Assume the rope remains steady.

   (c) At the center of his walk, Blondin stops and lowers a rope to lift up a drink bottle from the *Maid of the Mist*, a tour boat in the water below. If the bottle falls from a point 30 m above the water level, calculate the time it takes the bottle to to the water.

   (d) If the bottle has a mass of 0.5 kg, calculate its kinetic energy just before it strikes the water.

   (e) Fortunately, Blondin never fell during one of his daring walks. To reduce the chance of a fatal fall, Blondin carried a 40-foot-long pole. For maximum effectiveness, the pole should have weighted ends hanging below the tightrope walker's feet. Explain in detail why carrying such a pole can reduce a tightrope walker's chance of falling.

GO ON TO THE NEXT PAGE

2. Students are conducting an experiment to measure the electric charge on two small, identical pith balls. The balls are charged following an identical procedure, so the students assume that the two balls will have the same charge. The balls are then suspended in an inverted "V" shape from equal-length strings, as shown in the diagram.

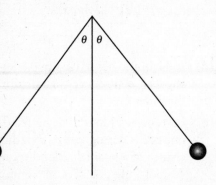

(a) Draw a diagram showing the electrical, gravitational, and any other forces acting on the left pith ball.

(b) In a previously abandoned experimental trial, the students noticed that the pith balls did not make symmetric angles with the vertical. How should they have interpreted this observation?

(c) Assuming the students do not use the trial in part (b) and that the balls hang at the same angle from the vertical, derive an expression for the charge, q, on the balls in terms of the mass, m; the length, l, of the strings; and the angle, θ, the pith balls hang from the vertical. Assume that m and q are the same for both balls. Show that

$$q = \left(\frac{(4\,mg\,l^2\,\tan\theta\,\sin^2\theta)}{k} \right)^{\frac{1}{2}}$$

(d) Calculate the charge, q, if the mass, m, is 0.005 kg; the length, l, is 0.1 m; and the angle, θ, is 20°.

(e) How many electrons does this charge correspond to? Is this result physically reasonable?

GO ON TO THE NEXT PAGE

IF YOU FINISH BEFORE TIME IS CALLED, YOU MAY CHECK YOUR WORK ON THIS SECTION ONLY. DO NOT TURN TO ANY OTHER SECTION IN THE TEST.

STOP

ANSWERS AND EXPLANATIONS

SECTION I

1. D

The acceleration must be in the opposite direction of the velocity for the car to slow down. Hence, if the car is traveling toward the north, the acceleration must be toward the south. To compute the value of the acceleration, apply the kinematic equation

$$v = v_0 + at$$

Use $v = 0$, $v_0 = 20$ m/s, and $t = 2$ min $= 120$ s. Solving for a gives $a = \dfrac{0 - 20 \text{ m/s}}{120 \text{ s}} = -0.17$ m/s^2. The minus sign indicates that the acceleration is in the opposite direction of the velocity. Note that applying the definition of acceleration, $a = \dfrac{\Delta v}{\Delta t}$, is equivalent.

2. C

Start by drawing a free-body diagram.

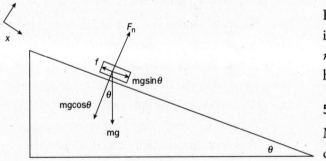

Just before the block starts to slide down the incline, it is not accelerating. Hence, the forces in both the x and y directions sum to zero: $\Sigma F_x = 0$ and $\Sigma F_y = 0$. Choose the positive x-axis to be down the incline and the positive y-axis to be perpendicular to the incline and upward. Summing the x and y forces gives

$$\Sigma F_x = 0 \qquad \text{and} \qquad \Sigma F_y = 0$$
$$mg \sin \theta - f_s = 0 \qquad \qquad F_n - mg \cos \theta = 0$$

Now the frictional force at this critical angle is $f_s = \mu_s F_n = \mu_s mg \cos \theta$, where F_n is the normal force, so $\mu_s mg \cos \theta = mg \sin \theta$ and $\mu_s = \dfrac{\sin \theta}{\cos \theta} = \tan \theta$.

3. E

Use energy considerations. Regardless of the angle at which the ball is thrown, the heights of the level ground and building are the same, so the potential energy change is not affected in any way by the angle. Because energy is a scalar rather than a vector quantity, the kinetic energy at the top is also independent of the angle at which the ball was thrown, so the kinetic energy when the ball hits the ground is completely independent of this angle. Because $K = \dfrac{1}{2} mv^2$, the velocity at which the ball hits the ground is also independent of the initial angle.

4. E

Because the object is moving at a constant velocity, its acceleration is zero. By Newton's second law, $\Sigma \mathbf{F} = ma$, any object with a zero acceleration must also have a zero net force acting on it.

5. C

Momentum, $p = mv$, is conserved. Because the car and truck stick together after the collision, the kinetic energy is not conserved. In addition, both the car and the truck have the same velocity after the collision. Both the car and the truck have a momentum of 15,000 kg · m/s ($p = mv$). However, momentum is a vector quantity, and direction matters, so they have opposite signs. Thus, both the initial and final momenta total zero, and the final velocity must be zero.

6.   D

The rock is moving in a uniform circular motion. The tension in the rope supplies the needed centripetal force. The centripetal force, F_c, is given by

$$F_c = \frac{mv^2}{r}$$

where the mass of the rock, m, is 5 kg; the speed, v, is 2 m/s; and the radius, r, of the circular path is 2 m. Putting the values into the formula gives

$$F_c = \frac{(5 \text{ kg})(2 \text{ m/s})^2}{(2 \text{ m})}$$

$$F_c = 10 \text{ N}$$

7.   A

The centripetal force required to keep the planet in a circular orbit around the star is supplied by the gravitational force between the planet and the star. Hence,

$$F_g = F_c$$

$$\frac{GMm}{R^2} = \frac{mv^2}{R}$$

$$\frac{GM}{R} = v^2$$

$$v = \left(\frac{GM}{R}\right)^{\frac{1}{2}}$$

where G is the universal gravitational constant, M is the mass of the star, m is the mass of the planet, R is the distance between the planet and the star, and v is the orbital speed of the planet.

8.   A

The buoyant force equals the weight of the water displaced. The volume of the water displaced is 75% of the volume of the block of wood. The wood must

therefore have a density that is 75% the density of water, or 75% of 1,000 kg/m^3.

9.   B

Air is a good approximation of an ideal gas; according to the ideal gas law, $PV = nRT$. In the ideal gas law, P is the pressure, V is the volume, n is the number of moles of gas, R is the universal gas constant, and T is the temperature. In this case n, R, and T remain constant, so V and P are inversely proportional. One-third the volume therefore yields three times the pressure.

10. A

By Gauss's law, a spherical charge distribution has the same electric field outside the charge distribution that a point charge of the same value located at the center of the distribution would have. The electric field at a distance r from a point charge Q is given by $E = \frac{kQ}{r^2}$.

11. C

Apply Coulomb's law for the force, F, between two charges, q_1 and q_2, a distance, r, apart: $F = \frac{kq_1q_2}{r^2}$. Note that q_2 is positive and q_3 is negative, so they produce forces in the opposite direction on charge q_1. Further note that the distance from charge q_1 to charge q_3 is twice the distance from charge q_1 to charge q_2 and that q_3 has four times the magnitude of q_2. Because Coulomb's law is an inverse square force, these two effects exactly cancel. Hence, the forces on charge q_1 produced by charges q_2 and q_3 are numerically equal but in the opposite direction. The net force on charge q_1 is therefore zero.

12. B

The capacitance for a parallel plate capacitor with no dielectric is given by $C = \frac{\varepsilon_0 A}{d}$, where A is the area of the plates and d is the distance between them. Taking the ratio of C_1 and C_2 will cause A and ε_0 to cancel out, so that $\dfrac{C_1}{C_2} = \dfrac{d_1}{d_2}$. If $d_1 = 2d_2$, then $C_1 = \dfrac{1}{2}C_2$. If the capacitors have identical dielectrics, the dielectric constant will also cancel out in the ratio.

13. D

Computing the current, I, from the power, P, and voltage, V, using $P = VI$, gives $1{,}200\text{ W} = (120\text{ V})I$ and $I = 10\text{ A}$. (Note that in the formula, V refers to the potential difference or voltage, but when it occurs after the number, V refers to the unit, volts.) Now apply Ohm's law, $V = IR$, to compute the resistance, R: $120\text{ V} = (10\text{ A})\,R$, so $R = 12\ \Omega$.

14. A

First find the equivalent resistance of the two resistors in series using

$$R_{eq} = R_1 + R_2 = 2\ \Omega + 4\ \Omega = 6\ \Omega$$

Then use $V = IR$ to find the current:

$$12\text{ V} = I\,(6\ \Omega) \text{ gives } I = 2\text{ A.}$$

15. C

Apply the right-hand rule for finding the direction of a force on a moving charge in a magnetic field. Curl the fingers of your right hand from the velocity vector to the magnetic field vector. Your thumb will point in the direction of the force on a positive charge, which is to the left in this case. The force on a negative charge will be in the opposite direction, or to the right in this case.

16. E

Magnetic fields are produced by electric fields that are changing in some way. If a charged particle, such as an electron, is moving, the electric field the particle produces will change. Hence, a moving electron will create a magnetic field.

17. B

The index of refraction, n, of a medium is the speed of light in a vacuum, c, divided by the speed of light in the medium, v. So

$$v = \frac{c}{n} = \frac{3 \times 10^8 \text{ m/s}}{1.5} = 2 \times 10^8 \text{ m/s}$$

18. A

Find the image distance, s_i, using: $\dfrac{1}{s_i} + \dfrac{1}{s_o} = \dfrac{1}{f}$. Using the focal length $f = 0.5$ m and the object distance $s_o = 1$ m gives $s_i = 1$ m. Hence, the image is formed 1m after the lens. Because the light rays actually cross at this point, the image is real. Finding the magnification, M, from $M = \dfrac{-s_i}{s_o}$ gives $M = -1$, so the image is its normal size but inverted. One can also answer this question by drawing a ray diagram.

19. D

The mass that is lost in each nuclear reaction is converted to energy according to the equation $E = mc^2$. $c = 3 \times 10^8$ m/s, so multiply 5×10^{-29} kg by 9×10^{16} m^2/s^2 to get 4.5×10^{-12} J. Because the mass lost is slightly less than 5×10^{-29} kg, the energy generated is slightly less than 4.5×10^{-12} J. Therefore, choose 4×10^{-12} J.

20. B

The energy, E, is directly proportional to the frequency, v, using $E = hv$, and the frequency is inversely proportional to the wavelength, λ, using $c = v\lambda$. Hence, the energy is inversely proportional to the wavelength, $E = \dfrac{hc}{\lambda}$. Doubling the wavelength will therefore produce half the energy.

SECTION II

1. For any problems involving forces acting on an object, always start with a free-body diagram to help visualize the situation. Then apply Newton's first or second law as appropriate. For the object in free fall, apply the appropriate kinematic equations. Again, a drawing of the situation may help you to visualize the situation. It is a good idea to define clearly the x and y directions used for the solution.

(a)

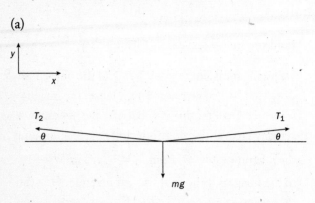

(b) Since there is no acceleration in this steady state, the forces sum to zero in both the x and y directions. Consider the x problem first.

$$\sum F_x = 0$$

$T_1 \cos\theta - T_2 \cos\theta = 0 \rightarrow$ Note, the angles on both sides are equal.

$T_1 = T_2 \rightarrow$ Therefore, drop the subscripts and simply use T.

Some people will understand intuitively that this result must be true from the symmetry of the situation.

Now consider the y forces.

$$\sum F_y = 0$$

$$T_1 \sin\theta + T_2 \sin\theta - mg = 0$$

$$2T \sin\theta = mg$$

$$T = \frac{mg}{2 \sin\theta}$$

$$T = \frac{(80 \text{ kg})(10 \text{ m/s}^2)}{2(\sin 2°)}$$

$$T = 11{,}000 \text{ N}$$

The required tension is so much greater than Blondin's weight because the angle is so close to horizontal that only a small component of the tension acts upward.

(c) Use the kinematic equations and neglect air resistance. First find the time.

$$y - y_0 = v_0 t + \frac{1}{2} a t^2$$

Take the downward direction as positive. Use $y_0 = 0$, $y = 30$ m, $v_0 = 0$, and $a = g$.

$$y = \frac{1}{2} g t^2$$

$$t^2 = \frac{2y}{g}$$

$$t^2 = \frac{2(30 \text{ m})}{(10 \text{ m/s}^2)}$$

$$t = 2.4 \text{ s}$$

(d) To calculate the kinetic energy, use

$$K = \frac{1}{2} m v^2$$

$$v = at + v_0 = (10 \text{ m/s}^2)(2.4 \text{ s}) + 0$$

$$= 24 \text{ m/s}$$

$$K = \frac{1}{2}(0.5 \text{ kg})(24 \text{ m/s})^2 = 140 \text{ J}$$

(e) The pole helps the tightrope walker in two ways. First, it increases the rotational inertia of the artist by distributing weight further from the center of mass, thereby slowing the rate the walker would rotate around the wire. Second, having weighted ends that hang down lowers the center of mass of the walker.

2. Start with a free-body diagram to analyze the electrical and gravitational forces on the pith balls that are balanced by the tension in the strings. This is an equilibrium situation, so the forces sum to zero. Set up and solve the equations for the charge.

(a) To avoid overcrowding the drawing, the forces and their components are shown on the right half and the distances are shown on the left half. Both sides are, however, symmetric.

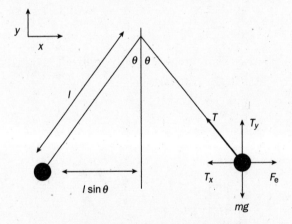

(b) A common first impulse is to guess that if the balls are hanging at unequal angles, there must be unequal charges on the pith balls. However, analysis of the free-body diagram in light of Coulomb's law and Newton's third law suggests that the electrical forces on each pith ball will still be the same if the charges on each ball are unequal. Hence, the gravitational force on each ball must be different if they are suspended at unequal angles. The balls are not the same mass.

(c) Both the x and y forces sum to zero. First consider the y forces on one of the balls. By symmetry, the result will be the same for both balls.

$$\sum F_y = 0$$

$$T_y - mg = 0 \rightarrow \text{use } T_y = T \cos \theta$$

$$T \cos \theta = mg$$

$$T = \frac{mg}{\cos \theta}$$

Now consider the x forces.

$$\sum F_x = 0$$

$$F_e - T_x = 0 \quad \text{use} \quad F_e = \frac{kq^2}{(2l \sin \theta)^2}, \quad T_x = T \sin \theta$$

Substituting gives

$$\frac{kq^2}{(2l \sin \theta)^2} = T \sin \theta = \left(\frac{mg}{\cos \theta} \right) \sin \theta$$

Solving for q^2 gives

$$q^2 = \frac{(4 \, mg \, l^2 \tan \theta \sin^2 \theta)}{k}$$

$$q = \left(\frac{(4 \, mg \, l^2 \tan \theta \sin^2 \theta)}{k} \right)^{\frac{1}{2}}$$

(d) Inserting the numbers into the expression gives

$$q^2 = \frac{(4)(0.005 \text{ kg})(10 \text{ m/s}^2)(0.1 \text{ m})^2 \, (\tan 20°) \, (\sin^2 20°)}{\left(\frac{9 \times 10^9 \text{ Nm}^2}{C^2} \right)}$$

$$q = 9.7 \times 10^{-8} \text{ C}$$

(e) Dividing the charge by the charge on a single electron or proton, $q_e = 1.6 \times 10^{-19}$ C, gives the total number of electrons or protons to which this charge corresponds:

$$n = \frac{(9.7 \times 10^{-8})}{(1.6 \times 10^{-19})}$$

$$n = 6 \times 10^{11}$$

This number looks like a very large number, but it is many orders of magnitude less than Avogadro's number, so it is a physically reasonable number of atoms that are ionized.

AP PHYSICS B DIAGNOSTIC TEST CORRELATION CHART

Use the following table to determine which Physics B topics you need to review most.

Area of Study	Question Number
Newtonian Mechanics	1, 2, 3, 4, 5, 6, 7
Fluid Mechanics and Thermal Physics	8, 9
Electricity and Magnetism	10, 11, 12, 13, 14, 15, 16
Waves and Optics	17, 18
Atomic and Nuclear Physics	19, 20

AP PHYSICS C: DIAGNOSTIC TEST

The chapters that follow in Part Three of this book contain a wealth of information and review questions about all the main topics covered on the Physics C exam. Ideally, you will have the time to go through every chapter and try every review question while working at a steady pace. But one thing often prevents students from doing this—the real world. The fact is, many students have schedules that are already chock-full of activities. Finding large chunks of time to devote to studying for a test—one that isn't even part of your regular schoolwork—isn't just difficult. It's next to impossible.

If this is the case with you, take the following 20-question Diagnostic Test. The questions in this Diagnostic are designed to cover most of the topics you will encounter on the AP Physics C Exam. After you take it, you can use the results to give yourself a broad idea of what subjects you are strong in and what topics you need to review more. You can use this information to tailor your approach to the review chapters in Part Three. Ideally you'll still have time to read all the chapters, but if pressed, you can start with the chapters and subjects you know you need to work on.

The Physics C Exam has two parts: Newtonian Mechanics and Electricity and Magnetism. You can take one or both parts. Give yourself about 13 minutes for each 10-question multiple-choice section and about 13 minutes for each free-response question. The whole Diagnostic should take about 50 minutes, and either of the two parts should take 25 minutes. Time yourself and take the entire test without interruption.

Be sure to read the explanations for all questions, even those you answered correctly. (This is something you should do on the Physics C Practice Test as well.) Even if you got a problem right, reading another person's answer can give you insights that will prove helpful on the real exam.

Good luck on the Diagnostic!

HOW TO COMPUTE YOUR SCORE

To compute your score for this Diagnostic Test, calculate the number of questions you got right, then divide by 20 to get the percentage of questions that you answered correctly.

The approximate score range is as follows:

5 = 80–100% (extremely well qualified)

4 = 60–79% (well qualified)

3 = 50–59% (qualified)

2 = 40–49% (possibly qualified)

1 = 0–39% (no recommendation)

A score of 49% is a 2, so in this case, you could definitely do better. If your score is low, keep on studying to improve your chances of getting credit for the AP Physics C exam.

AP PHYSICS C DIAGNOSTIC TEST ANSWER GRID

1. Ⓐ Ⓑ Ⓒ Ⓓ Ⓔ
2. Ⓐ Ⓑ Ⓒ Ⓓ Ⓔ
3. Ⓐ Ⓑ Ⓒ Ⓓ Ⓔ
4. Ⓐ Ⓑ Ⓒ Ⓓ Ⓔ
5. Ⓐ Ⓑ Ⓒ Ⓓ Ⓔ
6. Ⓐ Ⓑ Ⓒ Ⓓ Ⓔ
7. Ⓐ Ⓑ Ⓒ Ⓓ Ⓔ
8. Ⓐ Ⓑ Ⓒ Ⓓ Ⓔ
9. Ⓐ Ⓑ Ⓒ Ⓓ Ⓔ
10. Ⓐ Ⓑ Ⓒ Ⓓ Ⓔ

11. Ⓐ Ⓑ Ⓒ Ⓓ Ⓔ
12. Ⓐ Ⓑ Ⓒ Ⓓ Ⓔ
13. Ⓐ Ⓑ Ⓒ Ⓓ Ⓔ
14. Ⓐ Ⓑ Ⓒ Ⓓ Ⓔ
15. Ⓐ Ⓑ Ⓒ Ⓓ Ⓔ
16. Ⓐ Ⓑ Ⓒ Ⓓ Ⓔ
17. Ⓐ Ⓑ Ⓒ Ⓓ Ⓔ
18. Ⓐ Ⓑ Ⓒ Ⓓ Ⓔ
19. Ⓐ Ⓑ Ⓒ Ⓓ Ⓔ
20. Ⓐ Ⓑ Ⓒ Ⓓ Ⓔ

SECTION I: MECHANICS
Time—13 Minutes 10 Questions

Directions: Each of the questions or incomplete statements below is followed by five suggested answers or completions. Select the best choice in each case. Remember that you cannot use a calculator or the equation sheet for Section I.

Questions 1 and 2: A train travels a distance d (in a straight line) at a constant velocity v_1. It returns to the starting point along the same route, but at a constant velocity v_2, which is not equal in magnitude to v_1.

1. What is the train's average velocity over the trip?

   A $\dfrac{v_1 + v_2}{2}$

   B $\dfrac{v_1 - v_2}{2}$

   C 0

   D $v_1 + v_2$

   E $v_1 - v_2$

2. What is the average speed (not velocity) of the train over the trip?

   A $\dfrac{v_1 + v_2}{2}$

   B $\dfrac{v_1 - v_2}{2}$

   C 0

   D $2\dfrac{v_1 + v_2}{v_1 v_2}$

   E $2\dfrac{v_1 v_2}{v_1 + v_2}$

3. A ladder of mass m and length $2l$ leans against a frictionless wall making an angle of $60°$ with the ground. What does the minimum coefficent of friction at point C (where the ladder contacts the ground) have to be to keep the ladder from sliding down? Assume the ladder is initially motionless.

   A $\dfrac{N_2}{N_1}$

   B $\dfrac{1}{2}\sin 30°$

   C $\dfrac{1}{2}\cos 30°$

   D $\dfrac{1}{2}\tan 60°$

   E $\dfrac{1}{2}\tan 30°$

GO ON TO THE NEXT PAGE
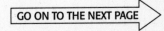

4. A rocket of mass m is launched vertically with an initial velocity v, with acceleration due to gravity at a constant magnitude g. What is the maximum height, h, that it reaches?

   (A) $\dfrac{v^2}{2g}$

   (B) $\dfrac{v^2}{g}$

   (C) $\dfrac{mv^2}{2g}$

   (D) $\dfrac{v}{2g}$

   (E) $\dfrac{2v^2}{g}$

5. An object of mass m, moving at a velocity of v, collides inelastically with an object at rest of mass M. What is the final velocity of M?

   (A) $\dfrac{m+M}{mv}$

   (B) $\dfrac{mv}{m+M}$

   (C) $\dfrac{mv^2}{2M}$

   (D) $\dfrac{mv}{M-m}$

   (E) $\dfrac{mv}{m-M}$

6. A small object of mass m is on a table. One edge is attached to a string that runs through a hole in the table and is connected to a mass M.

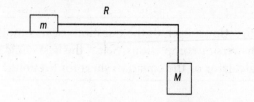

   m is smaller than M. If m moves in a circular path, what must its speed be to keep it moving at a constant radius R? If M is pulled down such that the new radius of circular motion is $\dfrac{R}{2}$, how does the angular momentum change?

   (A) $\sqrt{\dfrac{MgR}{2m}}$; angular momentum decreases by one-half.

   (B) $\sqrt{\dfrac{MgR}{2m}}$; angular momentum does not change.

   (C) $\sqrt{\dfrac{MgR}{m}}$; angular momentum doubles.

   (D) $\sqrt{\dfrac{MgR}{m}}$; angular momentum decreases by one-half.

   (E) $\sqrt{\dfrac{MgR}{m}}$; angular momentum does not change.

GO ON TO THE NEXT PAGE

7. A car of mass m and initial velocity v heads toward a brick wall at a distance d. What is the minimum coefficient of friction required between the car and the pavement to prevent a collision?

   (A) $\dfrac{g}{v^2}$

   (B) $\dfrac{2gd}{v^2}$

   (C) $\dfrac{v}{2gd}$

   (D) $\dfrac{v^2}{2mg}$

   (E) $\dfrac{v^2}{2gd}$

8. A thin ring of radius r rotates with speed v about a perpendicular axis through its center point. A brake is applied, slowing the ring at a constant rate until it stops in time t. What is the average power required to stop the ring from rotating?

   (A) $\dfrac{mv^2}{2}$

   (B) $\dfrac{mv^2}{rt}$

   (C) $\dfrac{mv^2}{2t}$

   (D) $\dfrac{mv^2}{rt^2}$

   (E) $\dfrac{mv^2 t}{r}$

9. A block of mass m is attached to a spring of spring constant k on a frictionless horizontal surface. The block is displaced a distance A from the equilibrium point x_0 and then released. Derive an expression, based on A, k, m, and x, for the velocity as a function of x, where x is the displacement from the equilibrium point.

   (A) $v(x) = \sqrt{\left(\dfrac{m}{k}\right)(A^2 - x^2)}$

   (B) $v(x) = \sqrt{\left(\dfrac{m}{k}\right)(A^2 + x^2)}$

   (C) $v(x) = \sqrt{\left(\dfrac{k}{m}\right)(A^2 - x^2)}$

   (D) $v(x) = \sqrt{\left(\dfrac{k}{m}\right)(A^2 + x^2)}$

   (E) $v(x) = \sqrt{\left(\dfrac{m}{k}\right)(A - x)}$

GO ON TO THE NEXT PAGE

10. A sphere has a radius R and density ρ, with a total mass M.

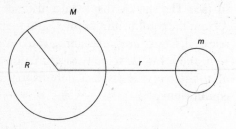

Which graph shows the correct magnitude of the gravitational force on a small mass m at a distance r from the center of the sphere?

(A)

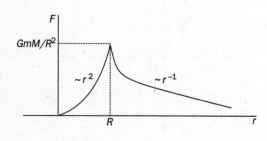

(B)

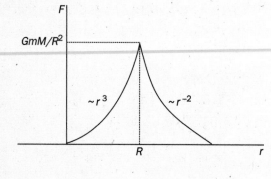

(C)

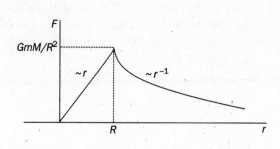

(D)

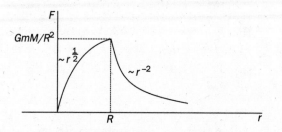

(E)

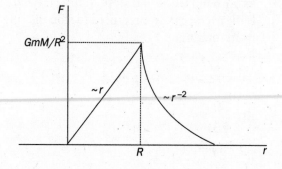

SECTION I: ELECTRICITY AND MAGNETISM
Time—13 Minutes 10 Questions

Directions: Each of the questions or incomplete statements below is followed by five suggested answers or completions. Select the best choice in each case. Remember that you cannot use a calculator or the equation sheet for Section I.

11. A dielectric rod with a total positive charge Q is brought close to two touching conducting spheres. At this point in time, what are the signs of the charges on the closer and farther sphere, respectively? Now, the rod is touched to the closer of the spheres, then removed. The spheres are then separated. What are the signs on the closer and farther spheres, respectively?

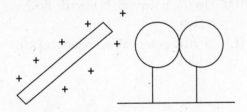

    (A) $-, +$ and $+, +$

    (B) $-, -$ and $+, +$

    (C) $-, -$ and $-, -$

    (D) $-, +$ and $-, +$

    (E) $+, -$ and $+, -$

12. A capacitor of area A, with a distance d between the two plates, is half filled with a substance with a dielectric constant k_1 and a substance with a dielectric constant k_2, as shown.

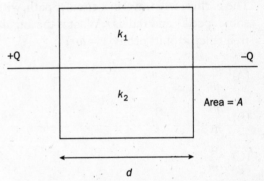

What is the total capacitance of the system?

    (A) $\dfrac{(k_1 - k_2)\varepsilon_0 A}{d}$

    (B) $\dfrac{(k_1 + k_2)\varepsilon_0 A}{d}$

    (C) $\dfrac{(k_1 - k_2)\varepsilon_0 A}{2d}$

    (D) $\dfrac{(k_1 + k_2)\varepsilon_0 A}{2d}$

    (E) $\dfrac{k_1 k_2 \varepsilon_0 A}{2d(k_1 + k_2)}$

GO ON TO THE NEXT PAGE

13. A particle of mass m and charge q travels in a uniform magnetic field that is pointed into the page, as shown.

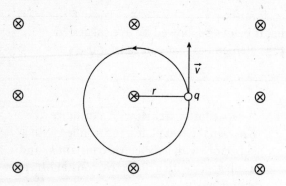

The particle will travel in a circular path, with some speed v and radius r. What is the angular frequency ω of its orbit? ($v = \omega r$)

(A) $\dfrac{m}{qB}$

(B) $\dfrac{qB}{m}$

(C) $\dfrac{vqB}{m}$

(D) $\dfrac{m}{vqB}$

(E) $\dfrac{mv^2}{2r}$

14. Two charges, $+q$ and $-q$, are located a distance d apart.

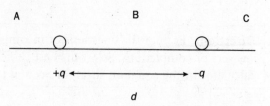

Along the axis through the charges, in what region(s) might one expect to have an electric potential of zero and an electric field of zero?

(A) Electric potential: A or C; electric field: B

(B) Electric potential: B; electric field: A or C

(C) Electric potential: no region; electric field: no region

(D) Electric potential: B; electric field: no region

(E) Electric potential: no region; electric field: B

GO ON TO THE NEXT PAGE

15. A solid conduction sphere carries a charge q.

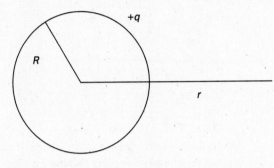

What plot accurately reflects the electric field at r?

(A)

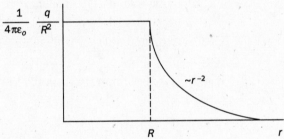

(B)

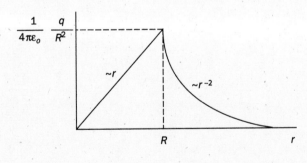

(C)

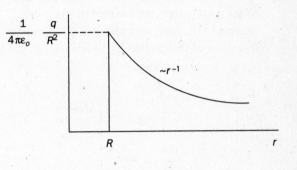

(D)

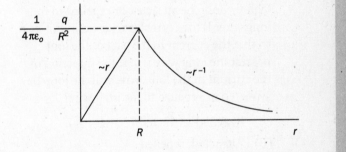

(E)

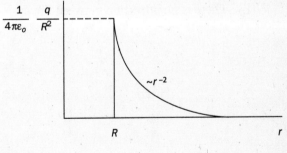

GO ON TO THE NEXT PAGE ⇒

16. A long wire carries a current I_1 in the direction
shown. A loop of wire also carries a current,
I_2, with the current in the part of the loop
nearest to the long wire going in the opposite
direction of the current in the long wire.

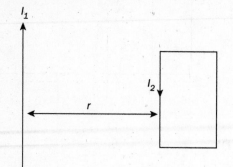

Will the loop be attracted or repelled to the
long wire? If the current is reversed in the loop
so that the current in the part of the loop
nearest the long wire is in same direction of
the current in the long wire, will the loop be
attracted or repelled to the long wire?

(A) Repelled, attracted

(B) Attracted, repelled

(C) Attracted, attracted

(D) Repelled, repelled

(E) There is no attraction between the loop
and the wire.

GO ON TO THE NEXT PAGE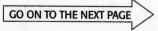

17. A cross section of two conducting tubes
is shown below. Both tubes carry an equal
current I. The current of the interior tube is
directed into the page, while the current of the
exterior tube is directed out of the page.

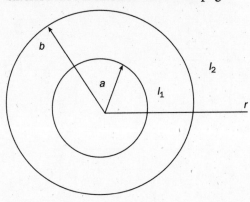

Which of the following is the correct graph
showing the magnetic field B as a function
of r?

(A)

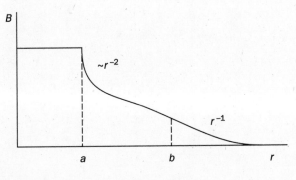

(B)

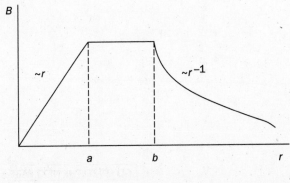

(C)

(D)

(E)

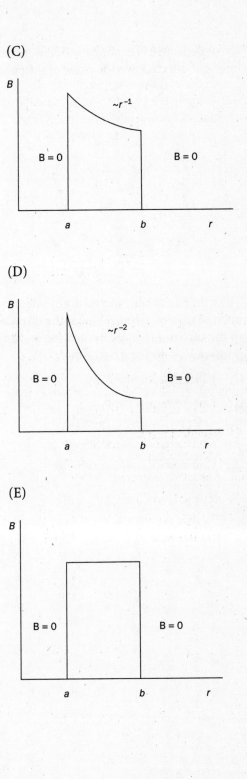

GO ON TO THE NEXT PAGE

18. A magnetic field is confined within a current loop and directed into the plane of the page.

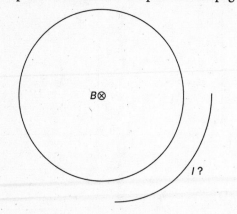

The field B is steadily increasing in intensity into the page over time. What is the direction of the current I induced in the wire, and does it increase or decrease over time?

(A) Clockwise, increase
(B) Clockwise, decrease
(C) Counterclockwise, increase
(D) Counterclockwise, decrease
(E) Counterclockwise, no change

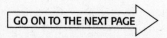

19. A resistor with a large resistance R is connected in parallel with a capacitor to a battery.

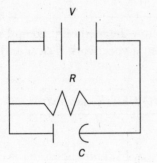

Which plot shows the correct amount of current flow over time, t, in the resistor?

(A)

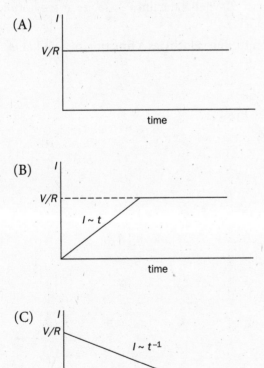

(B)

(C)

(D)

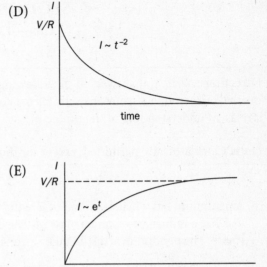

(E)

20. A circuit is connected to a battery of 10 volts as shown (ignore any internal resistance in the battery).

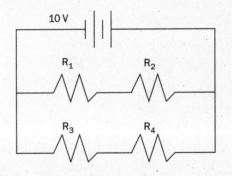

Each resistor has a resistance of 10 ohms. What is the current through each resistor, R_1 through R_4, respectively?

(A) 0.5, 0.5, 0.5, 0.5A

(B) 10, 10, 10, 10A

(C) 50, 50, 50, 50A

(D) 10, 5, 10, 5A

(E) 5, 10, 5, 10A

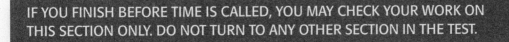

IF YOU FINISH BEFORE TIME IS CALLED, YOU MAY CHECK YOUR WORK ON THIS SECTION ONLY. DO NOT TURN TO ANY OTHER SECTION IN THE TEST.   STOP

SECTION II: MECHANICS
Time—13 Minutes 1 Question

Directions: Write out answers to the following question. Clearly show the method used and steps involved in arriving at your answers. Partial credit can only be given if your work is clear and demonstrates an understanding of the problem.

Note: Calculators are permitted, except for those with typewriter (QWERTY) keyboards.

1. A pendulum has a massless shaft of length L, with a disk attached by its center point to the end. The disk has a radius $R \leftarrow L$ and a mass M. The disk is rigidly attached to the shaft; that is, it is not free to rotate. The pendulum is displaced by an angle, θ, from the vertical.

    (a) Derive an expression for the velocity as a function of θ. What is the linear velocity at the bottom of the arc?

    (b) The disk is now free to rotate. Derive an expression for the velocity as a function of θ. What is the linear velocity at the bottom of the arc?

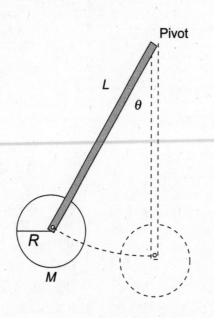

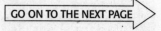
GO ON TO THE NEXT PAGE

_____

_____

_____

_____

_____

_____

_____

_____

_____

_____

_____

_____

_____

_____

_____

_____

_____

_____

_____

_____

_____

_____

IF YOU FINISH BEFORE TIME IS CALLED, YOU MAY CHECK YOUR WORK ON THIS SECTION ONLY. DO NOT TURN TO ANY OTHER SECTION IN THE TEST.

STOP

SECTION II: ELECTRICITY AND MAGNETISM
Time—13 Minutes 1 Question

Directions: Write out answers to the following question. Clearly show the method used and steps involved in arriving at your answers. Partial credit can only be given if your work is clear and demonstrates an understanding of the problem.

Note: Calculators are permitted, except for those with typewriter (QWERTY) keyboards.

2. Two spheres, both with dielectric constant ε, are embedded one inside of the other, as illustrated below. Both have charge densities of ρ, with the larger sphere containing a positive charge and the smaller sphere containing a negative charge. Their respective radii are R and r. A charge, $+q$, is placed at a distance d from the center of the larger sphere.

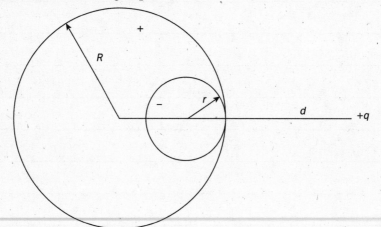

(a) What is the total charge of the two spheres?

(b) Write the equation for the electric potential at distance d, where $+q$ is situated. Do not take the time to reduce this equation to its simplest terms.

(c) Write the equation for the electric field felt by $+q$. Do not take the time to reduce this equation to its simplest terms.

Both spheres are now converted to conducting material, with no other changes being made (assume the spheres are insulated from one another).

(d) What is the electric potential at distance d, where $+q$ is situated? Do not take the time to reduce this equation to its simplest terms.

(e) What is the electric force felt by charge $+q$? Do not take the time to reduce this equation to its simplest terms.

GO ON TO THE NEXT PAGE

IF YOU FINISH BEFORE TIME IS CALLED, YOU MAY CHECK YOUR WORK ON
THIS SECTION ONLY. DO NOT TURN TO ANY OTHER SECTION IN THE TEST.

STOP

ANSWERS AND EXPLANATIONS

SECTION I: MECHANICS

1. C

Velocity is a vector. The average velocity is equal to the total change in distance (also a vector) divided by the total time of the trip: $\bar{v} = \dfrac{\Sigma d}{\Sigma t}$

Since the starting point is the same as the end point, the average velocity is zero. Therefore, $\dfrac{0}{t} = 0$.

2. E

Speed is scalar, with the average equal to the total distance (regardless of its direction) divided by the total time of the trip. The total distance is $2d$. The total time is $T = t_1 + t_2$. Since the magnitude of $d = vt$, $t = \dfrac{d}{v}$.

Thus, $t_1 = \dfrac{d}{v_1}$, and $t_2 = \dfrac{d}{v_2}$.

Doing the algebra, we get the average speed, which is $2\dfrac{v_1 v_2}{(v_1 + v_2)}$.

3. E

For the ladder to remain motionless, the total forces and torques, both of which are vectors, must be zero. Thus $\Sigma F_x = 0$, $\Sigma F_y = 0$, and $\Sigma \tau = 0$. Torque is equal to the vector product rF. It is immaterial about which point we measure the torque. We choose point C, since $r = 0$ for the frictional and N_2 forces.

For the x direction: $\Sigma F_x = 0$, so $N_1 = F_f = \mu N_2 = \mu mg$, where F_f is the force of friction and μ is the coefficient of frictional force.

For the y direction: $\Sigma F_y = 0$, so $N_2 = mg$.

The magnitude of the torque $= rF \sin \theta$.

Since the total torque has to be zero, we have

$lmg \cos \theta = 2l\, N_1 \sin \theta = 2l\mu mg \sin \theta.$

Doing the algebra yields

$$lmg \cos \theta = 2l\mu mg \sin \theta$$
$$\cos \theta = 2\mu \sin \theta$$
$$\mu = \frac{\cos \theta}{2 \sin \theta}$$
$$\mu = \frac{1}{2 \tan \theta}$$

for the minimum coefficient of friction with the ground that will prevent slipping. For $\theta = 60°$, $\dfrac{1}{\tan 60°} = \tan 30°$ (if this is not obvious, draw a 30-60-90 right triangle with sides a and b and solve for each). Therefore, $\mu = \dfrac{1}{2} \tan 30°$.

4. A

The change in kinetic energy of the rocket is equal to the work done against gravity on the rocket, which, in this case, is the change in potential energy:

$$\Delta K = -\int_0^h mg\, dy = -mgh$$

At the maximum height h, the velocity is zero, so we have

$$K(h) - K(0) = -mgh$$

or

$$0 - \frac{mv^2}{2} = -mgh$$

which gives

$$h = \frac{v^2}{2g}$$

for the maximum height h that the rocket can reach with an initial velocity of v.

5.   B

In an inelastic collision, momentum (but not kinetic energy) is conserved. $p = mv$, which is a vector. Also, since the collision is inelastic, the two masses will stick together after the collision.

So

$$p_{\text{before}} = p_{\text{after}}$$

which gives

$$mv + 0 = (m + M)V$$

Then $V = \dfrac{mv}{m + M}$, with V being in the same direction as v; that is, left to right.

6.   E

For m to remain unmoving, centripetal force (tension on m) must equal gravitational force from M (balanced by tension on M):

$$\frac{mv^2}{R} = Mg$$

So $v = \sqrt{\dfrac{MgR}{m}}$.

Since the force applied by pulling M down is at right angles to the angular momentum, the torque ($= rF \sin \theta$) is zero. Thus, angular momentum is conserved, so there is no change.

7.   E

The change in kinetic energy is equal to the work done by the frictional force on the car. Thus,

$$\Delta K = \int_0^d F_{\text{f}} \, dx = F_{\text{f}}$$

where F_{f} is the force of friction, which is equal to

$$F_{\text{f}} = \mu mg$$

where μ is the coefficient of friction between the car and the pavement. When the car is stopped, the velocity is zero, so we have

$$\frac{1}{2} mv^2 - 0 = F_{\text{f}} \, d = \mu mgd$$

Thus,

$$\mu = \frac{v^2}{2gd}$$

8.   C

By conservation of energy, change in kinetic energy = work done.

So $K(0) = \left(\dfrac{1}{2}\right) I\omega^2$, and $v = \omega R$, with $I = mr^2$.

Since $K(t) = 0$, the change in K is $\dfrac{1}{2} mv^2$, which is the work done to stop the wheel.

Work = Power $\times$ Time: $W = Pt$. So average power = $\dfrac{\text{total work}}{\text{total time}} = \dfrac{mv^2}{2t}$.

9.   C

The total energy for a simple harmonic system like the one described is

$$E_{\text{tot}} = \left(\frac{1}{2}\right) kA^2$$

where A is the maximum displacement of the mass from the equilibrium position x_0. The total energy is the total of the kinetic and potential energy, which are given as

$$K(x) = \left(\frac{1}{2}\right) m[v(x)]^2, \text{ and}$$

$$U(x) = \left(\frac{1}{2}\right) kx^2$$

where x is the displacement of the block from the equilibrium position. As a result, we have

$$E_{tot} = K(x) + U(x)$$

or

$$\left(\frac{1}{2}\right)kA^2 = \left(\frac{1}{2}\right)m[v(x)]^2 + \left(\frac{1}{2}\right)kx^2$$

Cancel out the term $\left(\frac{1}{2}\right)$ from both sides of the equation and get

$$kA^2 - kx^2 = m[v(x)]^2$$

This gives us

$$v(x) = \sqrt{\left(\frac{k}{m}\right)(A^2 - x^2)}$$

10. E

For gravitational force for a sphere or a point

$$F = \frac{GmM}{r^2}$$

In this case, m is a small constant mass. M however, is the mass enclosed by a sphere of radius r.

So for $r > R$, we have

$$F = \frac{GmM}{r^2}$$

where M is the total mass of the sphere since it is entirely closed by a sphere of radius r.

For $r < R$,

$$M(r) = \rho\left(\frac{4}{3}\right)\pi r^3 = \frac{4}{3}\rho\pi r^3$$

where ρ is the density.

ρ is a constant equal to the total mass divided by the volume, which is

$$\rho = \frac{M}{\left(\frac{4}{3}\pi R^3\right)}$$

This works out to

$$\rho = \frac{3M}{4\pi R^3}$$

So for $r < R$, we have

$$F = \frac{GmM(r)}{r^2}$$

Now,

$$M(r) = \rho\left(\frac{4}{3}\right)\pi r^3$$

and

$$\rho = \frac{3M}{4\pi R^3}$$

So we have

$$F = \frac{Gm}{r^2}\left(\frac{4}{3}\rho\pi r^3\right)$$

when we put in the expression for $M(r)$.

Substituting for ρ in this equation, we get

$$F = \frac{Gm}{r^2}\left(\frac{4}{3}\right)\left(\frac{3M}{4\pi R^3}\right)\pi r^3$$

Canceling the same terms in the numerator and denominator gives us

$$F(r < R) = \left(\frac{GmM}{R^3}\right)r$$

Since $\frac{GmM}{R^3}$ is a constant, for $r < R$, the gravitational force increases linearly as r does.

For $r > R$, we have

$$F = \frac{GmM}{r^2}$$

So F decreases by r^{-2}, with the maximum value occurring at

$$F(r = R) = \frac{GmM}{R^2}$$

SECTION I: ELECTRICITY AND MAGNETISM

11. A

In the first part, the positive charge on the rod is fixed in place, since it's a dielectric. Since charges are free to move within conductors, the effect of the electric field about the rod is to drive positive charges away. Thus, the near sphere has a negative net charge, while the farther sphere has a positive charge, since charges are free to move within conductions.

When the rod is touched to the near sphere, some amount of positive charge is placed on the spheres. When the rod is removed, both spheres retain some excess positive charge.

12. D

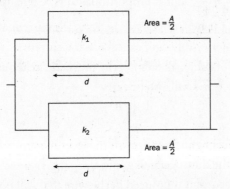

The trick here is to realize the capacitor can be treated as two capacitors in parallel, each with an area of $\dfrac{A}{2}$ and dielectric constants k_1 and k_2, respectively. Since in parallel, total capacitance is the sum of the individual capacitances, for a parallel plate capacitor of the type in the problem, it is generically $C = k\varepsilon_0 \dfrac{A}{d}$. We obtain

$$C_{\text{tot}} = \frac{(k_1 + k_2)\,\varepsilon_0}{2d}\,A.$$

13. B

The force on a charged particle is $F = qv \times B$, where $v \times B$ is the vector product of the particle's velocity and the magnetic field. The vector product follows the right-hand rule, so the force in this instance points to the center orbit of the particle. This inward force has to be balanced by the centrifugal force. Thus, we have $\dfrac{mv^2}{r} = qvB$. Since $v = \omega r$, this yields $\omega = \dfrac{qB}{m}$.

14. D

The electric potential is a scalar, equal to $\left(\dfrac{1}{4\pi}\varepsilon_0\right)\dfrac{q}{r}$. Since, q_1 and q_2 are of opposite signs, the potential midway between them will be zero. Thus, the answer for the first part is the B region. The electric field is a vector, with the magnitude equal to $\left(\dfrac{1}{4\pi}\varepsilon_0\right)\dfrac{q}{r^2}$ and the direction depending on the sign of the charge. In regions A and C, E_1 and E_2 would tend to cancel out but would never be zero, since the distance at a given point in these regions from the particles would never be equal. In region B, the electric fields are in the same direction (left to right) and so add up. Thus, the electric field is never zero along the axis of the charges.

15. E

In a conductor, all the charge resides on the surface.

For a spherical distribution of charge,

$$E = \left(\frac{1}{4}\pi\varepsilon_0\right)\frac{q}{r^2}$$

where q is the charge enclosed within r. Thus, inside the sphere, the electric field is zero since no charge is enclosed. Outside the sphere, the entire sphere is enclosed since the sphere defined by r encloses the whole conducting sphere, which holds a total charge q.

So we have the following:

For $(r < R)$, $E = 0$.

For $(r > R)$, $E = \left(\dfrac{1}{4}\pi\varepsilon_0\right)\dfrac{q}{r^2}$.

Since $\left(\dfrac{1}{4}\pi\varepsilon_0\right)q$ is a constant quantity, the electric field is proportional to $\dfrac{1}{r^2}$, or r^{-2}.

16. A

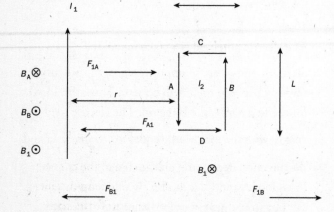

For a charge or current within a uniform field B, the force is $F = qv \times B$, or $F = IL \times B$, where $v \times B$ and $L \times B$ are vector products. (L is taken as having the same direction as I, the current). The magnitude of the field B from a long wire carrying a current is $B(r) = \dfrac{\mu_0 I}{2\pi r}$, with the direction of B into the page at the loop and stronger at wire A than at wire B.

Thus, the force between the long wire and wire B will be weaker than the force between the long wire and wire A. By the right-hand rule for force, we see that force on wire A is away from the long wire (field into page, current down page, force right). The forces between the long wire and wire B are in the opposite direction. Wires C and D contribute no net force, since by the right-hand rule the forces between

them and the long wire cancel out. Thus, the loop is repelled from the wire.

If the direction of the current in the loop is reversed, by the same reasoning as above, the loop is attracted to the wire.

17. C

For a long wire, $B(r) = \dfrac{\mu_0 I}{2\pi r}$, where I is the current enclosed within radius r. Since these are tubes, all of the current resides along the two surfaces, corresponding to r equal to a and b. Therefore, for $r < a$, there is no magnetic field. For $a < r < b$, there will be a magnetic field created by the interior but not the exterior tube (since this area is still inside of the exterior tube). This field will have strength $B(r) = \dfrac{\mu_0 I}{2\pi r}$, so it is proportional to r^{-1}. Finally, for $r > b$, the net current will be zero (the current of the interior tube will cancel out the current of the exterior tube). Therefore, the magnetic field outside of the tubes will also be zero.

18. E

The magnetic flux through the loop is increasing, with the direction being into the page. By Lenz's law, a current is induced in the wire that will oppose the increase in magnetic flux. Thus the current is counterclockwise. Because the flux changes constantly, Faraday's law predicts that there will be no change in the current.

19. E

At time $t = 0$, the charge on the capacitor is zero, so it acts as a short, and the current through the resistor is therefore zero. When the capacitor is fully charged, no current will flow through it, so the total current through the resistor is $I = \dfrac{V}{R}$. As the capacitor charges, since R is very large, the capacitor will charge very nearly as if R didn't exist, with the current through it decreasing proportionally to $e^{\frac{-t}{RC}}$. Thus, the current through R will increase roughly as $1 - e^{\frac{-t}{RC}}$.

20. A

In this circuit we have parallel and series components.

Within the two branches, the resistors are in parallel, so the total resistance in each branch is:

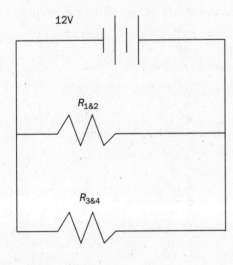

$$R_{1\&2} = R_1 + R_2$$
and
$$R_{3\&4} = R_3 + R_4$$

The circuit can now be thought of as $R_{1\&2}$ and $R_{3\&4}$ being in parallel. The formula for two resistances (a and b, for example) in parallel is

$$\frac{1}{R_{\text{tot}}} = \frac{1}{R_a} + \frac{1}{R_b}.$$

Finding a common denominator for the right side of the equation, we have

$$\frac{1}{R_{\text{tot}}} = \frac{R_a + R_b}{R_a R_b}.$$

Inverting the equation we have

$$R_{\text{tot}} = \frac{R_a R_b}{R_a + R_b}.$$

In this case, then we have

$$R_{\text{tot}} = \frac{R_{1\&2} R_{3\&4}}{R_{1\&2} + R_{3\&4}}.$$

Because each of the actual resistors is 20 ohms, this expression gives 10 ohms for the total resistance.

Now, $I_{\text{tot}} = \dfrac{V}{R_{\text{tot}}}$, so $I_{\text{tot}} = 1$ A. Because $R_{1\&2}$ is equal to $R_{3\&4}$, each parallel branch carries the same amount of current; that is, 0.5 A. Since the current in a set of resistors in series is the same for all the resistors, we see that each resistor in this circuit carries 0.5 A.

SECTION II: MECHANICS

1. (a) The change in kinetic energy is equal to the work done by gravity on the pendulum. In this case, the work done is Mgh, where $h = L(1 - \cos \theta)$. The kinetic energy K here is rotational kinetic energy, which has two parts: the K of the disk at the end of the shaft and the K of the disk as it rotates about its center of mass as the shaft moves. Recall that the disk is fixed, so as the shaft moves through an angle θ, the disk rotates about its center of mass by the same angle. Thus:

$$\left(\frac{1}{2}\right) I \omega^2 + \frac{1}{2} I_{\text{cm}} \, \omega_{\text{cm}^2} = MgL(1 - \cos \theta).$$

$I = ML^2$, and $\omega = \dfrac{v}{L}$, so the first term is equal to $\dfrac{1}{2} M v^2$.

I_{cm} about the center of mass for the disk is $\dfrac{MR^2}{2}$, while $\omega_{cm} = \dfrac{v}{L}$, so the second term is

$$\frac{1}{2}\left(\frac{MR^2}{2}\right)\left(\frac{v}{L}\right)^2 = \frac{1}{4}\,\frac{R^2}{L^2}\,Mv^2.$$

Therefore,

$$\frac{1}{2}\,Mv^2 + \frac{1}{4}\,\frac{R^2}{L^2}\,Mv^2 = MgL(1-\cos\theta)$$

$$(2L^2 + R^2)Mv^2 = 4MgL^3(1-\cos\theta)$$

$$v = \sqrt{\frac{4gL^3(1-\cos\theta)}{2L^2 + R^2}}$$

(b) Here, the shaft is free to rotate, so if the shaft is displaced by an angle θ, the disk doesn't rotate at all. Thus, the term containing I_{cm} goes to zero. We have

$$v(\theta) = \sqrt{2gL(1-\cos\theta)}\,.$$

SECTION II: ELECTRICITY AND MAGNETISM

2. (a) The total charge contained within a dielectric sphere is equal to the volume times the charge density: $Q = \left(\dfrac{4}{3}\right)\pi a^3 \rho$, where a is the radius of the sphere and ρ is the charge density. The sign on the charge will be positive or negative depending on the sign of the dielectric. Imagine only the larger sphere were present. Then the total charge of the sphere would be a positive charge equal to $Q_R = \left(\dfrac{4}{3}\right)\pi R^3 \rho$. Now replace the volume where the small sphere resides by a negatively charged

sphere with charge $-Q_r = \left(\dfrac{4}{3}\right)\pi r^3 \rho$. Removing the positively charged dielectric results in subtracting a charge Q_r, and putting in a negatively charged dielectric results in subtracting an additional charge Q_r. This results in the total charge equaling

$$Q_R - 2Q_r = \left(\frac{4}{3}\right)\pi\rho(R^3 - 2r^3).$$

(b) The electric potential for a charge q at a distance d from a point charge Q is $\left(\dfrac{1}{4}\pi\varepsilon_0\right)\dfrac{qQ}{d}$, where d is the distance from the center of the sphere, and q is the test charge. For the large sphere, q is located at distance d from the center, so the electric potential is $+\left(\dfrac{1}{4}\pi\varepsilon_0\right)\dfrac{Q_R}{d}$ if the small sphere did not exist. For the smaller sphere, the distance from the point where q is located and the center of the small sphere is $(d - R + r)$, and the charge is $-2Q_r$. The total electric potential (V_{tot}) is the sum of all the potentials, so $V_{tot} = V_R + V_r$:

$$V_{tot} = \left(\frac{1}{4}\pi\varepsilon_0\right)\frac{Q_R}{d} - 2\left(-\frac{1}{4}\pi\varepsilon_0\right)\frac{Q_r}{(d - R + r)}$$

One can substitute in for Q_R and Q_r and work out the expression in detail, but this is unnecessary.

(c) The electric field felt by a charge q is equal to q times the negative derivative of the electric potential. (Taking the derivative of $\dfrac{1}{d}$ gives $-\dfrac{1}{d^2}$.) Thus, the electric field felt by a charge q at a distance a from the center of a sphere of charge Q is qV:

$$E = \left(\frac{1}{4}\pi\varepsilon_0\right)\frac{qQ}{a^2}$$

where the direction of the electric field vector is generally determined by inspection of a diagram.

Thus, in this particular case, we have

$$E = q\left[\left(\frac{1}{4}\pi\varepsilon_0\right)\frac{Q_R}{d^2} - 2\left(-\frac{1}{4}\pi\varepsilon_0\right)\frac{Q_r}{(d-R+r)^2}\right],$$

where the direction of E would be toward the spheres.

(d) In conducting matter, all the charge is contained on the surface. If the object is a sphere, the total charge may be treated as residing at the central point of the sphere. In this case, $Q_{tot} = Q_R - 2Q_r$. Consequently, the electric potential at point d is $V = K_e\frac{Q_{tot}}{d}$.

Substitutions may be made for Q_{tot}, but this expression is sufficient.

(e) $F = qE$, and since $E = qV$, we have $E = qV = \frac{qQ_{tot}}{d^2}$ so $F = \frac{K_e qQ_{tot}}{d^2}$.

AP PHYSICS C DIAGNOSTIC TEST CORRELATION CHART

Use the following table to determine which Physics C topics you need to review most.

Area of Study	Question Number
Kinematics	1, 2
Newton's Laws of Motion	3, 6
Work, Energy, Power	4, 7
Systems of Particles, Linear Momentum	5
Circular Motion and Rotations	6, 8
Oscillations and Gravitation	9, 10
Electrostatics	11, 14
Conductors, Capacitors, Dielectrics	12, 15
Electric Circuits	19, 20
Magnetostatics	17
Electromagnetism	13, 16, 18

| Part Three |

AP PHYSICS REVIEW

HOW TO USE THE REVIEW SECTION

STEP 1: REVIEW THE CONCEPTS.

Each of the following review chapters begins by going over the main concepts that apply to the chapter's topic. The chapters will not include every shred of factual material related to physics, but instead will help you tie all the facts together to understand the topics and discuss briefly how they fit within the thematic emphases designed by the College Board.

STEP 2: ANSWER THE REVIEW QUESTIONS.

In each of the review chapters, you will be given sample questions that will help you learn and review the AP Physics course material. Questions from both sections of the AP Physics exams, multiple-choice and free-response, are included in each review chapter. The quantity and type of multiple-choice and free-response questions in the review chapters do not represent the exact material on the AP Physics exams, but they are plausible examples of topics covered.

STEP 3: REVIEW THE ANSWER EXPLANATIONS.

Following these questions are detailed answer explanations that explain how these questions address the course concepts. You will be given the correct answer for each question and be informed of the thought processes you should go through to reach a correct answer. Sometimes you will be given examples of how you might have reached the incorrect answer and how to avoid that problem in the future. Again, the emphasis is not just on repetition but on effective repetition. It's not only about learning information but about how to apply it to the task at hand—scoring well on the exams.

CHAPTER 3: KINEMATICS AND NEWTON'S LAWS OF MOTION

IF YOU MEMORIZE ONLY SIX EQUATIONS IN THIS CHAPTER . . .

Constant Acceleration:

$$v = at + v_0$$

$$s = s_0 + v_0 t + \frac{1}{2} at^2$$

$$v^2 = v_0{}^2 + 2a(s - s_0)$$

$$\overline{\sum F} = \overline{F_{net}} = \overline{ma} = \overline{F_1} + \overline{F_2} + \overline{F_3} + \cdots$$

$$a_c = \frac{v^2}{r}$$

$$F_{fric} \leq \mu F_N$$

GETTING STARTED IN THE RIGHT DIRECTION: VECTORS

The information in this chapter makes up a significant part of the AP Physics exams, accounting for about 16 percent of the B exam questions and about 19 percent of the C exam questions. However, vectors rarely show up on the test as isolated problems. Instead, they are embedded within more complex topics throughout the exam. Understood this way, vectors are the basis for nearly all mechanics problems and are essential in electromagnetism, optics, and atomic physics. Therefore, it is extremely important that you develop a thorough understanding of vector algebra and are able to apply these concepts with confidence.

BASIC DEFINITION AND RULES OF VECTORS

Direction is the element that distinguishes a **vector** quantity from a **scalar** quantity. A vector has both a magnitude and a direction. The **magnitude** is the length of the vector; this means that the magnitude is always positive. In physics, magnitude is often notated with the same mathematical symbol as absolute value:

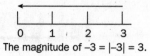

The magnitude of $-3 = |-3| = 3$.

The **direction** of a vector can be given in several formats. Examples include positive and negative values (in one dimension), angles, north, south, forward, up, down, coordinates, or any term useful for the physical situation.

In print, vector notation is often presented in bold font. The magnitude is typically represented by the same letter in italics (not bold). In handwritten work, it is common to place an arrow over the top of the vector.

$$\text{vector } v = \mathbf{v} = \vec{v}$$

AP Physics exam questions are written so that you will know if they are referring to a vector or to a scalar. Make sure to be equally clear in your free-response answers.

Examples of vector quantities include displacement, velocity, acceleration, force, and linear momentum. You should also be familiar with the terms **position vector** and **displacement vector**. The position vector is the vector that points from the origin of the coordinate axes to a point or particle: $\mathbf{r} = x\,\hat{\mathbf{i}} + y\,\hat{\mathbf{j}}$. The displacement vector is the resultant vector of the difference between two position vectors: $\Delta\mathbf{r} = \mathbf{r}_1 - \mathbf{r}_2$.

ADDITION AND SUBTRACTION OF VECTORS

One can visualize the addition of vectors graphically by adding them "tail to tip."

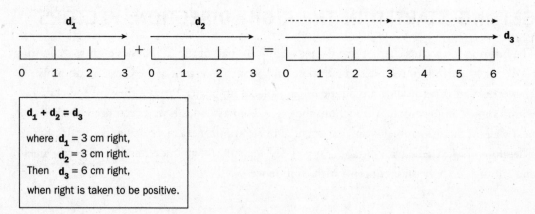

In one dimension, if the vectors are in the same direction, the magnitudes simply add directly and the direction remains the same. If the two vectors being added are in opposite directions, the magnitudes are simply subtracted. Vector addition is commutative (a + b = b + a) and associative [a + b + c = a + (b + c) = (a + b) + c]. This means that vector addition can be done in the order that makes the calculations simplest.

Here is a graphic representation of the addition of vectors in opposite directions.

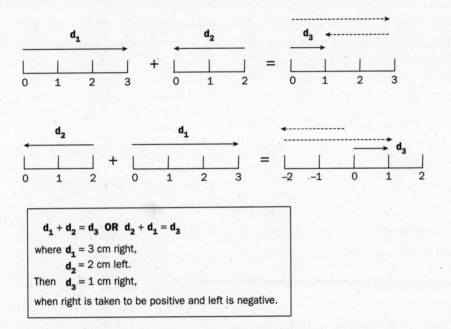

When working with more than one dimension, we must use a coordinate system. For simplicity, we will use rectangular coordinates in this section; additional systems are discussed later.

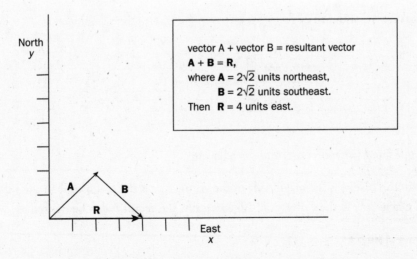

You should be familiar with rectangular coordinates in two dimensions (*x*, *y*) and in three dimensions (*x*, *y*, *z*). The "tail to tip" method can be used to find a **resultant**, or the "sum vector," of the addition of multiple vectors in multiple dimensions. Simply connect the vectors being added from tip to tail, then draw the resultant from the tail of the first to the tip of the last.

Often, numbers can be calculated much more accurately by using **components**. Remember that vectors have a magnitude and direction but they have no set place when being added. This means that we can move them anywhere within the coordinate system (or create a new set of coordinates around the vector) as long as they are kept parallel to the original. We can also separate the vector into components by making a **projection** onto any of the axes. These projections can be added separately, then used to find the resultant.

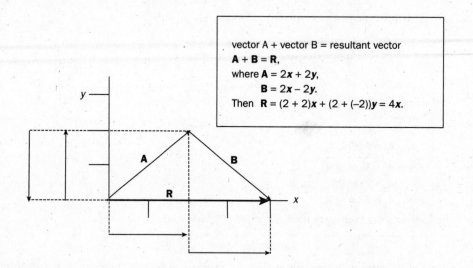

vector A + vector B = resultant vector
A + **B** = **R**,
where **A** = 2*x* + 2*y*,
   B = 2*x* − 2*y*.
Then **R** = (2 + 2)*x* + (2 + (−2))*y* = 4*x*.

Generally, it is easiest to find the components using trigonometry. Recall the following:

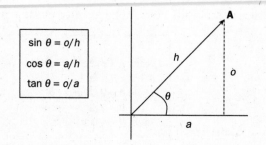

$$\sin \theta = o/h$$
$$\cos \theta = a/h$$
$$\tan \theta = o/a$$

where *o* = opposite, *h* = hypotenuse, and *a* = adjacent.

Notice that in this case, *h* is the magnitude of the vector A, *o* is the magnitude in the *y* direction, and *a* is the magnitude in the *x* direction. Algebraically, the vector A can be written as

$$\mathbf{A} = A \cos \theta \, x + A \sin \theta \, y.$$

Applying this method to the previous example we find the same result as before.

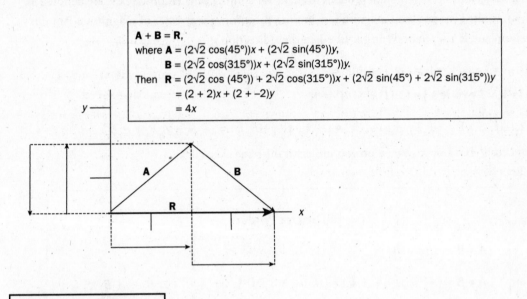

A + B = R,
where **A** = $(2\sqrt{2} \cos(45°))x + (2\sqrt{2} \sin(45°))y$,
      B = $(2\sqrt{2} \cos(315°))x + (2\sqrt{2} \sin(315°))y$.
Then **R** = $(2\sqrt{2} \cos(45°)) + 2\sqrt{2} \cos(315°))x + (2\sqrt{2} \sin(45°) + 2\sqrt{2} \sin(315°))y$
      = $(2 + 2)x + (2 + -2)y$
      = $4x$

FOR C EXAM ONLY

MULTIPLICATION OF VECTORS

There are three types of vector multiplication:

1. The multiplication of a vector by a scalar

2. The dot product

3. The cross product

The first type simply follows the distributive property and is much like addition. To find the product of a vector and a scalar, multiply the scalar by the magnitude and keep the direction the same.

Example: $3 \times 5x = 15x$

If there is more than one dimension, distribute the scalar to the magnitude of each component.

Example: $3(5x + 2y + 4z) = 15x + 6y + 12z$

The **dot product** is often used in physics problems associated with work and power. It describes the projection of one vector onto another and results in a **scalar product**.

A • **B** = $AB \cos \Phi$, where Φ = the acute angle between **A** and **B** and AB is the product of the magnitudes of the vectors.

The dot product is both commutative and distributive (see table below). Cartesian **unit vectors** are used in conjunction with the dot product to make for simpler calculations. These are denoted as $(\hat{\imath}, \hat{\jmath}, \hat{k})$, where $\hat{\imath}$ is a vector of magnitude 1 in the x direction, $\hat{\jmath}$ is a vector of magnitude 1 in the y direction, and $\hat{k}$ is a vector of magnitude 1 in the z direction.

$$\hat{\imath} \bullet \hat{\imath} = \hat{\jmath} \bullet \hat{\jmath} = \hat{k} \bullet \hat{k} = (1)(1)\cos 0° = 1$$

$$\hat{\imath} \bullet \hat{k} = \hat{\imath} \bullet \hat{\jmath} = \hat{k} \bullet \hat{\jmath} = (1)(1)\cos 90° = 0$$

Therefore, $\mathbf{A} \bullet \mathbf{B} = (A_x\hat{\imath} + A_y\hat{\jmath} + A_z\hat{k}) \bullet (B_x\hat{\imath} + B_y\hat{\jmath} + B_z\hat{k}) = (A_xB_x + A_yB_y + A_zB_z)$, where $A_x = A\cos\theta_x$ because $\cos\theta = \dfrac{A_x}{A}$ if you project it onto the axis.

The Dot Product

$$\mathbf{A} \bullet \mathbf{B} = AB\cos\Phi$$

$$\mathbf{A} \bullet \mathbf{B} = (A_x\hat{\imath} + A_y\hat{\jmath} + A_z\hat{k}) \bullet (B_x\hat{\imath} + B_y\hat{\jmath} + B_z\hat{k}) = (A_xB_x + A_yB_y + A_zB_z)$$

$$\mathbf{A} \bullet \mathbf{B} = \mathbf{B} \bullet \mathbf{A} \qquad \text{[commutative property]}$$

$$(\mathbf{A} + \mathbf{B}) \bullet \mathbf{C} = \mathbf{A} \bullet \mathbf{C} + \mathbf{B} \bullet \mathbf{C} \qquad \text{[distributive property]}$$

If $\mathbf{A} \perp \mathbf{B}$, then $\mathbf{A} \bullet \mathbf{B} = 0$ [because $\Phi = 90°$, $\cos 90° = 0$]

and if $\mathbf{A} \bullet \mathbf{B} = 0$, $\mathbf{A}$ and $\mathbf{B}$ must be perpendicular.

If $\mathbf{A} \| \mathbf{B}$, then $\mathbf{A} \bullet \mathbf{B} = AB$ [because $\Phi = 0°$, $\cos 0° = 1$]

so $\mathbf{A} \bullet \mathbf{A} = A^2$.

The **cross product** is commonly used in angular momentum and electromagnetism. It often describes two force fields and how they affect one another; it results in a **vector product**.

$\mathbf{A} \times \mathbf{B} = \mathbf{C}$, where the magnitude is $|\mathbf{C}| = C = AB\sin\Phi$ and the direction is perpendicular to the AB plane using the right-hand rule.

To find the direction, point your fingers in the direction of $\mathbf{A}$, then curl them in the direction of $\mathbf{B}$. Your thumb will be pointing in the direction of $\mathbf{C}$. Remember, if you cross $\mathbf{B}$ with $\mathbf{A}$, the direction changes. **The cross product is not commutative**.

Properties of the Cross Product

$\mathbf{A} \times \mathbf{B} = -\mathbf{B} \times \mathbf{A}$ \qquad [noncommutative]

If $\mathbf{A} \perp \mathbf{B}$, \qquad then $|\mathbf{A} \times \mathbf{B}| = AB$ \qquad [because $\Phi = 90°$, $\sin 90° = 1$]

If $\mathbf{A} \| \mathbf{B}$, \qquad then $|\mathbf{A} \times \mathbf{B}| = 0$ \qquad [because $\Phi = 0°$, $\sin 0° = 0$]

So $\mathbf{A} \times \mathbf{A} = 0$

and if $\mathbf{A} \times \mathbf{B} = 0$, $\mathbf{A}$ and $\mathbf{B}$ must be parallel.

If you try out the unit vectors, you will find that

$$\hat{\imath} \times \hat{\imath} = \hat{\jmath} \times \hat{\jmath} = \hat{k} \times \hat{k} = 0$$

$$\hat{\imath} \times \hat{\jmath} = -\hat{\jmath} \times \hat{\imath} = \hat{k}$$

$$\hat{\jmath} \times \hat{k} = -\hat{k} \times \hat{\jmath} = \hat{\imath}$$

$$\hat{k} \times \hat{\imath} = -\hat{\imath} \times \hat{k} = \hat{\jmath}$$

Therefore, $\mathbf{C} = \mathbf{A} \times \mathbf{B} = (A_y B_z - A_z B_y)\,\hat{\imath} + (A_z B_x - A_x B_z)\,\hat{\jmath} + (A_x B_y - A_y B_x)\hat{k}.$

You should recognize this as the determinant of

$$\mathbf{A} \times \mathbf{B} = \begin{vmatrix} \hat{\imath} & \hat{\jmath} & \hat{k} \\ A_x & A_y & A_z \\ B_x & B_y & B_z \end{vmatrix}$$

This is the quickest way to find a cross product, so review how to solve a 3×3 determinant using either the diagonal line method (rule of Sarrus) or the sum of three 2×2 determinants method.

COORDINATE SYSTEMS

Any physics problem must be worked by defining a convenient point of space as the origin, O. The position is then measured relative to that origin. The majority of the AP Physics exam deals with the **Cartesian coordinate system**, also known as the **rectangular coordinate system**. A few problems, however, are more easily addressed using other systems. You should be familiar with the basics of each coordinate system.

Cartesian coordinates are defined by three mutually perpendicular axes. You can describe a point in space by noting how far you must move parallel to each of these axes. The exam sticks to the more common right-handed coordinates (x, y, z), as opposed to left-handed coordinates in which x and y are interchanged.

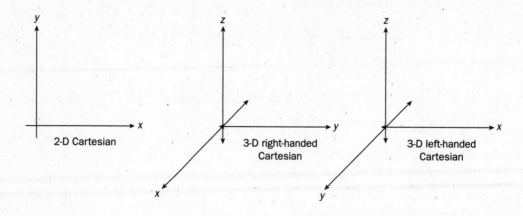

2-D Cartesian

3-D right-handed Cartesian

3-D left-handed Cartesian

Polar coordinates are two-dimensional (r, θ). The distance from the origin to the point, or radial distance, is denoted as r. The angular coordinate, sometimes called the polar angle, is the measure of the angle from the x-axis counterclockwise to the point; it is denoted as θ.

In terms of x and y, $r = \sqrt{(x^2 + y^2)}$ and $\theta = \tan^{-1}\left(\dfrac{y}{x}\right)$.

In Cartesian coordinates, $x = r \cos \theta$ and $y = r \sin \theta$.

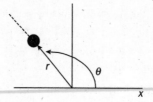

FOR C EXAM ONLY

Cylindrical coordinates are a three-dimensional extension of polar coordinates in which the third axis is simply z (r, θ, z). In physics problems, they are often given different symbols $(\rho, \Phi, z,$ respectively). The radial axis, r (or ρ) is the distance along the x-y plane. The azimuthal angle (also known as the longitude angle or bearing) is denoted as θ (or Φ). The height above the x-y plane is z.

In terms of x, y, and z: $r = \sqrt{(x^2 + y^2)}$ $\quad \varepsilon\,[0, \infty]$

$$\theta = \tan^{-1}\left(\frac{y}{x}\right) \quad \varepsilon\,[0, 2\pi]$$

$$z = z \quad \varepsilon\,(-\infty, \infty)$$

In Cartesian coordinates, $x = r\cos\theta$, $y = r\sin\theta$, and $z = z$.

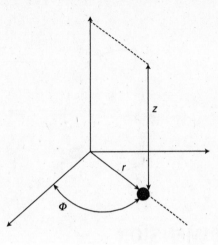

FOR C EXAM ONLY

Spherical coordinates define a point in space by giving a radial distance from the origin, a longitude angle, and a latitude angle. There is no one standard notation, so be aware of the order in which coordinates are given. Do *not* depend on the symbol used. As in cylindrical and polar coordinates, the first term is the radial coordinate, r (or ρ). The second term, θ (or Φ or λ) is the longitude angle (or azimuthal angle); it lies in the x-y plane and is taken counterclockwise from the x-axis. The third term, an angle in the y-z plane, is the trickiest. It is called the latitude angle, Δ, when it is measured from the y-axis. More commonly, this polar angle Φ (or θ) is measured from the z-axis. To clarify, co-latitude is $\Phi = 90° - \Delta$.

In short, spherical coordinates are (r, θ, Φ), (ρ, Φ, θ), or (r, λ, Δ), which describe radial distance, the angle within the x-y plane, and the angle out of the x-y plane, respectively. In the following formulas, we use the first definition.

In terms of x, y, and z: $r = \sqrt{(x^2 + y^2 + z^2)}$ $\varepsilon\ [0, \infty]$

$$\theta = \cos^{-1}\left(\frac{z}{r}\right) \qquad \varepsilon\ [0, \pi]$$

$$\Phi = \tan^{-1}\left(\frac{y}{x}\right) \qquad \varepsilon\ [0, 2\pi]$$

In Cartesian coordinates: $x = r\sin\theta\cos\Phi$

$$y = r\sin\theta\sin\Phi$$

$$z = r\cos\theta$$

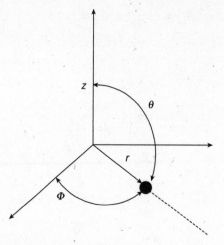

MOTION IN ONE DIMENSION

These are the simplest of the mechanics exam questions. For these and other exam problems, it is essential to understand and be able to use the kinematic equations. There are two approaches to deriving these equations: algebraic substitution and calculus. The first method is generally taught in AP Physics B, and calculus is generally used for AP Physics C. This chapter will demonstrate both, but correct use of the equations is more important than their derivation. Additionally, you will be given an equation sheet for the free-response portion of the exam.

Generally, the questions will give you information about a system, and you will be looking for the missing element. If you are familiar with the equations, you will simply be able to do the algebraic calculation.

DISPLACEMENT

Displacement is a vector form of distance; it is an object's change in position. Note that displacement follows a direct path and is *not* the distance traveled. For example, consider a person who runs one mile, doing four laps, on a track; her distance would be one mile, but her displacement would be zero because she ended in the same spot she started.

Displacement in one dimension is often presumed to be on the x-axis and is defined as

$$\Delta x = x_f - x_i = x_2 - x_1 = x - x_o$$

where $x_i = x_1 = x_o$ is the initial position and $x_f = x_2 = x$ is the final position.

Any of these forms are correct, but because displacement may not always be on the x-axis, the equation sheet for the exam defines displacement to be s:

$$\Delta s = s - s_o$$

SI units for displacement are meters (m).

VELOCITY

Average velocity is the vector form of speed and is defined as the displacement per time interval.

$$\overline{\mathbf{v}} \equiv \frac{\mathbf{s} - \mathbf{s_o}}{t - t_o} = \frac{\Delta \mathbf{s}}{\Delta t} = \frac{\text{displacement}}{\text{time interval}}$$

The SI units for velocity are meters per second (m/s). You can see that if we let $s_o = 0$ and $t_o = 0$, then

$$\overline{\mathbf{v}} = \frac{\mathbf{s}}{t} \quad \text{or} \quad \mathbf{s} = \overline{\mathbf{v}}t$$

This can be represented by the slope of the graph s versus t, from which the linear best fit is taken because v is an average.

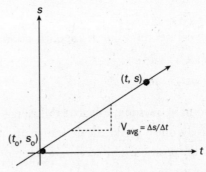

While average velocity is a measurement taken over the entire interval, **instantaneous velocity** is a measurement taken at a particular instant in time using calculus notation.

$$\mathbf{v} = \lim_{\Delta t \to 0} \frac{\Delta \mathbf{s}}{\Delta t} = \frac{d\mathbf{s}}{dt}$$

It describes how fast an object is traveling during a very small interval of time. This limit is actually the first derivative of the position function. Visually, it is represented by the slope of the tangent to the s versus t curve at a point.

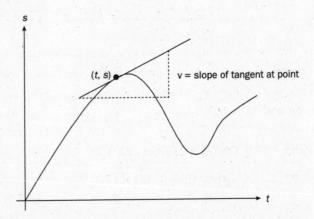

When reading the questions on the AP test, make sure you don't confuse *displacement* and *distance*, or *velocity* and *speed*.

Speed is a scalar quantity and is equal to the magnitude of velocity. The steeper the curve of the graph, the greater the magnitude of the velocity. If the slope is negative, velocity is negative as well, which implies that the object is moving in the negative direction.

ACCELERATION

Acceleration describes the change in velocity within a time interval; the units for acceleration are meters per second per second (m/s^2). If you are interested in a large time interval, you will want to find the **average acceleration**.

$$\bar{a} = \frac{v - v_o}{t - t_o} = \frac{\Delta v}{\Delta t}$$

The **instantaneous acceleration** defines this value at a specific point in time.

$$a = \lim_{\Delta t \to 0} \frac{\Delta v}{\Delta t} = \frac{dv}{dt} = \frac{d}{dt}\left(\frac{ds}{dt}\right) = \frac{d^2 s}{dt^2}$$

To approach a limit about the point in question, calculate the average acceleration for successively smaller points in time. Alternatively, one can take the derivative of the velocity with respect to time, or the second derivative of the position. The acceleration is represented graphically by the slope of the line or curve of *v* versus *t*.

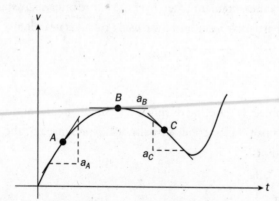

At point *A*, the slope of the velocity curve is positive, so a_A, the acceleration at point *A*, is positive and v_A, the velocity at that point, is increasing. At point *B*, the slope of the velocity curve is horizontal and equal to 0, so $a_B = 0$ and v_B is constant. At point *C*, the slope of the velocity curve is negative, so a_C is negative and v_C is decreasing.

KINEMATIC EQUATIONS FOR CONSTANT ACCELERATION

Some very useful equations can be derived if (and only if) acceleration is constant.

WITH CALCULUS

$v = \int a\, dt = at + c_1$ since $a = \dfrac{dv}{dt}$

if $t_0 = 0$, $v_0 = 0 + c_1$, so $\boxed{v = at + v_0}$ $\quad$ *Equation A*

Similarly, $v = \dfrac{ds}{dt} \Rightarrow s = \int v\, dt = \int (at + v_0)\, dt = \dfrac{1}{2}at^2 + v_0 t + c_2$

When $t_0 = 0$, $c_2 = s_0$, so $\boxed{s = s_0 + v_0 t + \dfrac{1}{2}at^2}$ $\quad$ *Equation B*

Solve Equation *A for t*: $\quad t = \dfrac{v - v_0}{a}$

Insert this into Equation *B*: $s = v_0 \left(\dfrac{v - v_0}{a} \right) + \dfrac{1}{2}a\dfrac{(v - v_0)^2}{a^2}$

$\Rightarrow 2as = 2v_0 v - 2v_0^2 + v^2 - 2v_0 v + v_0^2 = v^2 - v_0^2$

So $\boxed{v^2 = v_0^2 + 2as}$ $\quad$ *Equation C*

WITHOUT CALCULUS

$\bar{v} = \dfrac{v_0 + v}{2} = \dfrac{\Delta s}{\Delta t} \Rightarrow \boxed{\Delta s = \dfrac{1}{2}(v_0 + v)\Delta t}$ $\quad$ *Equation D*

$a = \dfrac{v - v_0}{t - t_0} = \dfrac{v - v_0}{\Delta t} \Rightarrow \boxed{v = v_0 + a\Delta t}$ $\quad$ *Equation E*

Substitute Equation *E* into Equation *D*: $\Delta s = \dfrac{1}{2}(v_0 + v_0 + a\Delta t)\Delta t$

$\qquad\qquad\qquad\qquad\qquad\qquad = \dfrac{1}{2}(2v_0 \Delta t + a\Delta t^2)$

$\Rightarrow \boxed{\Delta s = v_0 \Delta t + \dfrac{1}{2}a\Delta t^2}$ $\quad$ *Equation F*

Solve Equation *D* for Δt: $\Delta t = \dfrac{2\Delta s}{v_0 + v}$

Substitute this into Equation *E*: $v = v_0 + a\left(\dfrac{2\Delta s}{v_0 + v} \right) \Rightarrow \boxed{v^2 = v_0^2 + 2a\Delta s}$ $\quad$ *Equation G*

KINEMATIC EQUATIONS FOR CONSTANT ACCELERATION

$s = \bar{v}t \qquad\qquad s = s_0 + v_0 t + \dfrac{1}{2}at^2$

$v = v_0 + at \qquad v^2 = v_0^2 + 2as$

AP EXPERT TIP

It is common for the AP exams to ask about the motion graphs representing constant velocity and uniform acceleration. Make sure to familiarize yourself with them.

Always simplify these equations as much as you can. For example:

$s_o = 0$ if you start at the origin.

$v_o = 0$ if you start from rest.

$a = 0$ if v is constant.

GRAVITY AND FREE FALL

These equations can be applied to any object in free fall. First, we must ignore any outside forces such as air resistance. The only force acting on the object is assumed to be gravity, as is true in a vacuum. Displacement is in the vertical direction, which we will call y. The acceleration of gravity is constant and has been experimentally found to be $g = -9.81$ m/s^2. We can substitute these values into the kinematic equations for constant acceleration.

$$\mathbf{a} = -g, \quad \mathbf{s} = y; \text{ so } \mathbf{v} = \mathbf{v_o} - gt, \text{ and } \mathbf{y} = \mathbf{y_o} + \mathbf{v_o}t - \frac{1}{2}gt^2$$

From these equations, we can conclude two objects with different masses have the same acceleration ($a = -g = $ constant), the same speed ($v = v_o - gt$), and the same position ($y = y_o + v_o t - \frac{1}{2}gt^2$) at any point in time after being dropped simultaneously.

MOTION IN TWO DIMENSIONS

The same rules apply for two- and three-dimensional situations. The key difference is that we are now working with vectors. Use the same kinematic equations but allow displacement, velocity, and acceleration to be vectors.

$$\mathbf{s} = \bar{\mathbf{v}}t \qquad\qquad \mathbf{s} = \mathbf{s_o} + \mathbf{v_o}t + \frac{1}{2}\mathbf{a}t^2$$

$$\mathbf{v} = \mathbf{v_o} + \mathbf{a}t \qquad\qquad v^2 = v_o^2 + 2\mathbf{as}$$

Notice that t is scalar. Therefore, v must be a vector in the direction of Δs.

$$\bar{\mathbf{v}} = \frac{\mathbf{s} - \mathbf{s_o}}{t - t_o} = \frac{\Delta \mathbf{s}}{\Delta t}$$

Similarly, a must be a vector in the direction of Δv.

$$\mathbf{a} = \frac{\mathbf{v} - \mathbf{v_o}}{t - t_o} = \frac{\Delta \mathbf{v}}{\Delta t}$$

This implies that acceleration could be due to a change in speed *or* direction.

These vectors can be resolved into components. The resolution of s yields the position vector, in which x represents the horizontal distance from the origin and y represents the vertical distance from the origin.

$$s = x\hat{i} + y\hat{j}$$

FOR C EXAM ONLY

The components of velocity and acceleration can be defined as derivatives or limiting cases. Either way, think of each component as the velocity (or acceleration) along a particular axis.

$$\mathbf{v} = v_x\hat{i} + v_y\hat{j} = \frac{ds}{dt} \quad \text{or} \quad \lim_{\Delta t \to 0} \frac{\Delta s}{\Delta t} \qquad \text{where } v_x = \frac{dx}{dt} \quad \text{or} \quad v_x = \lim_{\Delta t \to 0} \frac{\Delta x}{\Delta t}$$

$$v_y = \frac{dy}{dt} \quad \text{or} \quad v_y = \lim_{\Delta t \to 0} \frac{\Delta y}{\Delta t}$$

$$\mathbf{a} = a_x\hat{i} + a_y\hat{j} = \frac{d\mathbf{v}}{dt} = \frac{d^2s}{dt^2} \quad \text{or} \quad \lim_{\Delta t \to 0} \frac{\Delta \mathbf{v}}{\Delta t} \qquad \text{where } a_x = \frac{dv_x}{dt} = \frac{d^2x}{dt^2} \quad \text{or} \quad \lim_{\Delta t \to 0} \frac{\Delta v_x}{\Delta t}$$

$$a_y = \frac{dv_y}{dt} = \frac{d^2y}{dt^2} \quad \text{or} \quad \lim_{\Delta t \to 0} \frac{\Delta v_y}{\Delta t}$$

Refer to the section on vectors earlier in this chapter if you need help working with these components.

If we use these vectors in conjunction with the kinematic equations, we will have two components for each equation because we are in two dimensions.

$$\mathbf{s} = \mathbf{s_o} + \mathbf{v_o}t + \frac{1}{2}\mathbf{a}t^2: \qquad x = x_o + v_{ox}t + \frac{1}{2}a_xt^2, \qquad y = y_o + v_{oy}t + \frac{1}{2}a_yt^2$$

Often, these component equations can be solved simultaneously for unknowns. The first step in any problem is to note the information that you are given, determine what information you need, and identify and use the appropriate equation.

PROJECTILE MOTION

Projectile motion is an excellent example of motion in two dimensions. Sometimes called ballistic motion, **projectile motion** describes the movement of an object that has been launched with any initial velocity and allowed to follow its trajectory without any outside forces working upon it other than gravity. The object will have no acceleration and, therefore, has a constant velocity in the horizontal direction. It will experience the acceleration of gravity in the vertical direction, just as would an object in free fall.

Consider an object launched at an angle θ with initial velocity v_o. Let y be the vertical axis and x be the horizontal axis. The first step in assessing a projectile motion problem is to sketch a diagram and resolve v_o into components.

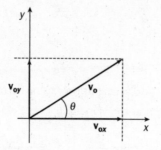

Use trigonometry to find values for v_{ox} and v_{oy}.

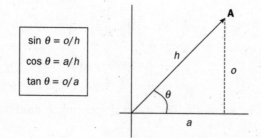

$$\sin\theta = o/h$$
$$\cos\theta = a/h$$
$$\tan\theta = o/a$$

$\mathbf{v_x} = v_o\cos\theta = \text{constant}$ $\qquad$ $\mathbf{v_y} = \mathbf{v_{oy}} + \mathbf{a}_y t = v_o\sin\theta - gt$

$\mathbf{x} = \mathbf{v_x}t = v_o(\cos\theta)t$ $\qquad$ $\mathbf{y} = \mathbf{v_{oy}}t + \dfrac{1}{2}\mathbf{a}_y t^2 = v_o(\sin\theta)t - \dfrac{1}{2}gt^2$

$\mathbf{a_x} = 0$ $\qquad$ $\mathbf{a}_y = -g$ $\qquad$ $\mathbf{v_y}^2 = \mathbf{v_{oy}}^2 + 2\mathbf{a}_y y = v_o^2(\sin^2\theta) - 2gy$

You can now substitute these values into the kinematic equations.

If you solve x for t and put this value into the equation for y, you will see that it is an equation for a parabola; it has the form $y = ax^2 + bx + c$.

$$y = (\tan\theta)x - \left(\frac{g}{2v_o^2(\cos^2\theta)}\right)x^2$$

This is illustrated in the following figure.

Notice that the vertical magnitudes of v_y are the same at equal heights. Also notice that the horizontal vector v_x is always the same.

AP EXPERT TIP

When performing a projectile calculation, your horizontal and vertical values should never appear in the same equation. Using subscripts is a great way to avoid getting mixed up.

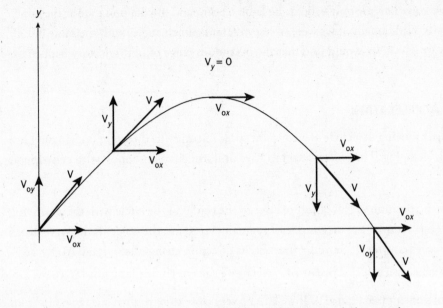

Projectile motion always occurs in a symmetrical parabola (provided we neglect air resistance). Thus, we can make several assumptions. First, notice that at the highest point the object will reach, the velocity in the y direction changes direction from up to down. This means that at the maximum height, $v_y = 0$.

Then $t = \dfrac{v_{oy}}{g}$ and $y_{\max} = \dfrac{v_{oy}^2}{2g}$

To find the time of flight of the object, we will assume that it lands at the same height from which it was launched, $y_o = y_f = 0$. Solving for t, we find that there are two solutions; $t = 0$ is the initial time, so the other solution must be the time of flight:

$$t_{\text{flight}} = 2\,\frac{v_{oy}}{g}$$

Notice that this is twice the time of the maximum height. We can use this time to find the range, or total distance traveled in the horizontal direction, by substituting this value of t into $x = v_x t$.

projectile motion when $\qquad v_o = v_f$

maximum height: $\qquad y_{\max} = \dfrac{\mathbf{v_{oy}}^2}{2g} = \dfrac{v_o(\sin^2\theta)}{2g}$

time of flight: $\qquad y_{\text{flight}} = \dfrac{2\mathbf{v_{oy}}}{g} = \dfrac{2v_o\sin\theta}{g}$

range: $\qquad y_{\max} = \dfrac{2\mathbf{v_{oy}}v_x}{g} = \dfrac{2v_o(\sin\theta)v_o(\cos\theta)}{g} = \dfrac{v_o^2\sin 2\theta}{g}$

AP EXPERT TIP

Only the vertical component of velocity changes (due to gravity). The horizontal component stays the same throughout flight.

Remember, these equations are only valid if the launch point and impact point are at the same height, $y_o = y_f = 0$. Additionally, if we were to test different launch angles with the same initial velocity where $y_o = y_f = 0$, we would find that the maximum range of our trajectory happens at 45°.

CENTRIPETAL ACCELERATION

Consider an object moving in a circle at a constant speed. Because its direction is changing, its velocity vector is changing. The object is said to have uniform circular motion with **centripetal acceleration**.

We can describe the position of this object relative to the center of the circle with the position vector r. As the object moves, the magnitude of r remains constant; the object is the same distance from the center as it was when it started. However, its direction changes (see figure A). Recall that the average velocity is $v = \dfrac{\Delta r}{\Delta t}$. Because t is scalar, we can conclude that v must have the same direction as displacement (see figure B). If we let t be very close to zero, then v is tangential to the circle. Because the speed is not changing, the length or magnitude of v remains constant.

Similarly, $a_c = \dfrac{\Delta v}{\Delta t}$, and therefore a_c must have the same direction as Δv (toward the center of the circle) (see figure C).

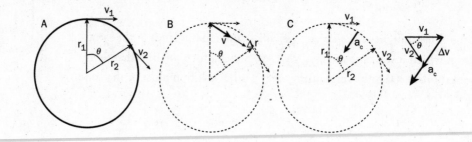

Because of similarity of triangles in the figure, we can solve for the magnitude of acceleration.

$$\frac{\Delta r}{r} = \frac{\Delta v}{v} \qquad\qquad \text{by symmetry of triangles}$$

$$\Rightarrow \frac{\Delta r}{r\Delta t} = \frac{\Delta v}{v\Delta t} \qquad\qquad \text{divide both sides by } \Delta t$$

$$\Rightarrow \frac{v}{r} = \frac{a_c}{v} \qquad\qquad \text{since } v = \frac{\Delta r}{\Delta t} \text{ and } a_c = \frac{\Delta v}{\Delta t}$$

$$\text{then } a_c = \frac{v^2}{r} \qquad\qquad \text{solve for centripetal acceleration}$$

To find the velocity, it is convenient to find the period, or time it takes to complete one revolution. The distance of one revolution is the circumference of the circle, $2\pi r$.

$$v = \frac{\Delta r}{\Delta t} = \frac{2\pi r}{T} = \frac{\text{circumference}}{\text{period}}$$

so $\quad a_c = \frac{v^2}{r} = \left(\frac{2\pi r}{T}\right)^2 \frac{1}{r} = \frac{4\pi^2 r}{T^2}$

Another approach is to find the tangential speed. The angular displacement of the object is defined as arc length per radius, $\Delta\theta = \frac{\Delta s}{r}$. The angular speed is defined as $\omega_{avg} = \frac{\Delta\theta}{\Delta t}$. If we let the time interval become infinitely close to 0, the velocity becomes tangential, or perpendicular, to r.

$$\frac{\Delta\theta}{\Delta t} = \frac{\Delta s}{r\Delta t} \qquad\qquad \text{let } \Delta t \rightarrow 0$$

$$\omega_{avg} = \frac{1}{r} v_t \Rightarrow r\omega = v_t$$

This equation is only true if the angle is measured in radians.

We can use this speed in the equation for the magnitude of centripetal acceleration to come up with our third form of a_c.

$$a_c = \frac{v^2}{r} = \frac{4\pi^2 r}{T^2} = r\omega^2, \text{ where } a_c \text{ is always directed toward the center of the circle.}$$

This will be applied to Newton's second law.

NEWTON'S LAWS

FORCE

Dynamics is the study of forces and their effects on the motion of objects. The cause of any change in an object's velocity is a **force**; the push or pull exerted on an object causes acceleration. There are three key examples of force. The first is the force that causes a stationary object to move; an example of this is throwing a baseball. The second type of force causes a moving object to stop; this is demonstrated when a person catches a ball. The third example of force is not a change in the magnitude of velocity but is rather a change in the direction of velocity; this can be seen when a batter hits a pitch.

The SI unit for force is the newton (N), the force it takes to accelerate 1 kg of mass at a rate of 1 m/s in one second.

$$N = kg \times m/s^2$$

Forces can be **contact forces** in which the force is caused by actual contact, or they can be **field forces** in which the force is caused without visible contact. Examples of each type of force can be seen in the table below.

Common Force	Type	Definition	Direction
friction	contact	acts to oppose sliding motion between surfaces	parallel to surface, opposite direction of motion
normal	contact	exerted by surface of object	perpendicular to and away from surface
spring	contact	restoring force from push or pull of a spring	opposite the displacement
tension	contact	pull exerted by a taut cable attached to object	parallel to cable, away from object
weight	field	due to gravitational attraction exerted by Earth on object	toward large object, usually Earth
electrostatic	field	strength increases with proximity to source	toward source, radial

AP EXPERT TIP

It's always a good idea to draw a quick free-body diagram before trying to solve any force problems.

Forces are vectors and so must be combined using vector addition. This is more easily visualized with the use of free-body diagrams. A **free-body diagram** shows all forces acting on a single object with arrows; the length of the arrows represents magnitude, and the arrows point in the direction of the vectors. The arrows are drawn radially from the center of the object. A complete, clear diagram will let you easily convert the vectors into components along a convenient axis.

The sum of all vector forces acting on an object is called the **net external force**. Consider a game of tug-of-war. There are three people on the left, each pulling with a force of 50 N, and two people on the right who are each pulling with a force of 65 N. Let left be negative and right be positive. We can draw a free-body diagram to help us visualize this situation.

$$\longleftarrow \quad \mathbf{F}_{left} = -150\text{N} \quad \bullet \quad \mathbf{F}_{right} = 130\text{N} \quad \longrightarrow$$

The net external force on the rope is thus

$$F_{net} = -50\text{N} + (-50\text{N}) + (-50\text{N}) + 65\text{N} + 65\text{N} = -20\text{N} = 20\text{N to the left}$$

THE NORMAL FORCE AND FRICTION

Although it may often be neglected to simplify basic physics problems, friction is a very important force because it is present in nearly every physical situation. **Friction** is the opposing force between two surfaces that are sliding or rolling parallel to one another. It opposes the motion of the object. The resistance of any fluid, such as air or water, is a form of friction. The magnitude of friction depends upon the surfaces involved. Conveniently, the force of friction is proportional to the normal force between the surfaces in contact.

The **normal force** is the force that is exerted by any object in contact with another object. It is what keeps us from falling through our chairs. If a 100 lb person were to sit on a certain chair, she would exert a force of 100 lb on that chair. The chair would react by exerting a force of 100 lb on the person sitting in it. If a 150 lb person were to sit on a certain chair, he would exert a force of 150 lbs on that chair, and it would react by exerting a force of 150 lb on the person sitting in it (see the discussion of Newton's third law later in this chapter for more on reaction forces).

The proportionality of frictional force to normal force is found by defining a dimensionless constant, the coefficient of friction, μ. There are two special cases: static friction and kinetic friction.

Static friction is the case in which there is no relative motion between the two surfaces, as in the case of a book sitting on an incline.

As the normal force increases, so too does the maximum force of friction.

$$F_s \leq \mu_s F_N$$

If we were to consider a case in which a force $F > \mu_s F_N$ acts on the object, the situation would no longer be static. The point just before the object begins to move is the maximum friction.

$$F_{s,max} = \mu_s F_N$$

Kinetic friction occurs when there is relative motion between the two surfaces, as in a book sliding across a table.

$$F_k = \mu_k F_N$$

Because it is easier for an object to continue to move than for it to begin to move,

$$\mu_s > \mu_k$$

AP EXPERT TIP

Have you ever pushed a couch across a carpet? It's always harder to start the motion than to continue it, because static friction is stronger than kinetic.

NEWTON'S FIRST LAW: STATIC EQUILIBRIUM

Newton's First Law: "An object at rest will remain at rest, and an object in motion continues in motion with a constant velocity unless it experiences a net external force."

Newton's first law is often called the inertial law. **Inertia** is the tendency of an object to resist a change in motion. **Mass** is the measurement of inertia. If an object has a large mass, it is more difficult to change its velocity, whether it is at rest or is already in motion.

Mathematically, if the net external force $F_{net} = 0$, then $a = 0$ and $v = $ constant.

This scenario may seem to contradict common knowledge. If someone kicks a ball, it will eventually stop moving. However, this happens because the force of friction is acting on the ball, so the net force is not zero.

NEWTON'S SECOND LAW: DYNAMICS OF A SINGLE PARTICLE

Newton's Second Law: "The magnitude of the acceleration of an object is directly proportional to the resultant force acting on it and inversely proportional to its mass. The direction of the resultant acceleration is in the direction of the resultant force."

In other words, $F = ma$. If a net force acts on an object, it will cause an acceleration of the object. This relationship can be shown experimentally by applying identical forces to differing masses. One will find that $\dfrac{m_1}{m_2} = \dfrac{a_2}{a_1}$. Therefore, mass is simply a constant of proportionality.

There is often more than one force acting on an object. We employ the **principle of superposition** to find the net force. Two or more forces acting on an object accelerate it as if the object were acted upon by a single force equal to the vector sum of the individual forces.

$$\sum F = F_{net} = ma = F_1 + F_2 + F_3 + \dots$$

Because this is a vector equation, we must remember to resolve it into scalar components.

$$F = ma: F_x = ma_x, F_y = ma_y$$

It is useful to note how this may look in some common situations. For example, **weight** is defined as the force of gravity on an object. Therefore, $w = mg$.

Also recall that centripetal acceleration is $a = \dfrac{v^2}{r}$. For any object to be moving in a circular motion at constant speed, it must have some net force changing its direction. This net force is known as **centripetal force**. A centripetal force is a *net force*—it is not an individual force that shows up on a free-body diagram. It is always the resultant of forces such as tension, friction, or gravity.

The net force that maintains circular motion is equal to mass times centripetal acceleration. Recall the various forms of centripetal acceleration. If we substitute any of these forms into the equation for net force, we will get a valid equation for centripetal force.

$$a_c = \frac{v^2}{r} = \frac{4\pi^2 r}{T^2} = r\omega^2, \text{ where } a_c \text{ is always directed toward the center of the circle.}$$

$$F_c = ma_c = \frac{mv^2}{r} = \frac{m2\pi v}{T} = \frac{m4\pi^2 r}{T^2} = mr\omega^2$$

NEWTON'S THIRD LAW: SYSTEMS OF TWO OR MORE BODIES

Newton's Third Law: "If two objects interact, the magnitude of the force exerted on object A by object B is equal to the magnitude of the force simultaneously exerted on object B by object A. These forces are opposite in direction."

Forces always exist in pairs that are called **action-reaction pairs**. For every action force, there is a simultaneous equal and opposite reaction force. Mathematically, if A exerts a force F_{AB} on B, and B exerts a force F_{BA} on A, then $F_{AB} = -F_{BA}$.

It is important to notice that in this action-reaction pair, each force is acting on a different object. Therefore, the net force is not necessarily zero, and acceleration can occur. Consider a hammer driving a nail. When the hammer hits the head of the nail, it exerts a force, $F_{h \text{ on } n}$, on the nail. The nail also exerts a force on the hammer—$F_{n \text{ on } h}$—of equal magnitude and opposite direction. $F_{h \text{ on } n}$ causes the nail to be driven into the wood, while $F_{n \text{ on } h}$ causes the hammer to bounce back.

Field forces are also found in pairs. Consider the force of gravity. We know that the Earth exerts a force on an object causing it to accelerate toward the ground. But the object also exerts a force on the Earth. This means that the Earth must be accelerating toward the object. We do not see this acceleration because it is so small.

$$\mathbf{F_g} = m_{obj}\mathbf{a_{obj}} \qquad \mathbf{F}_{object \text{ on } earth} = m_e\mathbf{a_e}$$

$$\mathbf{F_g} = -\mathbf{F_{obj}} \Rightarrow m_{obj}\mathbf{a_{obj}} = m_e\mathbf{a_e}$$

$$\text{sm} \times \text{lg} = \text{lg} \times \text{sm}$$

The Earth's mass is much larger than the mass of the object. Therefore, when we set the magnitudes equal, we can see that the acceleration of the Earth must be very tiny compared to the acceleration of the object.

REVIEW QUESTIONS

1. A truck is stopped at a stoplight. When the light turns green, the truck accelerates at 2.5 m/s². At the same instant, a car drives past the truck going 15 m/s. How long does it take for the truck to catch up with the car?

   (A) 15 s
   (B) 12 s
   (C) 1.2 s
   (D) 1.5 s
   (E) 24 s

2. A weather balloon is released and rises at a constant speed of 15 m/s relative to the ground. The wind blows eastward at 6.5 m/s. What is the velocity of the balloon?

   (A) 1.6 m/s, 66° above the horizontal axis
   (B) 160 m/s, 24° above the horizontal axis
   (C) 16 m/s, 66° east of the vertical axis
   (D) 16 m/s, 24° east of the vertical axis
   (E) 16 m/s, 24° west of the vertical axis

3. During a softball game, a batter hits a pop fly. If the ball remains in the air for 6 seconds, how high did it rise?

   (A) 44 m
   (B) 36 m
   (C) 22 m
   (D) 50 m
   (E) 74 m

4. The wind is blowing with a force of 250 N southeast. If a 150 kg person is trying to jog with an acceleration of 2.00 m/s² north, what force must the jogger exert? (You will need a calculator for this problem.)

   (A) 582 N, 45 degrees east of north
   (B) 123 N, 45 degrees west of north
   (C) 550 N, due north
   (D) 508 N, 20 degrees west of north
   (E) 300 N, due north

5. A lead block with a mass of 2.0 kg is placed on a horizontal cellar door. The hinged door is slowly lifted until the block begins to slide, which happens when the door is lifted 30 degrees from the ground. What is the coefficient of static friction between the lead block and the cellar door?

   (A) 0.58
   (B) 0.27
   (C) 0.17
   (D) 0.65
   (E) 0.29

6. A boy holds a metal sign that weighs 100 N against a wall by pushing on it horizontally. If the coefficient of static friction between the sign and his hands is 0.6, and the coefficient of static friction between the sign and the wall is also 0.6, with what force must the boy push to keep the sign in place?

   (A) 75 N
   (B) 92 N
   (C) 83 N
   (D) 167 N
   (E) 42 N

7. A child is being pulled in a wagon at an angle of 45 degrees from the horizontal; the child weighs 100 N, and the wagon weighs 50 N. Assume the coefficient of kinetic friction is 0.20. What is the approximate force of friction if the wagon is to travel at a constant speed?

   (A) 25 N

   (B) 50 N

   (C) 75 N

   (D) 100 N

   (E) 125 N

8. A girl twirls a lasso above her head in a constant, counterclockwise, horizontal circular motion, describing a circle with a diameter of 1.50 meters. The loop experiences a centripetal acceleration of 1.50 m/s². If the girl releases the lasso when it is directly in front of her, at what velocity will the rope project?

   (A) 1.13 m/s, forward

   (B) 2.25 m/s, forward

   (C) 1.50 m/s, left

   (D) 1.06 m/s, left

   (E) 2.13 m/s, left

9. A box weighing 300 N is pushed across a floor by a force of 500 N coming downward at an angle of 35 degrees with the floor. If the coefficient of kinetic friction between the box and the floor is 0.57, how long does it take the box to move 4.00 meters if it starts from rest?

   (A) 1.8 s

   (B) 2.5 s

   (C) 7.9 s

   (D) 1.0 s

   (E) 1.1 s

Use the following figure to answer question 10.

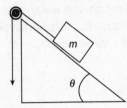

10. In the figure above, what minimum force is needed on the rope to keep the object in static equilibrium? Assume friction is present.

   (A) $\dfrac{(\mu_s \cos\theta + \sin\theta)}{mg}$

   (B) $mg(\cos\theta - \mu_s \sin\theta)$

   (C) $mg(\sin\theta - \mu_s \cos\theta)$

   (D) $mg(\mu_s \cos\theta + \sin\theta)$

   (E) $\dfrac{mg}{(\mu_s \cos\theta + \sin\theta)}$

11. A block is given an initial velocity of v in the horizontal direction. After a displacement d, it comes to rest due to friction. What must the coefficient of kinetic friction be between the block and the plane on which it slides?

   A $\dfrac{v}{2dg}$

   B $\dfrac{v^2}{2gd}$

   C $2vgd$

   D $\dfrac{d^2}{2gv}$

   E $\dfrac{1}{2}v^2gd$

12. A small 50 kg object is tied to the end of a cord and whirled in a horizontal circle of radius 2.0 meters. If it makes one revolution in 1.5 seconds, what force must be exerted to maintain the movement?

(A) 1,750 N

(B) 1,000 N

(C) 850 N

(D) 2,650 N

(E) 550 N

13. An object moves clockwise around a circular path at a constant speed. Which figure below accurately portrays its acceleration and velocity?

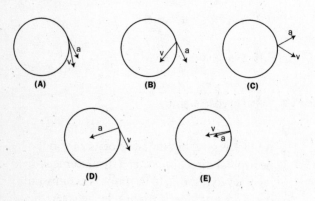

14. A projectile with a mass of 100 kg is fired at an angle of 45° above horizontal and lands at the same altitude on the other side of a canyon 15.0 km from its launch point. What must its initial speed have been? Neglect air resistance.

(A) 456 m/s

(B) 12.1 m/s

(C) 384 m/s

(D) 14.4 m/s

(E) 208 m/s

15. If an object is moving west, in what direction is its acceleration?

(A) West

(B) East

(C) South

(D) North

(E) More information is needed to determine the direction of acceleration.

FREE-RESPONSE QUESTION

The following graph shows the position of a roller coaster as it changes with time.

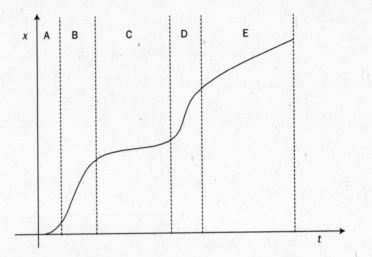

(a)   At which segment(s) is the velocity the greatest?

(b)   Where might the peaks of the coaster's hills be located?

(c)   Graph the velocity versus time.

(d)   What does the slope of the velocity versus time graph tell you?

(e)   What is happening in section E?

ANSWERS AND EXPLANATIONS

1. B

You are given a value for the acceleration of the truck; its initial velocity must be zero because it is stopped. The car has a constant velocity, which is given, and therefore its acceleration must be zero.

$$a_{\text{truck}} = 2.5 \text{ m/s}^2 \quad v_{\text{o,truck}} = 0 \quad\quad a_{\text{car}} = 0 \quad\quad v_{\text{car}} = 15 \text{ m/s}$$

Also, let the initial position be zero along the x-axis. We want to find the time when the distance from this origin is the same for both the car and the truck. Recall the kinematic equation $x = x_o + v_o t + \frac{1}{2} at^2$. You know all of the variables except x and t for both vehicles. Set $x_{\text{truck}} = x_{\text{car}}$.

$$x_{\text{car}} = v_{\text{o,car}} t + \frac{1}{2} a_{\text{car}} t^2 = (15 \text{ m/s})t$$

$$= x_{\text{truck}} = v_{\text{o,truck}} t + \frac{1}{2} a_{\text{truck}} t^2 = \frac{1}{2} (2.5 \text{ m/s}^2)t^2$$

$$\Rightarrow 15 \text{ m/s} = \frac{1}{2} (2.5 \text{ m/s}^2)t \Rightarrow t = 12 \text{ s}$$

Notice that if you needed to figure out *where* (i.e., the distance at which) the truck caught up with the car, you could put this value for t into either equation and solve for x.

2. D

This problem is asking you to add vectors to find a resultant. The first step is to create a diagram and assign axes.

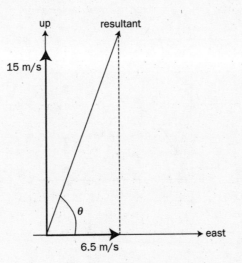

To find the magnitude, use the Pythagorean theorem, $a^2 + b^2 = c^2$, where $a = 15$ m/s, $b = 6.5$ m/s, and $c =$ resultant.

$$a^2 + b^2 = c^2$$
$$15^2 + 6.5^2 = 267.25 = (magnitude)^2$$
$$\Rightarrow magnitude = \sqrt{267.25} = 16.35 \text{ m/s}$$

To find the direction, use trigonometry.

$$\tan\theta = \frac{\text{opposite}}{\text{adjacent}} = \frac{15}{6.5} = 2.3$$
$$\Rightarrow \theta = \tan^{-1}(2.3) = 66°$$

There are only two significant figures, so the magnitude is 16 m/s. Notice that the angle you found is measured from the horizontal. This angle measured from the vertical would be $90° - \theta = 24°$.

3. A

This is an example of projectile motion in which the path of the ball is a parabola that has the same initial and final heights. The question tells you that the total time of flight is 6 seconds. You can also infer that acceleration is due only to gravity because the ball is in free fall. The ball flies in a symmetrical parabola, so it must hit its maximum height halfway through the flight, at a time of 3 seconds.

$$t_0 = 0, \quad t_{\text{flight}} = 6 \text{ seconds}, \quad a = -g, \quad t_{\text{top}} = 3 \text{ seconds}$$

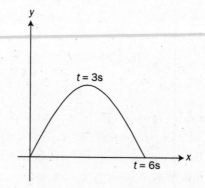

There is not enough information to plug into one of the kinematic equations and simply solve for the maximum height. It would be helpful to know the initial velocity. Recall $y = y_o + v_o t - \frac{1}{2} g t^2$. Let the initial height equal zero, which implies that the final height is also zero.

$$y_o = 0, \ y_f = 0$$

$$y_f = y_o + v_o t_f - \frac{1}{2} g t^2_f = 0 + v_o (6s) - \frac{1}{2}(9.81\,m/s^2)(6s)^2 = 0$$

$$\Rightarrow v_o = 29.43\,m/s^2$$

Now use the same equation to solve for the height at the top, halfway through the flight.

$$y_{top} = y_o + v_o t_{top} - \frac{1}{2} g t_{top}^2$$

$$y_{max} = 0 + (29.43\,m/s)\,3s - \frac{1}{2}(9.81\,m/s^2)(3s)^2 = 44\,m$$

4. D

A free-body diagram illustrates this problem best.

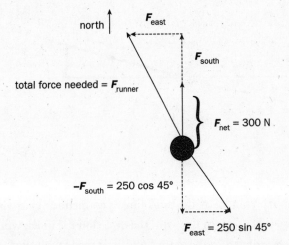

To find the amount of force needed to combat the wind, first resolve the wind into its south and east components. Since the wind is blowing exactly 45 degrees east of south, $F_{south} = F_{east} = 250 \cos 45° = 176.778$ N. To accelerate due north at the desired acceleration, the runner must counteract the wind force. To find the amount of force needed to accelerate a 150 kg person at $a = 2$ m/s² north, invoke Newton's second law.

$\mathbf{F} = m\mathbf{a},$ where $\mathbf{a} = 2\,\text{m/s}^2$ and $m = 150\,\text{kg}$

$\mathbf{F}_{\text{net (N/S)}} = (150\,\text{kg})(2\,\text{m/s}^2) = 300\,\text{N north}$

The total northern component of the force is $300 + 176.778 = 476.778$ N. The total western component of the force is 176.778 N. Therefore, the total force is

$F_{\text{net}} = \sqrt{(300)^2 + (176.778)^2} = 508$ N. The direction $\theta = \arctan\left(\dfrac{176.778}{476.778}\right) = 20°$ west of north.

5. A

Begin by drawing a picture, then create a free-body diagram from your sketch.

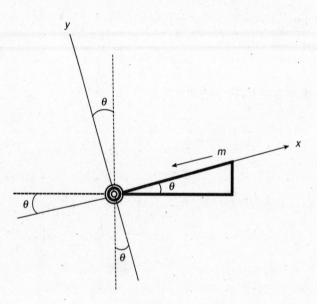

In this picture, the door, represented by the thick line, is an incline plane with an angle θ at the hinge. The axis is arbitrary; here it is in line with the direction the block will slide. By using the geometric rules for opposite angles, you can find the angle made with the force of gravity.

In this free-body diagram, treat the block as a point mass. The axes have been moved, but they are parallel to those in the picture. This allows you to find the components of the force from gravity.

You want to find the coefficient of static friction as the block starts to move. This is defined as $\mu_{max} = \dfrac{F_s}{F_N}$. Therefore, you first must find both the normal force and the force of static friction. The normal force is the force that is perpendicular to the surface. In other words, it opposes the y component of the force of gravity. The force of static friction is the force needed to hold the object in place, opposing the x component of gravity.

$$F_g = -mg, \quad F_{g,y} = -mg\cos\theta, \quad F_{g,x} = -mg\sin\theta$$

$$\text{where } m = 2 \text{ kg}, \quad g = 9.81 \text{ m/s}^2, \quad \theta = 30°$$

$$F_N = -F_{g,y} = mg\cos\theta, \quad F_s = -F_{g,x} = mg\sin\theta$$

$$\Rightarrow \mu_s = \frac{F_s}{F_N} = \frac{mg\sin\theta}{mg\cos\theta} = \tan\theta = 0.577$$

6. C

When drawing the free-body diagram for this problem, recall Newton's third law: For every action there is an equal and opposite reaction. This means that as the boy pushes on the sign, the sign is responding with the same magnitude of force. More importantly, these two forces do not cancel out because they are on different objects. Therefore, the sign must also be pushing on the wall with the same force as the boy, and the wall will be pushing back toward the sign. Because the coefficients of friction are the same, the frictional forces needed to hold the sign up are split equally between the boy and the wall. Only the y component will actually counteract the weight of the sign.

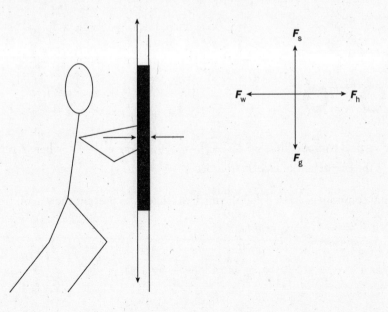

The total force from the wall and the boy's hands must be equal to the normal force. You can use the definition of frictional force, which must be equal and opposite to the weight of the sign, to find the normal force.

$$F_N = \frac{F_s}{\mu_s}, \quad \text{where } F_s = F_g = 100 \text{ N} \quad \text{and } \mu_s = 0.6$$

$$F_N = \frac{100 \text{ N}}{0.6} = 166.667 \text{ N}$$

Because the force from the boy's hands must be equal to the force from the wall, and the total of the two is the normal force, the boy must push with exactly half of the normal force.

$$F_N = F_h + F_w, \quad F_h = F_w \quad \Rightarrow F_h = \frac{1}{2} F_N = 83.3 \text{ N}$$

7.   A

Because the child is being pulled at a constant velocity, there is no acceleration. This means that you are dealing with an equilibrium situation.

In order for the wagon not to be lifted off the ground, the vertical component of the force plus the normal force F_N exerted by the ground on the wagon must equal the downward force exerted by the child and wagon (150 N).

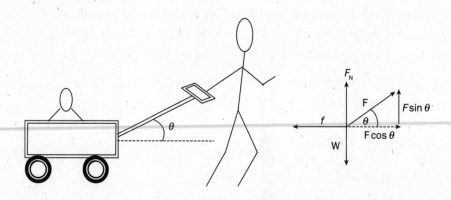

Solve for the vertical component: $F\sin 45° + F_N = 150$, where $F_N = \frac{f}{\mu}$, where f is the frictional force and μ is the coefficient of kinetic friction.

The horizontal component must equal the force of friction to prevent the wagon from accelerating:

$f = F\cos 45°$ or $F = \frac{f}{\cos 45°}$. Combine:

$$\frac{f}{\cos 45°} \sin 45° + \frac{f}{\mu} = 150 \Rightarrow f\left(\tan 45° + \frac{1}{\mu} \right) = 150$$

Since $\tan 45° = 1$ and $\dfrac{1}{\mu} = 5$, you get $6f = 150$ or $f = 25$ N.

8. D

Visualize this problem, then draw a free-body diagram for the loop. The path of the lasso creates a circle with a diameter of 1.50 m, so its radius must be half this value. Because this is a constant motion, the tangential velocity must be constant. Recall Newton's first law: if there are no forces acting upon an object, it will continue to move at a constant velocity. When the girl releases the rope, there will not be any more centripetal acceleration, so the velocity will be the same magnitude but in a straight line tangential to the circle. This limits you to choices (C) and (D), because the rope will head left if released in front.

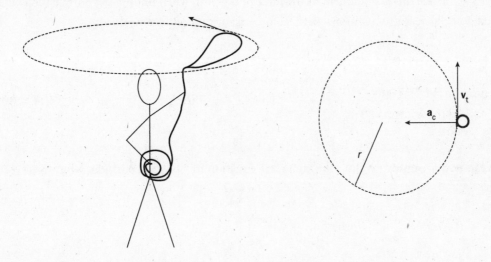

Now you need only to find the magnitude of the velocity by using the equation for centripetal acceleration.

$$a_c = \frac{v_t^2}{r} \Rightarrow v_t = \sqrt{a_c r}, \quad \text{where } a_c = 1.50 \ \text{m/s}^2, r = 0.750 \ \text{m}$$

$$v_t = \sqrt{1.50 \ \text{m/s}^2 \times 0.750 \ \text{m}} \quad = 1.06 \ \text{m/s}$$

9. A

To find the time it takes to move a certain distance, you must first find the acceleration. You can do this by using Newton's second law, $\Sigma F = ma$, along the x-axis.

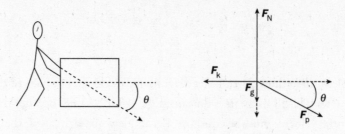

$$\Sigma F_x = m\mathbf{a} = F_{p,x} - F_k = F_p \cos \theta - F_k$$

To use the definition for kinetic friction, $F_k = \mu_k F_N$, we must first find F_N.

Be sure to include the y-component of the force of the push when finding the normal force; the normal force increases as additional vertical force against the surface is added.

$$F_N = mg + F_p \sin \theta, \qquad mg = 300 \text{ N} \Rightarrow m = \frac{300}{9.81} \text{kg}$$

Now, $\mathbf{a} = \dfrac{F_p \cos \theta - \mu_k (mg + F_p \sin \theta)}{m}$, where $F_p = 500$ N, $\theta = 35°$, $\mu_k = 0.57$, $m = 30.6$g

$\mathbf{a} = 2.454 \text{m/s}^2$

This can now be substituted into the kinematic equation for constant position, which is solved for time.

$$x = x_o + v_o t + \frac{1}{2}at^2, \quad \text{where } x_o = 0, v_o = 0$$

$$\Rightarrow t^2 = \frac{2x}{a}, t = \sqrt{\frac{2x}{a}}, \quad \text{where } x = 4.00 \text{ m}, a = 2.454 \text{ m/s}^2$$

So $t = 1.80$ s.

10. C

The rope is being pulled down, but the force will be directed up the slope because of the pulley. Therefore, you merely need to find the force along the slope needed to hold the block in equilibrium. A free-body diagram illustrates that this force must counteract the force of gravity on the object but it has the help of static friction.

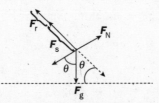

$$\Sigma F_{\text{slope}} = 0 = F_{\text{r}} + F_{\text{s}} - F_{\text{g}} \sin \theta, \quad \text{where} \quad F_{\text{s}} = \mu_{\text{s}} F_{\text{g}} \cos \theta \text{ and } F_{\text{g}} = mg$$

$$\Rightarrow F_{\text{r}} = mg \sin \theta - \mu_{\text{s}} mg \cos \theta = mg (\sin \theta - \mu_{\text{s}} \cos \theta)$$

There will be several problems on the exam where you need only to derive a formula. When there are numbers, they are often simple calculations. The key is to know the concepts at work in the problem.

11. B

The information given is $s = d$ and $v_{\text{o}} = v$. You are looking for the coefficient of kinetic friction, $F_{\text{k}} = \mu_{\text{k}} F_{\text{N}}$, so you will need to find the normal force. Because the only force working on the block is gravity, $F_{\text{N}} = mg$. The only force working in the horizontal direction is friction, so $F_{\text{k}} = \mu_{\text{k}} F_{\text{N}} = ma = -\mu_{\text{k}} mg \Rightarrow a = -\mu_{\text{k}} g$. Now you can use the kinematic equation for constant acceleration.

$$v^2 = v_{\text{o}}^2 + 2as, \quad \text{at rest}, \quad v = 0$$

$$\Rightarrow 0 = v^2 - 2\mu_{\text{k}} gd \Rightarrow \mu_{\text{k}} = \frac{v^2}{2gd}$$

12. A

This is a simple centripetal force problem. You need only to list the given information and decide which formula to use.

Given $r = 2$ m and $T = 1.5$ s, you can use

$$F_{\text{c}} = ma_{\text{c}} = \frac{m 4\pi^2 r}{T^2} = 1{,}754.7 \text{ N}$$

13. D

Centripetal acceleration always points toward the center of the circle, and velocity is always tangential to the circle. You will find many questions that use simple diagrams. There is essentially nothing to solve; you are just demonstrating your grasp of the concepts. Be careful to look at each figure and to pay attention to the labels.

14. C

This problem is another projectile motion problem in which you can use symmetry because the initial and final heights are the same. It can be derived that the range of a projectile

$$x_{\text{max}} = \frac{2 v_{\text{o}}^2 \sin \theta \cos \theta}{g} = \frac{v_{\text{o}}^2 \sin 2\theta}{g}$$

You are given the launch angle and range, so the only unknown variable is the initial velocity. Be sure to convert all units so that they are the same. This is easiest when done before you start plugging values into equations. You can see that the answer choices are in m/s, so we know we want our range in meters.

$$x_{max} = 15 \text{ km} = 15,000 \text{ m}, \qquad \theta = 45°$$

Solve for initial velocity.

$$v_0^2 = \frac{gx}{\sin 2\theta} = \frac{9.81 \text{m/s}^2 \times 15,000 \text{ m}}{\sin 90°} = 147,150 \text{ m}^2/\text{s}^2$$

$$\Rightarrow v_0 = \sqrt{147,150 \text{ m}^2/\text{s}^2} = 383.86 \text{ m/s}$$

15. E

Be prepared for questions that require you to think critically about the answer choices. For this question, it is easy to choose west without thinking. However, think of different situations. The object could be slowing down, in which case its acceleration is east. The object could also be turning, in which case the acceleration would be perpendicular to the direction of motion. The object could also be traveling west at a constant velocity, in which case the acceleration would be zero and would have no direction.

FREE-RESPONSE QUESTION

(a) Sections B and D have the greatest velocity. You know this because $v = \dfrac{\Delta x}{\Delta t}$, which is the slope of the graph. The slope is the steepest in these two areas.

(b) There may be a peak of a roller coaster hill in section C. Here, the slope levels out and shows that the velocity is very near zero.

(c)

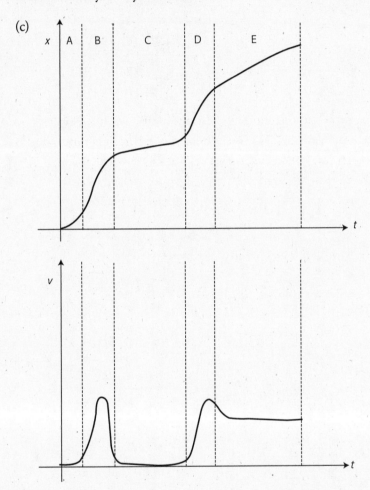

(d) The slope of this graph describes acceleration, $a = \dfrac{\Delta v}{\Delta t}$. By looking at the steepness of the slope, you can see that the car is accelerating in the first half of sections B and D and decelerating at the end of B and D. This reinforces the possibility of a peak in section C.

(e) The roller coaster must level out at this point because there is a constant velocity. If you account for friction, the angle of the hill must be such that the component of the roller coaster's weight along the track must be equal to the kinetic frictional force.

CHAPTER 4:   WORK, ENERGY, POWER, AND SYSTEMS OF PARTICLES

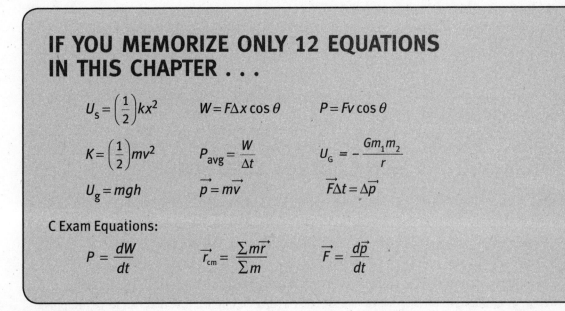

IF YOU MEMORIZE ONLY 12 EQUATIONS IN THIS CHAPTER . . .

$$U_s = \left(\frac{1}{2}\right)kx^2 \qquad W = F\Delta x \cos\theta \qquad P = Fv\cos\theta$$

$$K = \left(\frac{1}{2}\right)mv^2 \qquad P_{avg} = \frac{W}{\Delta t} \qquad U_G = -\frac{Gm_1 m_2}{r}$$

$$U_g = mgh \qquad \vec{p} = m\vec{v} \qquad \vec{F}\Delta t = \Delta\vec{p}$$

C Exam Equations:

$$P = \frac{dW}{dt} \qquad \vec{r}_{cm} = \frac{\sum m\vec{r}}{\sum m} \qquad \vec{F} = \frac{d\vec{p}}{dt}$$

INTRODUCTION

The concepts in this chapter all relate in some way to two important conservation laws: the law of conservation of energy and the law of conservation of momentum. These conservation laws provide very powerful methods for solving problems. Physicists' faith in the inviolable nature of these laws was the basis of Wolfgang Pauli's 1931 prediction that subatomic particles called neutrinos existed. Pauli had no direct evidence for neutrinos, but in some nuclear reactions, it was observed that energy and momentum did not appear to be conserved even though mass and charge were conserved. This led Pauli to predict that there must be particles with no mass or charge that carried off the extra momentum and energy. Despite a complete lack of evidence for nearly 30 years (until neutrinos were discovered in 1959), physicists preferred to believe that elusive, massless particles with a high amount of momentum and energy existed rather than to believe that these laws of physics could be violated.

The topics surrounding these two laws are quite important to physics; out of 70 multiple-choice questions, you can expect to see 6–7 questions on the B exam and 9–10 questions on the C exam relating to them. The exam questions covering these topics will be divided as follows:

ENERGY, WORK AND POWER

B exam	3 to 4 questions
C exam	5 questions

MOMENTUM AND ITS CONSERVATION

B exam	2 to 3 questions
C exam	4 to 5 questions

It cannot be stressed enough: You need to be familiar with the topics in this chapter in order to receive a high score on either exam.

WORK AND THE WORK ENERGY THEOREM

Work is one of those words that has a very specific meaning in physics, a meaning that differs from the way we understand the term in everyday life. For example, if you try to hold a heavy weight completely motionless above your head for an extended time, you will soon become very tired. You may start to sweat, and you will feel like you are working very hard. However, according to the way work is defined in physics, you are doing absolutely no work when you hold this weight motionless. In physics, **work** requires that a force be applied *and* that this force be applied over some displacement. Holding even a very heavy weight motionless is not work under this definition, because the displacement is zero. It doesn't matter how large the force is or how tired you are: if the displacement is zero, the work done will also be zero. On the other hand, pushing a flea a very small distance involves both a force and a displacement. Both are very small—but not zero—which means that, in this case, work is done. From the perspective of physics, more work is done when you flick a flea off your arm than when you hold a 100-pound weight motionless over your head for an hour.

To calculate work, you must multiply the **force** applied, F, by the **displacement**, s, over which the force is applied. The force and displacement must be parallel. If they are not, use the component of the force that is parallel to the displacement or the component of the displacement that is parallel to the force. To find the parallel component of either one, multiply by the cosine of the angle, θ, between the force and displacement vectors. Thus, work is given by

$$W = Fs \cos \theta$$

The unit in which work and energy are measured is the **joule**, in which $1\ J = 1\ N \cdot m$. Be very careful about the angle you choose as θ. It should be the angle between the force and displacement vectors. As in the figures below, draw the extended vectors so that the arrows from both vectors point away from the vertex of the angle and then use the angle between these vectors as θ. If you are not sure which angle to use, your drawing will tell you which component of the force is parallel to the displacement.

> **AP EXPERT TIP**
>
> Notice that since $\cos 90° = 0$, there is no work done if the force and displacement are perpendicular. In particular, **centripetal forces do no work** on an object.

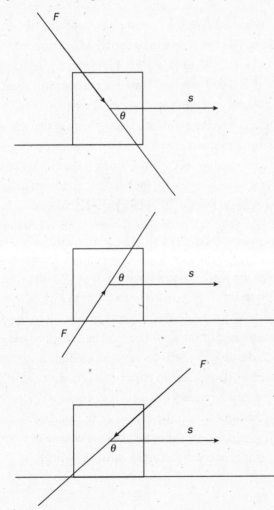

Notice that both force and displacement are vectors. Work, however, is a **scalar**. There are two ways to multiply vectors. The first way, called the **cross product**, gives a vector as an answer. Work is an example of the other way of multiplying vectors, the **dot product**. The dot product of two vectors gives a scalar as the answer. The dot product of two vectors is defined as the product of the magnitudes of each of the vectors and the cosine of the angle between the vectors. Work is a dot product of the force and displacement vectors. It is given by

$$W = \mathbf{F} \cdot \mathbf{s} = Fs \cos \theta$$

Because work is a scalar, the direction does not matter, but the sign is very important. Think again about holding a heavy weight over your head and letting it down gently when you get too tired to continue holding it up. You are applying an upward force, but the displacement is downward. The angle θ is therefore 180°, and $\cos \theta = -1$. In this case, the work you do on the weight is negative. It is easy to make sign mistakes in problems involving work and energy, but it is important to get the signs right.

Questions of this sort can be both tricky and subtle. For example, what is the net work done on a 20 N object if it is lifted a distance of 10 m at a constant velocity? Many students' first impulse will be to multiply 20 N by 10 m and give 200 J as an answer. However, this response is incorrect. Whatever lifted the object did indeed do 200 J of work. However the question asked for the **net** work, which is the total work done by all the forces acting on the system (i.e., the work done by the net force). The fact that the object is lifted at a constant velocity means that the acceleration is zero; hence, the net force is zero. If the net force is zero, the net work must also be zero. When the object is being lifted, there are two forces acting on the object: the upward lifting force and Earth's downward gravitational force. If the acceleration is zero, then these forces must balance. The lifting force is 20 N upwards, and the gravitational force is 20 N downwards. Using $W = Fs \cos \theta$, the work done by the upward lifting force is $W = (20 \text{ N})(10 \text{ m})(\cos 0°) = 200 \text{ J}$. The work done by the Earth is, however, $W = (20 \text{ N})(10 \text{ m})(\cos 180°) = -200 \text{ J}$. The net work is therefore zero.

This counterintuitive result relates to the **work energy theorem**. If work is done on an object, some, but not all, of the work results in the displacement or position change of the object. Some of the work also results in a change in the object's motion. The energy relating to the object's position is called potential energy, which is discussed in the next section. The energy relating to the object's motion is called kinetic energy, K. If the object's speed changes, its kinetic energy also changes. Because energy is conserved, the total work done on an object must equal the object's total energy change during the process. Because potential energy is the energy of position, any change in potential energy is taken care of by the force of opposing movement, usually either gravity or friction. Therefore the net work, W_{net}, done on an object equals the change in kinetic energy, ΔK, of the object. This statement is the work energy theorem. In equation form, it is

$$W_{net} = \Delta K = K - K_0$$

Kinetic energy is the energy of the object's motion. It includes both the mass, m, and the speed, v, of the object. It is given by the equation

$$K = \frac{1}{2}mv^2$$

Notice that this equation gives the same units, joules, for both kinetic energy and work. Kinetic energy, or the energy an object has by virtue of its motion, is a very important concept that comes up in other places besides the work energy theorem.

Be careful when you apply the work energy theorem to solve problems. The work in this theorem is the net, or total, work done on an object. If you are trying to find the net work, then this theorem is the tool you need. If, however, you are trying to find the work done by just one of the forces acting on an object (i.e., not the net work), the work energy theorem will give the wrong answer. The work energy theorem is a powerful but subtle principle. Make sure that you understand its subtleties, or you are likely to use it incorrectly.

CONSERVATIVE FORCES AND POTENTIAL ENERGY

As mentioned in the previous section, potential energy is the energy that an object has by virtue of its position. Potential and kinetic energy are intertwined; because energy is conserved, it can change forms between kinetic and potential energy as the object changes its position and speed. Potential energy is also intimately connected with **conservative forces**. All conservative forces are associated with a form of potential energy. **Nonconservative forces** are not.

To understand conservative forces, it helps to begin with an example of a force that is not conservative: friction. When work is done against friction, some of the energy is converted to heat energy. This energy is no longer available for doing useful work. For example, when you slide a book across a table, you do some work against friction and convert this energy to heat. If you then return the book to its original position, you again do work against friction and convert still more energy to heat. Friction is a nonconservative force.

Gravity, on the other hand, is a conservative force. If you lift a book from the floor to the table, you must do some work to overcome gravity. In this case, however, none of the energy is converted to heat. It is still available for useful work in the form of potential energy. If after lifting the book you let go, it will fall to the floor, increasing its speed as it falls. The work you did against gravity is not converted to heat and is recoverable as useful work, as is true for all conservative forces. The work you did against gravity gave the book gravitational potential energy, which was converted to kinetic energy when the book returned to its original position.

In general, work done against nonconservative forces is converted to heat or some other nonrecoverable type of energy. Work done against conservative forces, however, becomes potential energy. It is not lost.

Potential energy is fully recoverable and can easily be converted to useful work or other forms of energy.

All conservative forces have an associated potential energy. The potential energy of an object at a certain position is the work done against the conservative force to get the object into its current position. For example, the gravitational potential energy of the book is the work done against gravity while lifting the book. Remember that potential energy is a relative quantity rather than an absolute quantity. An object doesn't have an absolute amount of potential energy at a given position. Rather, there is a difference between the potential energy in that position and in a reference position, where the potential energy is taken to be zero. The choice of this zero-point reference position for gravitational potential energy near the surface of the Earth is completely arbitrary. But there are conventions about this choice for other types of potential energy or for gravitational potential energy when not near the surface of the Earth. When working an appropriate problem, choose the zero point at any position that is convenient for the particular problem. In the previous example about lifting a book, for example, if the book is lifted above the floor, one would probably choose the floor level as the zero point. If the book is lifted above a table, choose the table level as the zero point for the potential energy. Because the zero point is arbitrary, you should not spend time worrying about where the zero point will be. Instead, focus on defining your chosen level clearly and using that level consistently.

Near Earth's surface, the gravitational force on an object is simply its weight, $w = mg$. If the object is lifted some height, h, then the work done against gravity is the weight times the height. So the gravitational potential energy, U_g, is given by

$$U_g = mgh$$

where m is the object's mass and h is its height above or below the arbitrary reference level where the gravitational potential energy is taken as zero. You should note that this equation is a special case that applies only near the surface of the Earth. For the more general case, the force used to find the work done against gravity is drawn from Newton's law of gravity for the force between two objects. Applying this variable force over a distance gives the general form for the gravitational potential energy:

$$U_G = -\frac{Gm_1m_2}{r}$$

Here, the reference point at which the gravitational potential energy is zero is taken to be at infinity. This potential energy then represents the amount of work required

---

AP EXPERT TIP

In problems near the Earth's surface, it is useful to choose the lowest point in the problem as the zero point of potential energy. That way, potential energy will never be negative.

to bring the masses, m_1 and m_2, from an infinite distance apart to a distance r from each other.

Gravity is not the only conservative force; therefore, gravitational potential energy is not the only type of potential energy. Electrical forces are also conservative, and the electrical potential energy, U_E, between two charges is given by

$$U_E = -\frac{q_1 q_2}{4\pi\varepsilon_0 r}$$

Here, ε_0 is the electrical permittivity of free space, and q_1 and q_2 are the charges separated by a distance r. You should notice that this equation has a mathematical form similar to the equation for gravitational potential energy. That is because both the gravitational and electrical forces have similar inverse-square mathematical forms for the force laws. Electrical potential energy is also related to the familiar concept of voltage. Voltage, more properly called the potential difference, is the potential energy per unit charge.

For an ideal spring, **Hooke's law** is also a conservative force, and thus there is a potential energy associated with an ideal spring. This potential energy, U_s, is given by

$$U_s = \frac{1}{2}kx^2$$

in which k is the spring constant and x is the distance that the spring is either stretched or compressed from its equilibrium position. Unlike an ideal spring, a real spring will experience some frictional loss as it oscillates. However, the loss is usually small enough that it can be ignored over a small number of oscillations.

CONSERVATION OF ENERGY

Energy is *always* conserved. The conservation of energy is a fundamental law of physics that has never been violated, as far as modern scientists know. Energy can change forms, but it can never be created or destroyed. When Einstein published his famous equation $E = mc^2$, he modified our understanding of energy to include the energy equivalent of mass. Mass is simply another form of energy. If we include the energy equivalent of the total mass, the total energy in the universe remains constant.

Because the energy of an isolated system is always conserved, energy conservation provides a very powerful tool for solving problems. Since energy is conserved regardless of the exact process that occurs, we don't need to know anything about exactly what happens to use the conservation of energy to solve a problem. The details don't matter. For example, if you throw a ball from the top of a building, you can predict the speed at which it will hit the ground using projectile motion

AP EXPERT TIP

Throughout the AP Physics exams, if you truly have no idea how to begin a problem, energy conservation should be a top choice as a starting point.

kinematics. Most students find these problems challenging. You can also find the speed much more easily using conservation of energy. Potential energy is converted to kinetic energy. Both are scalars, so no components or directional information is needed, and the problem becomes much easier to solve.

The total kinetic energy plus the total potential energy of a system is called its mechanical energy. In the absence of nonconservative forces, especially friction, the total mechanical energy of a system is conserved. But if energy is *always* conserved, why is it necessary to specify that there is no friction? This specification is necessary because heat energy is not included in the definition of mechanical energy, so mechanical energy is conserved only when there is no friction to convert some of the energy into heat. When friction is present, some of the mechanical energy is converted to heat (i.e., takes a different form). Note that heat energy is simply the random kinetic energy of individual molecules, which move faster if the temperature is higher. Nonetheless, this random kinetic (heat) energy is not included in a system's mechanical energy.

When applied to mechanical energy in the absence of friction, the law of conservation of energy states that the total kinetic energy, K, plus the total potential energy, U, remains constant. Therefore,

$$K + U = \text{constant}$$

or

$$K_i + U_i = K_f + U_f$$

The i and f subscripts refer to the initial and final states of the system. For both states, the total kinetic energy, K, must include the kinetic energy of all the objects in the system. Likewise, the total potential energy, U, must include all the objects in the system and all the sources of potential energy (e.g., gravitational potential energy and/or spring potential energy, etc.). This principle may seem straightforward, but in a complex system containing several objects or multiple forms of potential energy, the application of this principle can become rather involved.

To solve problems using the conservation of mechanical energy, first determine whether the principle applies. For example, there should be no friction at work in the problem; if there is, this is not the right way to solve it. Next draw pictures of the system showing the initial and final stages of the system, as in the following illustration.

AP EXPERT TIP

If there *is* friction in the problem, then the work done by friction will equal the change in mechanical energy.

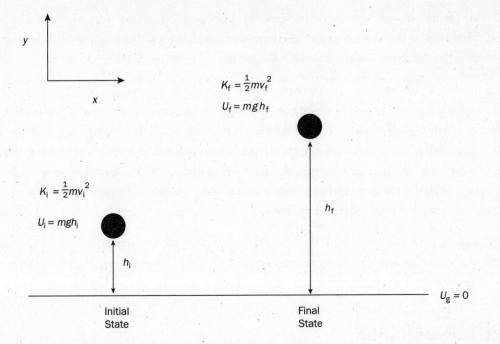

On your drawings, be sure to indicate clearly your zero reference level for gravitational potential energy and the directions of your x and y coordinates. Doing this on your diagram will help ensure that you do not forget the level that you selected and inadvertently change it in the middle of the problem. Now indicate the total kinetic and potential energies for the initial and final stages of the system. Substitute these energies into the conservation of energy equation. Finally, substitute the known numerical values into your equation and solve it for the unknown values.

POWER

Power is our way of measuring the rate at which work is done. Just as velocity—a measurement of the rate at which position changes—is change in position divided by time interval, power is simply work (or change in energy) divided by time interval. Therefore, the average power is given by

$$P_{avg} = \frac{W}{\Delta t}$$

The units of power are joules per second, which are defined as **watts**: 1 J/s = 1 W.

Recalling that work is a force times a distance and that distance divided by time is velocity provides another useful formula for power. Making these substitutions produces

$$P_{avg} = Fv\cos\theta$$

Just as it did in the equation for work, the $\cos\theta$ factor gives the component of the force that is parallel to the velocity.

Watts are the unit in which power is measured, but horsepower is another common unit that measures power. As the name suggests, the horsepower derives from the rate at which a horse can do work. The conversion is 1 horsepower = 746 watts.

Keep the distinction between work or energy and power clear in your mind. Power is the rate at which work is done or the rate at which energy is used. If the work is done rapidly, it is possible to have a relatively high power even if the total work is relatively small. For example, a crane may be able to lift a heavy load 100 m vertically in as little as 10 seconds. However, if the load is instead lifted to the same height in 100 seconds, the crane did as much total work at the slow setting as it did at the fast. It did the same work in a much longer time, though, so its power output at the slow setting was much less than the power output at the fast setting.

Now suppose the crane needs to lift the load 200 m but must go even slower to be safe. The crane at the fast setting has a much greater power output than at the slower setting. However, because the distance is doubled, it must do more total work to lift the load 200 m.

CENTER OF MASS

Most of us have tried balancing a small object such as a pen or a stick on the end of a finger. This game is an attempt to find the object's **center of mass**, because the object will balance only if the center of mass is above the support point. The center of mass concept has many familiar applications, so many people have an intuitive understanding of some of the ideas even before they study the concept formally in a physics class. For example, sport-utility vehicles have a higher incidence of rollover accidents because they are taller than other cars, and their higher center of mass makes them less stable. Football players, wrestlers, and other people expecting to receive heavy blows do not stand tall with their feet together. Instead, they spread their feet apart and bend their knees. The wider base created by this posture makes it harder to push them over, because the wide base makes it difficult to move their centers of mass beyond their support points. Lowering the center of mass by bending one's knees also increases stability.

The concept of center of mass also has many useful scientific applications. For example, we often say that the moon orbits the Earth, but in reality both the Earth and moon orbit the center of mass of the Earth-moon system. For an extended object, we can often simplify calculations by assuming that all the gravitational force acts on one point: the center of gravity. While this is not always true, on the AP exam this point can always be assumed to be at the same place as the center of mass.

For two objects of the same mass, the center of mass of the system will be halfway between the centers of mass of the two objects. If one object is more massive than the other, then the center of mass of the system is closer to the center of mass of the more massive object. For example, if one object's mass is twice that of the other object, the center of mass of the system will be one-third of the way from the center of mass of the more massive object to the center of mass of the less massive object. For more complex systems of point particles, it is not possible to find the center of mass by intuition alone, so there is an equation for computing the center of mass. This equation is simply an average mass weighted by the positions of the masses. The position of the center of mass must be computed for each dimension: x, y, and z. The x position of the center of mass, x_{cm}, is given by

$$x_{cm} = \frac{(\sum x_i \, m_i)}{(\sum m_i)} = \frac{(\sum x_i \, m_i)}{M}$$

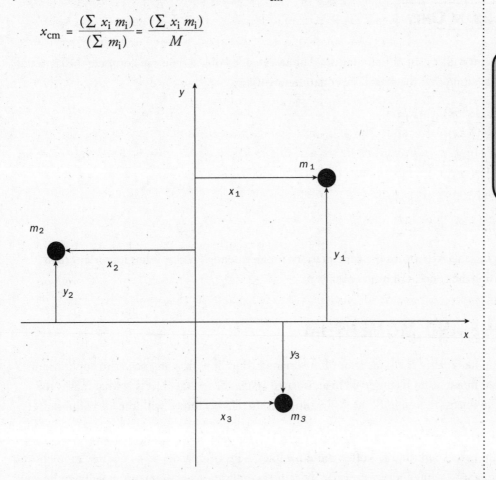

> **AP EXPERT TIP**
>
> The Physics B exam has not recently used center of mass calculations, but all students should still understand the concept.

As shown in the above figure, x_i and m_i represent the x position and mass of the ith point mass in the system. The total mass of all the masses is M. The summations indicated are over all the point masses in the system. Essentially, this equation means that you must multiply the mass times the x coordinate of that mass for all the masses in the system. Then you must sum all these products and divide the sum by the total mass of the system. To find the y and z positions of the center of mass, apply the same equation, using the y or z coordinates:

$$y_{cm} = \frac{(\sum y_i \, m_i)}{(\sum m_i)} = \frac{(\sum y_i \, m_i)}{M}$$

$$z_{cm} = \frac{(\sum z_i \, m_i)}{(\sum m_i)} = \frac{(\sum z_i \, m_i)}{M}$$

FOR C EXAM ONLY

Not all systems will consist of point masses. For an extended object, you can compute the center of mass by integrating over the mass. The equations are these:

$$x_{cm} = \frac{(\int x \, dm)}{(\int dm)} = \frac{(\int x \, dm)}{M}$$

$$y_{cm} = \frac{(\int y \, dm)}{(\int dm)} = \frac{(\int y \, dm)}{M}$$

$$z_{cm} = \frac{(\int z \, dm)}{(\int dm)} = \frac{(\int z \, dm)}{M}$$

Actually being able to perform these integrations for an extended object is less important than understanding the concept of center of mass.

IMPULSE AND MOMENTUM

Momentum, like energy, is an important fundamental concept with many powerful applications. The fact that momentum is conserved in an isolated system of several objects is what makes the concept of momentum so significant. Momentum is also closely linked with the idea of impulse.

Applying momentum conservation is conceptually similar to applying energy conservation. However, the actual application is often more complex. This complexity arises because momentum is a vector quantity, while energy is a scalar. Therefore, applications of momentum will require you to divide the momentum into x and y (and rarely z) components and solve both the x and the y problems. Never forget that momentum is a vector quantity. A very common mistake is to forget to divide a two-dimensional momentum problem into x and y components. If you forget to do this, you will get the wrong answer.

To understand the relationship between impulse and momentum, begin with Newton's second law:

$$\mathbf{F}_{net} = m\mathbf{a}$$

The bold type (**F** and **a**) indicates vector quantities. Now recall the definition of average acceleration:

$$\mathbf{a} = \frac{\Delta \mathbf{v}}{\Delta t}$$

Substituting this definition into Newton's second law and multiplying both sides by Δt gives

$$\mathbf{F}_{net} \Delta t = m\Delta \mathbf{v} = \Delta(m\mathbf{v})$$

This equation is the impulse-momentum equation; this equation applies only if the mass involved is constant, which may not be the case on the C exam. The quantity on the left-hand side, $\mathbf{F}_{net} \Delta t$, is the impulse. **Impulse**, sometimes represented by the variable j, is simply the net force multiplied by the time over which the force is applied. The quantity on the right-hand side, $\Delta(m\mathbf{v})$, is the change in momentum. **Momentum**, p, is the mass times the velocity:

$$\mathbf{p} = m\mathbf{v}$$

Notice that because momentum is defined in terms of velocity—a vector quantity—momentum is also a vector quantity.

The impulse is equal to the change in momentum. Note that this impulse-momentum equation simply states that the force is equal to the rate of change in momentum, which is how Newton originally stated his second law. The impulse-momentum equation is thus simply a restatement of Newton's second law.

CONSERVATION OF LINEAR MOMENTUM AND COLLISIONS

Look at the impulse-momentum equation and consider its implications for the case when there is a force in a constant direction but zero before and after the interaction. The equation is

$$\mathbf{F}_{net} \Delta t = m\Delta \mathbf{v} = \Delta(m\mathbf{v}), \text{ or in integral form, } \mathbf{p}_f - \mathbf{p}_i = \int_{t_i}^{t_f} \mathbf{F}dt, \text{ where i refers to the initial}$$

condition and f refers to the final condition.

If the net external force, $\mathbf{F}_{net}$, is zero, then the impulse is also zero, and the change in momentum, $\Delta(m\mathbf{v})$, must be zero. In equation form:

$$\mathbf{F}_{net} \Delta t = 0 \quad \Rightarrow \quad \Delta\mathbf{p} = \Delta(m\mathbf{v}) = 0$$
$$\Rightarrow \quad \mathbf{p} = m\mathbf{v} = \text{constant}$$

This equation is the law of conservation of momentum. It states that the total momentum of an isolated system with no net external forces is always conserved. Like the law of conservation of energy, the law of conservation of momentum is a fundamental conservation law that has no known exceptions. It is also an extremely powerful tool for solving problems.

Imagine a bat hitting a baseball. While the ball is thrown, an external force (gravity) is acting on it, while another external force (friction) is acting on both the bat and the ball when contact is made. You might be tempted to think that conservation of momentum entails that momentum is conserved only if these external forces were brought into the internal bat-and-ball system in order to conserve the momentum of the ball. While this may be true, it usually can be neglected. In this case, the collision of the bat and ball lasts for only a fraction of a second, which means the average impulse is very large compared to gravity or friction.

The trick to applying the law of conservation of momentum is to select an isolated system with no net external forces acting on it so that the momentum of the system will be conserved. When you throw the ball, you and the ball constitute an isolated system. According to the law of conservation of momentum, if you are at rest with zero momentum before you throw the ball, then the total momentum of the system consisting of you and the ball must also be zero after you throw the ball. You will therefore have a backward momentum equal to the ball's forward momentum. You do not, however, recoil backward, because the friction between you and the ground imparts an impulse to you that offsets the recoil momentum (if you stood on a frictionless or near-frictionless surface, like an ice rink, you would possibly recoil from the act of throwing). If you tried to throw a ball that was as massive as you are, you would feel the backward recoil. This principle is what causes a rifle to recoil when someone fires it. The forward momentum of the bullet is sufficient to impart a backward velocity to the rifle.

Remember, you can *only* use the law of conservation of momentum if the net external force is zero. Keep this in mind on the AP Physics exam so you don't waste any time with these kinds of problems.

Also, when you apply the law of conservation of momentum, remember the *entire* law. It applies only in the case of an isolated system, which is a system with no net external forces acting on it. When solving problems, be sure that you select an isolated system so that momentum will be conserved for your system.

Problems involving collisions commonly require students to apply the law of conservation of momentum. Usually, these problems present a collision between two bodies, but the principle will still apply to collisions involving more than two bodies. The two objects colliding form an isolated system, so momentum is conserved. Think about two balls colliding in midair. Even if we ignore air resistance, Earth's gravity acts on the balls. So the two balls do not really form an isolated system. We can, however, *treat* the balls as an isolated system if the collision occurs rapidly enough that there is no time for Earth's gravity to have a significant effect on the balls during the collision.

Before and after the collision, gravity will affect each ball's velocity. Hence, in problems that use the law of conservation of momentum, the velocities of the balls before and after the collision must be understood to be their velocities *immediately* before and *immediately* after the collision so there is not enough time for gravity (or any other external force) to have a measurable effect. With this caveat, we consider the objects involved in a collision to form an isolated system. Therefore, momentum is conserved in all collisions.

What about energy—is it conserved in collisions? It doesn't make sense to consider the potential energy in a collision, because the collision occurs at a specific position that does not change during the collision. Potential energy, which is the energy related to the position, therefore cannot tell us anything about the collision. It does, however, make sense to consider the kinetic energy before and after the collision. So the real question is whether kinetic energy is conserved in collisions.

The answer depends on the type of collision. Think about dropping a very bouncy ball on a concrete floor. As it bounces back up, it doesn't quite come back up to its original height, but let's pretend that this is an especially good, highly elastic ball. (Here we are considering an ideal case, just as we do when we ignore friction.) When you drop the ball, it has some potential energy, which is converted into kinetic energy as the ball falls. After the collision with the floor, the ball's kinetic energy is converted back to potential energy as the ball bounces up to its original height. If some of the kinetic energy had been lost in the collision, then the ball would not bounce up to its original height. So if the ball bounces back to its original height, as it does in the ideal case, the kinetic energy in the collision was conserved. If the ball does not quite reach its original height, as in the real case, then the kinetic energy was not conserved in the collision.

If the kinetic energy is conserved in a collision, we call the collision an *elastic* collision or a completely elastic collision. Most real collisions are not completely elastic, but many collisions are elastic to a very good approximation. In these collisions, we can use conservation of kinetic energy as well as conservation of momentum to solve the problem.

If after a collision the kinetic energy is not conserved, we refer to that case as an *inelastic* collision. Momentum is always conserved, so we can still use the conservation of momentum in these collisions, but we cannot use the conservation of kinetic energy. If the kinetic energy in an inelastic collision is not conserved, where did it go? The law of conservation of energy tells us that energy cannot simply vanish. In an inelastic collision, the kinetic energy goes into such things as producing a sound, heating the objects via the enhanced movement of the molecules due to impact with the other object, and changing the shape of the objects.

What if you drop a mud ball instead of a bouncy ball? The mud ball does not bounce; it has no elasticity at all. This type of collision is called a completely inelastic collision. A completely inelastic collision is one in which the objects remain together after the collision. Momentum is conserved, as it is for all collisions, but we have the additional information that the two objects have the same velocity after the collision.

A final type of collision is not really a collision but an explosion. Think about a firecracker exploding. If it is at rest, the initial momentum is zero, so the final momentum must also be zero. The pieces fly off in all directions, and the vector nature of momentum causes the momentum of the pieces flying in opposite directions to cancel. Hence, the total momentum after the collision is zero. Kinetic energy is, however, a scalar, so there is no directional cancellation. The initial kinetic energy is zero, but the final kinetic energy is greater than zero: the kinetic energy has increased. An explosive collision is one in which the kinetic energy increases. The energy usually comes from chemical energy stored in the material that explodes, such as gunpowder in a firecracker. More commonly, it can come from the potential energy stored in a cocked spring that is sprung during flight or even from the energy stored in our bodies, as when we throw a ball.

The first step in solving a collision problem is to identify the type of collision involved. Is it completely elastic, inelastic, or completely inelastic? Explosive collision problems are less common, but the approach is the same. Look for clues in the wording of the problem or in the physical situation described. Many problems will simply state that the collision is elastic or that you should assume the collision is elastic. If the problem says or implies that the objects remain together after the collision, then it is a completely inelastic collision. If the problem involves a bomb or something similar, then the collision is probably explosive. If the problem gives no information about the type of collision, it is usually best to assume that the collision is inelastic but not that it is completely inelastic.

The next step is to draw a diagram showing the situation. You actually need two diagrams to illustrate the situation, one depicting the scene before the collision and the other showing the objects after the collision. Be sure to define clearly your x- and y-axes and the positive direction of each axis. The diagram should include the masses and velocities of both objects before and after the collisions. Indicate the directions of the velocities if they are known. If they are not known, then guess. A guess in the positive direction is usually the least confusing. If your answer has a minus sign, the velocity is in the opposite direction of your guess. Indicate the known and unknown quantities on your diagram. In many problems, some of the velocities will be zero. Take these zero velocities into account to simplify your diagram. The following figures illustrate sample diagrams for both one- and two-dimension collisions.

AP EXPERT TIP

Whenever you see a problem with a collision or explosion, your first approach should always use conservation of momentum.

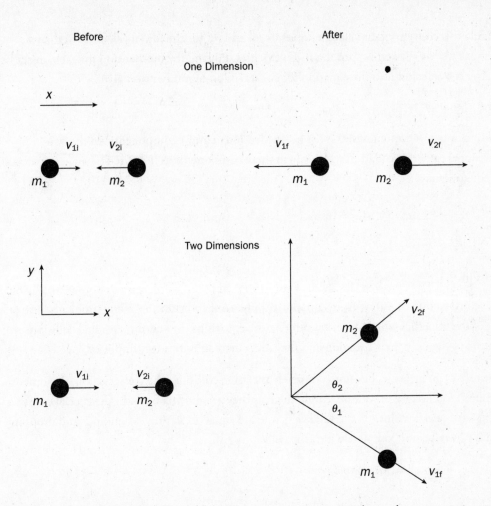

Momentum is conserved in all collisions, so your next step is to write down the conservation of momentum equation as it applies to the specific problem. If it is a one-dimensional problem, then you will have one equation. But remember that momentum is a vector and must be divided into *x* and *y* components, so a two-dimensional problem will have *two* conservation of momentum equations. The law of conservation of momentum tells us that the total momentum must be the same before and after the collision. Use an i subscript to indicate the initial state before the collision and an f subscript to indicate the final state after the collision. Conservation of momentum thus gives

$$p_i = p_f$$

Now you must divide this momentum into the momentum for each object before and after the collision. If we have a collision between two objects called object 1 and object 2, we can use 1 and 2 subscripts to indicate the object. So m_1 and m_2 indicate the masses of object 1 and object 2.

Applying this convention to velocities gives us v_{1i} and v_{2i} for the initial velocities of the two objects, as well as v_{1f} and v_{2f} for the final velocities. Putting the momentum for each object before and after the collision into the equation for conservation of momentum gives

$$m_1 v_{1i} + m_2 v_{2i} = m_1 v_{1f} + m_2 v_{2f}$$

Now remember that momentum must be divided into x and y components. In the two-dimensional case, this equation will apply to these components as indicated by an x or y subscript. The equations are

$$m_1 v_{1ix} + m_2 v_{2ix} = m_1 v_{1fx} + m_2 v_{2fx}$$

and

$$m_1 v_{1iy} + m_2 v_{2iy} = m_1 v_{1fy} + m_2 v_{2fy}$$

These conservation of momentum equations apply to all collisions. In many problems, one or more of the velocities will be zero, simplifying the equations. In one-dimensional collisions, you will need to use only the x equation and may drop the x subscript for simplicity.

In the case of an inelastic, but not completely inelastic, collision, the conservation of momentum equations are all that you will use. Apply them as they are shown above. In the special case of a completely inelastic collision, the final velocities are equal. Equating v_{1f} and v_{2f} and dropping the 1 and 2 subscripts for completely inelastic collisions gives:

$$m_1 v_{1ix} + m_2 v_{2ix} = (m_1 + m_2) v_{fx}$$

and

$$m_1 v_{1iy} + m_2 v_{2iy} = (m_1 + m_2) v_{fy}$$

Remember that the equations above apply *only* to completely inelastic collisions.

The conservation of momentum equations are not on the equation sheet provided with the AP Physics exam. You will need to be able to write them down by looking at your drawing and summing the momentum for each object before and after the collision. To do this step correctly, you must make sure that your drawings are complete. Otherwise you are likely to forget something and make an error.

In the case of a completely elastic collision, we know the additional fact that kinetic energy, K, is also conserved. The equation is:

$$K_i = K_f$$

As for momentum, add the kinetic energy for each object before and after the collision. Recalling that $K = \frac{1}{2}mv^2$ and canceling out the $\frac{1}{2}$ in each term gives

$$m_1 v_{1i}{}^2 + m_2 v_{2i}{}^2 = m_1 v_{1f}{}^2 + m_2 v_{2f}{}^2$$

As mentioned earlier, kinetic energy is a scalar, not a vector, quantity. So this equation does not need to be divided into components. There is only one conservation of kinetic energy equation, even for two-dimensional problems.

After writing down the appropriate equations for the specific problem, the final step in solving the problem is to insert the known values into the equations and solve for the unknown values. This step may be easy for some problems, but it can also be fairly complex. The worst-case scenario is a two-dimensional elastic collision problem with none of the velocities equal to zero. For a problem like this, you will have to be able to solve a system of three equations in three unknowns. The solution will often be quite messy and time consuming. The time constraints of the AP Physics exam will make such problems prohibitive without some simplifying assumptions. However, even if there are some simplifying assumptions, you will need to understand how to apply all the pieces so you can set up the problem.

Beware of the fact that many physics textbooks provide special equations that apply only to specific cases, such as when one of the objects is initially at rest or when two objects meet head-on in an elastic collision. These equations apply only to these specific cases. If you try to apply them to the wrong type of problem, you will get the wrong answer. Special-case equations like these will not be given to you on the AP Physics exam. You cannot possibly memorize all the special-case equations without risking confusion; most physics professors and professional physicists, even those with advanced degrees in physics, do not know all of these special equations. Instead, they know how to apply the fundamental principles to whatever situation is presented in a specific problem. So forget about those special equations. Focus on understanding the *general* equations outlined above and learning how to apply them. If you can do that, you will be able to solve the collision problems that appear on the exam.

REVIEW QUESTIONS

1. While mowing a level lawn, Bill applies a force of 20 N directed at an angle of 60° below the horizontal. How much work does he do pushing the mower a distance of 10 m? (Use cos 60° = 0.5 and sin 60° = 0.866.)

   (A) 100 J
   (B) 174 J
   (C) 200 J
   (D) 0 J
   (E) −200 J

2. A spring with a spring constant of 100 N/m is compressed a distance of 0.1 m horizontally. When the spring is released, it pushes a 0.25 kg mass that was initially at rest against the spring. What is the maximum speed attained by the mass?

   (A) 0 m/s
   (B) 1 m/s
   (C) 20 m/s
   (D) 0.1 m/s
   (E) 2 m/s

3. How much energy does a 100 W light bulb consume in an hour?

   (A) 100 J
   (B) 100 W
   (C) 360,000 J
   (D) 360,000 W
   (E) 60 J

4. A 4,000 kg elevator rises from the basement to the top of a 100 m tall building. Which of the following is closest to the net work done on the elevator as it rises through the middle 20 m of its trip at a constant velocity? Use $g = 10$ m/s^2.

   (A) 800,000 J
   (B) −800,000 J
   (C) 80,000 J
   (D) −80,000 J
   (E) 0 J

5. A 4,000 kg elevator rises to the top of a tall building. What is closest to the work done on the elevator by Earth's gravity as it rises a height of 20 m at a constant velocity near the middle part of its ascent? Use $g = 10$ m/s^2.

   (A) 800,000 J
   (B) −800,000 J
   (C) 80,000 J
   (D) −80,000 J
   (E) 0 J

6. A ball of mass m is thrown from a height h with an initial speed v_0 at an initial angle θ. Which of the following expressions gives the speed, v, at which the ball will strike the ground?

   (A) $v = (2gh)^{\frac{1}{2}}$
   (B) $v = (2gh + v_0^2 \sin \theta)^{\frac{1}{2}}$
   (C) $v = (2mgh + v_0^2 \sin \theta)^{\frac{1}{2}}$
   (D) $v = (2gh + v_0^2)^{\frac{1}{2}}$
   (E) $v = (2gh + v_0^2 \tan \theta)^{\frac{1}{2}}$

7. Imagine a conservative force with the force law $F = (aqz)$. The variables a, q, and z have no particular meaning here, but assume their product has units of force. Which expression might give the associated potential energy?

   (A) $U = (aqz)t$, where t represents a time
   (B) $U = (aqz)v$, where v represents a velocity
   (C) $U = (aqz)K$, where K represents a kinetic energy
   (D) $U = (aqz)V$, where V represents an electrical potential
   (E) $U = (aqz)x$, where x represents a distance

8. Two cars collide in a low-speed collision in an intersection. After the collision, the two bumpers are locked so that the cars are stuck together. Which of the following statements is true about this collision?

   (A) Momentum is conserved, but kinetic energy is not.
   (B) Kinetic energy is conserved, but momentum is not.
   (C) Both momentum and kinetic energy are conserved.
   (D) Neither momentum nor kinetic energy is conserved.
   (E) It is impossible to tell what might be conserved without knowing the masses and velocities of the vehicles.

9. A roller coaster at a theme park is traveling at a speed of 20 m/s just before it starts up a hill. If the roller coaster barely makes it over the hill, which of the following is closest to the maximum possible height of the hill? Use $g = 10$ m/s^2 and neglect friction.

   (A) 1 m
   (B) 400 m
   (C) 50 m
   (D) 20 m
   (E) To answer this question, one must know the mass of the roller coaster.

10. A glass vase falls off a table but lands on a thick carpet and does not break. The vase has a mass of 0.5 kg and falls a distance of 0.8 m before hitting the carpet. It is in contact with the carpet for 0.2 s before completely stopping. Which of the following is closest to the average stopping force that the carpet applies against the vase? Use $g = 10$ m/s^2.

    (A) 0.4 N
    (B) 5 N
    (C) 10 N
    (D) 40 N
    (E) 100 N

11. An astronaut of mass M throws a ball of mass m while floating weightless in space. They are both initially at rest. After it is thrown, the ball has a velocity v. What is the astronaut's velocity after the throw?

   (A) v in the same direction as the ball

   (B) v in the opposite direction from the ball

   (C) $\dfrac{mv}{M}$ in the same direction as the ball

   (D) $\dfrac{mv}{M}$ in the opposite direction from the ball

   (E) $\dfrac{Mv}{m}$ in the opposite direction from the ball

12. Which of the following does not have a conservation law associated with it?

   (A) Potential energy

   (B) Angular momentum

   (C) Momentum

   (D) Energy

   (E) Charge

13. A small ball of mass m is traveling with an initial velocity v in the positive x direction. It collides with a larger ball of mass M that is initially at rest. After the collision, the smaller ball is at rest. What is the final velocity V of the larger ball?

   (A) $V = v$

   (B) $V = -v$

   (C) $V = 0$

   (D) $V = \left(\dfrac{M}{m}\right)v$

   (E) $V = \left(\dfrac{m}{M}\right)v$

14. A person with a mass of 80 kg runs up a flight of stairs in 4 seconds. If the elevation gain is 4 m, what is the power the person expends against gravity?

   (A) 80 W

   (B) 320 W

   (C) 800 W

   (D) 3,200 W

   (E) 12,800 W

15. There is a potential energy associated with gravitational and spring forces. Is there also a potential energy associated with frictional forces?

   (A) No, because gravitational and spring forces are the only forces that have an associated potential energy.

   (B) Yes, because gravitational, spring, and frictional forces are all conservative.

   (C) No, because frictional forces are conservative, while gravitational and spring forces are not.

   (D) Yes, because all forces have an associated potential energy.

   (E) No, because gravitational and spring forces are conservative, while frictional forces are not.

FREE-RESPONSE QUESTION

A one-dimensional collision occurs between a small ball of mass m and a more massive ball of mass M. The more massive ball is initially at rest. The less massive ball has an initial velocity, v_i, in the positive direction. After the collision, the less massive ball has a final velocity, v_f, in the negative direction with the same magnitude as the initial velocity: $v_f = -v_i$.

(a) Derive a formula for the final velocity, V, of the more massive ball.

(b) If the more massive ball has twice the mass of the less massive ball (i.e., $M = 2m$), what is the final velocity, V, of the more massive ball?

(c) Compare the values of the balls' kinetic energy before and after this collision. Is the kinetic energy conserved?

(d) What can you conclude about this collision from the answer to part (c)? What type of collision is this?

ANSWERS AND EXPLANATIONS

1. A

This problem is a direct application of calculating the work using the equation $W = Fs \cos \theta$.

Here the force, F, is 20 N, and the distance, s, is 10 m. Bill is pushing downward and forward on the mower, so the force vector points forward at an angle 60° below the horizontal. The displacement vector points forward. Hence the angle, θ, between these two vectors is 60°. Substituting these numbers gives

$$W = (20\ \text{N})(10\ \text{m})(\cos 60°) = (20\ \text{N})(10\ \text{m})(0.5)$$
$$W = 100\ \text{J}$$

The correct answer is therefore 100 J, choice (A).

Note: In this problem, the angle given turned out to be θ in the equation for work. That won't always be the case. If, for example, the angle had been expressed as 30° from the vertical, using $\cos 30°$ would have given the wrong answer.

2. E

This question applies the concept of conservation of energy. The spring is horizontal, so gravitational potential energy does not change and does not enter into this problem. When the spring is compressed and the mass is at rest, all of the energy is spring potential energy; the kinetic energy at this point is zero. When the spring is released, this spring potential energy is converted into kinetic energy. The maximum speed occurs when the spring potential energy is zero and all the energy is kinetic energy. So conservation of energy gives

$$E_i = E_f$$
$$K_i + U_{si} = K_f + U_{sf}$$

As noted above, K_i and U_{sf} are both zero, so

$$U_{si} = K_f$$
$$\frac{1}{2} kx^2 = \frac{1}{2} mv^2$$
$$(100\ \text{N/m})\,(0.1\text{m})^2 = (0.25\ \text{kg})\,v^2$$

Solving this equation for v gives $v = 2$ m/s, choice (E).

3. C

This question requires you to understand the distinction between work and power. First, work or energy is measured in joules; power is measured in watts, which are joules per second. Thus, you should immediately eliminate (B) and (D) because neither of them is in the right units. Power is work divided by time: $P_{avg} = \dfrac{W}{\Delta t}$. The power consumption of a 100-watt light bulb is 100 W, and an hour is 3,600 seconds. Multiply the power by the time to get the total energy consumed.

$$W = P\,\Delta t$$
$$W = (100\ \text{W})(3,600\ \text{s})$$
$$W = 360,000\ \text{J}$$

The correct answer is therefore 360,000 J, choice (C).

4. E

This question applies the work energy theorem. Read very carefully; the question specifies the net work. The word *net* tells you that the work energy theorem applies in this case.

$$W_{net} = \Delta K = K - K_0$$

The net work, W_{net}, equals the change in kinetic energy, ΔK. Notice that during this part of the elevator's ascent, it is moving at a constant velocity.

That means that the kinetic energy does not change. According to the work energy theorem, if the kinetic energy does not change, the net work is zero. The correct answer is therefore 0 J.

5. B

On the surface, this question is similar to the previous question, but you should pay attention to the subtle difference. Subtle differences can make a big difference when you are trying to answer a question correctly! This question does *not* ask for the net work performed. Therefore, the work energy theorem does *not* apply here. Instead, this question asks for the work done by Earth's gravity on the elevator. You must use the equation for the work done by a force, which in this case is the Earth's gravitational force. Notice that the question does not ask for the work done by the motor on the elevator. This additional distinction makes a difference in the answer.

The work equation you need to apply is $W = \mathbf{F} \bullet \mathbf{s} = Fs \cos \theta$. The force, F, here is the gravitational force or weight, w, of the elevator (note the distinction between $W =$ work and $w =$ weight):

$$w = mg = (4{,}000 \text{ kg})(10 \text{ m/s}^2)$$

$$w = 40{,}000 \text{ N}$$

The displacement, s, is just 20 m, as given in the question stem.

Now comes the part that will trip you up if you are not careful. What is the angle θ? If you don't think about this carefully, you are likely just to say that $\theta = 0°$ and $\cos \theta = 1$ because the elevator is rising vertically. If you do so, you will miss the negative sign in the answer. Pay careful attention to the directions of the vectors.

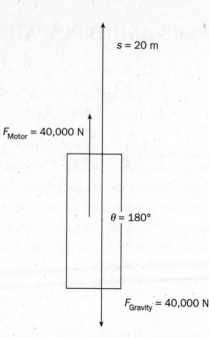

As shown in this figure, the elevator is rising. Therefore the displacement vector points upward. The gravitational force, as always, acts downward. Therefore the angle, θ, between these two vectors is 180° rather than 0°. So $\cos \theta = -1$. Now you are ready to substitute the numbers into the equation for work:

$$W = Fs \cos \theta$$

$$W = (40{,}000 \text{ N})(20 \text{ m})(\cos 180°)$$

$$W = -800{,}000 \text{ J}$$

The correct answer is therefore −800,000 J.

The incorrect choice (A), 800,000 J, is the work done by the motor on the elevator. The motor supplies the upward force, which is in the same direction as the displacement, $\theta = 0°$. Hence, the work done by the motor is 800,000 J. Notice that the sum of the work done by the Earth and the work done by the motor, which is the net work, is zero (800,000 J − 800,000 J = 0 J). When solving problems involving work, pay attention to exactly what work the question asks about.

Questions 4 and 5 show how important it is to read each question carefully and completely, noticing subtleties that can make a big difference in the answers. If you read the questions too quickly and miss these distinctions, you will be led astray.

6. D

On a problem like this, you should first look at the units. The question asks for a speed, so the answer should have the correct units for speed, which is m/s. Choice (C) does not have the correct units in the *mgh* term; it has the units for energy. You should therefore eliminate (C) right away.

There are two ways to work this problem: a hard way and an easy way. The hard way is to treat this problem as a two-dimensional projectile motion problem. It is possible to get the right answer that way, but you will have to solve two equations in two unknowns and one of the equations may be a quadratic equation. Doing the problem this way on the AP exam will up take valuable time.

Notice that the question asks only for the speed, not the velocity. If the question asked for the velocity, you would have no choice but to use the more difficult projectile motion approach described above. However, since you need only the speed (a scalar quantity) to solve the problem, you have the option of using the law of conservation of energy. Energy is also a scalar quantity, so you cannot use this law to arrive at an answer about a vector such as velocity. The fact that energy is a scalar also makes the angle at which you throw the ball irrelevant, so don't worry about this information.

To solve the problem this simpler way, apply the law of conservation of energy. At the start of its flight, the ball has potential energy and kinetic energy. At the end, this extra potential energy has been converted to even more kinetic energy, increasing

the speed of the ball. If you choose the ground level as the zero reference level for potential energy, then the potential energy of the ball when it strikes the ground is zero. Conservation of energy gives

$$E_i = E_f$$

$$K_i + U_{gi} = K_f + U_{gf}$$

Use the expressions

$$U_{gf} = 0$$

$$K_f = \frac{1}{2}mv^2$$

$$U_{gi} = mgh$$

$$K_i = \frac{1}{2}mv_0^2$$

Putting these expressions into the energy equation gives

$$\frac{1}{2}mv_0^2 + mgh = \frac{1}{2}mv^2$$

Now you need to solve this equation for v. Notice that the mass appears in every term, so it will cancel out. (Objects fall at the same rate regardless of their masses.) Now multiply both sides of the equation by 2 to get rid of the $\frac{1}{2}$. Doing this step will give an equation for v^2:

$$v^2 = v_0^2 + 2gh$$

Now just take the square root of both sides of the equation to get the correct answer:

$$v = \left(2gh + v_0^2\right)^{\frac{1}{2}}$$

If you work this problem correctly using projectile motion kinematics, you will get the same answer. Because that method is more difficult and time consuming, a quicker approach is preferable for a timed test. However, be careful that you don't go too far looking for shortcuts.

7. E

This question requires you to understand the relationship between a conservative force and potential energy. The potential energy is the amount of work required to move something against the associated conservative force. Recall that work is a force times a distance. Therefore, the answer to this question should involve force multiplied by some distance. All of the answer choices for this question are force multiplied by some quantity, so this fact alone is not enough to eliminate any options. However, choice (E) is the only option that multiplies the force by a distance, making $U = (aqz) x$ the correct answer. Notice that the question says "might," so (E) *might* represent the associated potential energy, but it might not. If the force is a constant, it will, but if the force varies in some way, there will likely be other numerical factors in the expression to account for the variable nature of the force. Because the question gives no information as to whether the force is constant or variable, we cannot find the exact expression.

8. A

To answer this question, you must to identify the type of collision and then recall what is conserved in each type of collision. The fact that the two cars are stuck together after the collision is important, because that statement should tell you that it is a completely inelastic collision. In a completely inelastic collision, momentum is conserved (as it is for all collisions). Kinetic energy, however, is conserved only in an elastic collision. Hence, the correct answer is that momentum is conserved but kinetic energy is not conserved.

9. D

Apply conservation of energy to answer this question. If you take the roller coaster's original level as the zero-point reference for gravitational potential energy, the roller coaster's initial energy is entirely potential energy. If the roller coaster barely makes it over the top of the hill, it has no speed and consequently no kinetic energy at the top of the hill. The final energy is therefore entirely gravitational potential energy. In this problem, the initial kinetic energy is converted to the final gravitational potential energy. So $E_i = E_f$ gives

$$K_i = U_{gf}$$

$$\frac{1}{2}mv_i^2 = mgh_f$$

Here the initial speed, v_i, is 20 m/s, and the final height, h_f, is the unknown. Notice that the mass, m, of the roller coaster cancels out, so you can eliminate (E). Think about the consequences if this were not the case: all roller coaster passengers would have to be weighed before getting on the ride to make sure it had the correct mass to climb the hills. If you train yourself to do this type of thinking (i.e., to consider what else would have to be true if a certain answer choice were true), it will often help you to eliminate physically impossible and highly unlikely choices.

Now solve the equation for h_f and insert the numerical values:

$$h_f = \frac{v_i^2}{(2g)}$$

$$h_f = \frac{(20 \text{ m/s})^2}{(2 \times 10 \text{ m/s}^2)}$$

$$h_f = 20 \text{ m}$$

The correct answer is therefore 20 m.

10. C

Answering this question requires putting multiple concepts together. First, you must find the speed at which the vase hits the carpet. You can do this part either by using conservation of energy or by using kinematics. Next, you must recognize that this velocity corresponds to a certain momentum, and use the impulse momentum equation to find the impulse. Finally, knowing the time that the stopping force acts on the vase allows you to find the average force acting on the vase from the value of the impulse.

First, find the velocity. If you use the conservation of energy approach to this problem, the initial potential energy before the vase falls is converted to kinetic energy. Choose the floor level as the zero point for potential energy so the final potential energy is zero. Energy conservation gives

$$E_i = E_f$$

$$0 + U_{gi} = K_f + 0$$

$$mgh = \frac{1}{2}mv^2$$

$$v^2 = 2gh$$

$$v^2 = (2)(10 \text{ m/s}^2)(0.8 \text{ m}) = 16 \text{ m}^2/\text{s}^2$$

$$v = 4 \text{ m/s}$$

You can also get this same velocity using kinematics. Either way, the vase strikes the carpet at 4 m/s.

Now think about what happens when the vase strikes the carpet: the carpet compresses for a short time, 0.2 s. As the carpet is compressing, it is applying an upward force to the vase, causing it to stop. After the vase stops, its momentum is zero. The impulse equation relates the average force, F, the time, Δt, and the change in momentum, Δp. This equation in one dimension is

$$F_{net} \, \Delta t = \Delta p$$

Solving for the force gives

$$F_{net} = \frac{\Delta p}{\Delta t}$$

$$F_{net} = \frac{(0 - 0.5 \text{kg})(4 \text{m/s})}{(0.2 \text{ s})}$$

$$F_{net} = -10 \text{ N}$$

The negative sign simply indicates that the direction of the force is opposite to the direction of the velocity. Since the question asks nothing about the direction of the force, you can ignore the negative sign. Therefore, the correct answer is that the average force on the vase is 10 N.

11. D

This question applies conservation of momentum. The astronaut and ball are both initially at rest, so the initial momentum is zero, and the final momentum must thus also be zero. For the final momentum to be zero, the final momentum of the ball and the astronaut must be in opposite directions. You can therefore eliminate (A) and (C), in which the astronaut's velocity is in the same direction as the ball's. You should also look at the units. In some cases, a choice will be presented in incorrect units, which will allow you to eliminate it. In this case, however, all the choices have the correct units for velocity, so none of them can be eliminated immediately.

The conservation of momentum equation, in one dimension, tells us that

$$0 = p_i = p_f$$

$$0 = MV - mv$$

where M and V are the mass and velocity of the astronaut. The mass and velocity of the ball are m and v. Solving the equation for V gives

$$V = \frac{mv}{M}$$

The correct answer for the astronaut's velocity is therefore $\frac{mv}{M}$ in the opposite direction of the ball's movement.

12. A

This question requires that you recall the conservation laws. As discussed in this chapter, momentum and total energy are conserved. From the discussion of conservation of energy in this chapter, you should know that energy can change form but the total energy (including the energy equivalent of mass) is conserved. Because the law of conservation of energy allows energy to change form, specific types of energy, such as potential energy, are not conserved. Thus potential energy does not have an associated conservation law. Conservation of charge and angular momentum are discussed elsewhere in this book.

13. E

This is an example of a one-dimensional collision problem. As in all collision problems, momentum is conserved. The problem does not specify that this is an elastic collision, so you should not assume that kinetic energy is conserved. The balls do not remain together after the collision, so it is not a completely inelastic collision, either. Thus, the only option for solving this problem is to use momentum conservation. Notice that all motion occurs along the x-axis; this means that you can use the one-dimensional form of momentum conservation.

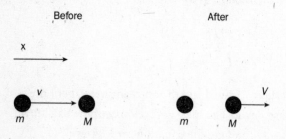

You should start by drawing a picture of the situation before and after the collision, as shown in the above figure. This problem has simplifying aspects: both the initial velocity of the larger ball and the final velocity of the smaller ball are zero. So momentum conservation gives

$$p_i = p_f$$

Because this is a one-dimensional problem, this momentum equation does not have to be divided into x and y components. The equation gives

$$mv + 0 = 0 + MV$$

Solving for V gives

$$V = \left(\frac{m}{M} \right) v$$

The correct answer for the final velocity of the larger ball is therefore $\left(\frac{m}{M} \right) v$.

14. C

If you recall that 1 horsepower is equal to 746 watts, then you know that choices (D) and (E) are high numbers of horsepower and would therefore be quite unrealistic as a description of the amount of power a person expends climbing stairs. You cannot, however, eliminate the smaller values so readily.

To solve the problem, use the fact that power is defined as work divided by time. Compute the work the person does against gravity and divide by the time. The work done against gravity is the potential energy change, mgh. We have for the power, P,

$$P = \frac{W}{\Delta t}$$

The work, W, is in this case mgh, so

$$P = \frac{mgh}{\Delta t}$$

$$P = \frac{(80 \text{ kg})(10 \text{ m/s}^2)(4 \text{ m})}{4 \text{ s}}$$

$$P = 800 \text{ W}$$

The correct answer for the person's power output against gravity is 800 W.

15. E

This question requires that you understand the relationship between conservative forces and potential energy. All conservative forces have an associated potential energy. Nonconservative forces do not have an associated potential energy, because these forces do not result in energy that is stored in a recoverable way. Gravitational and spring forces are examples of conservative forces. Frictional forces are not conservative; when friction acts, energy is "lost" as heat rather than being stored as potential energy. Total energy is still conserved but not in the same way that it is when a conservative force acts.

FREE-RESPONSE QUESTION

This is a collision question, and you are not explicitly told what type of collision it is. Instead, you are asked to solve the problem and then to examine the kinetic energies and deduce what type of collision has occurred. To solve the problem, you must apply momentum conservation. It is a one-dimensional collision, so you do not need to divide the momenta into x and y components. Assume that all motion occurs in the x direction.

As with all collision problems, you should start with a drawing showing the situation before and after the collision to help you visualize what is happening.

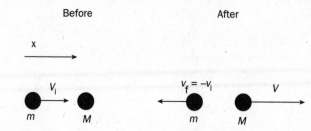

(a) To derive a formula for the final velocity, V, of the more massive ball, you need to apply the momentum conservation equation:

$$p_i = p_f$$

From the drawing, add the momenta of each ball before and after the collision. The equation becomes

$$mv_i + 0 = mv_f + MV$$

Now recall that $v_f = -v_i$, and substitute it into the equation above:

$$mv_i = -mv_i + MV$$

Solving this equation for V gives

$$V = \frac{2mv_i}{M}$$

(b) Now if $M = 2m$, we can substitute this value into the result for part (a). This substitution gives

$$V = \frac{2mv_i}{2m}$$

$$V = v_i$$

After the collision, the more massive ball has the same velocity (magnitude and direction) as the less massive ball's initial velocity.

(c) For this part, compute the initial, K_i, and final, K_f, kinetic energies and compare them. Before the collision, the more massive ball is at rest and has no kinetic energy. The initial kinetic energy is therefore

$$K_i = \frac{1}{2}mv_i^2$$

The final kinetic energy includes both balls:

$$K_f = \frac{1}{2}mv_i^2 + \frac{1}{2}MV^2$$

Now use the results from the previous part of this problem.

$V = v_i$ and $M = 2m$ give

$$K_f = \frac{1}{2}mv_i^2 + \frac{1}{2}2mv_i^2$$
$$K_f = \frac{3}{2}mv_i^2$$

Comparing the two kinetic energies gives

$$K_f = 3K_i$$

The kinetic energy is not conserved in this collision, because it has *increased*. Note that when a quantity is conserved, it can neither increase nor decrease; the quantity must remain unchanged. Therefore, even though kinetic energy is not lost in this collision, we say that it is not conserved.

(d) The kinetic energy increases in this collision. It is, therefore, an explosive collision. For the collision to occur in this way, one of the balls would have to have some type of explosive material attached to it or perhaps a cocked spring.

CHAPTER 5: CIRCULAR MOTION, OSCILLATION, AND GRAVITATION

IF YOU MEMORIZE ONLY 16 EQUATIONS IN THIS CHAPTER . . .

$$T_p = 2\pi \sqrt{\frac{\ell}{g}} \qquad F_G = \frac{-Gm_1m_2}{r^2} \qquad \tau = rF\sin\theta \qquad \vec{F}_s = -k\vec{x} \qquad T_s = 2\pi\sqrt{\frac{m}{k}} \qquad T = \frac{1}{f}$$

C Exam Equations:

$$a_c = \omega^2 r \qquad \theta = \theta_0 + \omega_0 t + \frac{1}{2}\alpha t^2 \qquad \vec{\tau} = \vec{r} \times \vec{F}$$

$$\omega = \frac{v}{r} \qquad I = \int r^2 dm = \sum mr^2 \qquad \vec{L} = \vec{r} \times \vec{p} = I\vec{\omega}$$

$$\omega = \omega_0 + \alpha t \qquad K = \frac{1}{2}I\omega^2 \qquad \vec{\tau}_{net} = I\vec{\alpha} \qquad T = \frac{2\pi}{\omega}$$

INTRODUCTION

All of the topics in this chapter relate to rotation or circular motion. Out of the 70 multiple-choice questions on the AP Physics exams, you can expect to see about 7 questions related to these topics on the B exam and about 12 to 13 questions related to these topics on the C exam. The exam questions covering these topics will be divided as follows:

CIRCULAR AND ROTATIONAL MOTION

B exam	2 to 3 questions
C exam	6 to 7 questions

SIMPLE HARMONIC MOTION AND GRAVITATION

B exam	4 to 5 questions
C exam	6 to 7 questions

If you have a good understanding of translational motion (covered in chapters 3 and 4), then you should be able to master circular or rotational motion fairly easily. To understand rotational motion, it often helps to think about the analogies between translational motion and rotational motion. Translational motion quantities all have analogous rotational motion quantities. For example, in rotational motion, the angular velocity is analogous to the translational quantity of velocity. Furthermore, just as momentum is conserved in the absence of external forces, angular momentum must be conserved in the absence of external torques. Once you understand how this analogous relationship works, you can use it as a memory aid; you can recall equations relating to circular or rotational motion if you remember the analogous translational equation. The table below shows the analogies between circular/rotational motion and translational motion.

TRANSLATIONAL-ROTATIONAL MOTION ANALOGUES

Translational Motion	Rotational Motion
Displacement x, y, s	Angle θ (remember: in radians)
Velocity v	Angular velocity ω
Acceleration a	Angular acceleration α
Force F	Torque τ
Mass m	Moment of inertia I
Newton's second law $F = ma$	$\tau = I\alpha$
Momentum $p = mv$	Angular momentum $L = I\omega$
Kinetic energy $K = \frac{1}{2}mv^2$	Rotational kinetic energy $K_R = \frac{1}{2}I\omega^2$

UNIFORM CIRCULAR MOTION

You should recall from your studies and from the previous chapters in this book that velocity is a vector quantity. Think about what that means: direction matters. Among other things, this entails that an object moving in a circular path *cannot* have a constant velocity. The direction is constantly changing, so the velocity must also be constantly changing—even if it is moving at a constant speed. Any object moving at a constant speed in a circular path is said to be in a state of uniform circular motion.

To be in uniform circular motion, an object must change its velocity continuously as it moves around the circular path. However, this change affects the direction of the velocity but not the speed. To force the object into a circular path at constant speed, the acceleration must constantly be perpendicular to the velocity and directed toward the center of the circular path, as shown in the following figure.

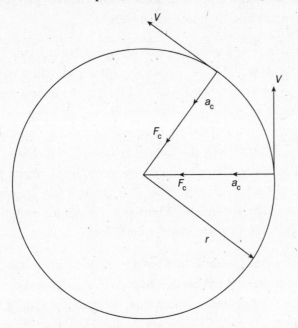

It is called the centripetal acceleration, a_c, and is given by

$$a_c = \frac{v^2}{r}$$

When the speed is constant, $v = \dfrac{d}{t} = \dfrac{2\pi r}{T}$, where T is the period of rotation and $2\pi r$ is the circumference of the circle.

Very often you will be given ω, the angular velocity, rather than v. These two are related by the equation $v = \omega r$. Substituting this into our equation above for centripetal acceleration gives us

$$a_c = \frac{v^2}{r} = \omega^2 r$$

From Newton's second law, a force is required to produce an acceleration. Because $F = ma$, this centripetal force, F_c, is given by

$$F_c = ma_c = \frac{mv^2}{r} = m\omega^2 r$$

AP EXPERT TIP

It is critical to know which way the force, acceleration, and velocity vectors point. Memorize this. Don't forget: Velocity is NOT constant in uniform circular motion!

The mass, m, represents the mass of the object following the circular path. Just as the centripetal acceleration is constantly changing direction, the centripetal force constantly changes direction so that it is always perpendicular to the velocity and directed toward the center of the circle. This centripetal force can be supplied in a variety of ways. In the case of a ball twirling on a string, the tension in the string supplies the centripetal force. If the string were to break, the ball would continue to travel in a straight line, as required by Newton's inertial law. In the case of a planet orbiting the sun, the centripetal force is the gravitational force between the sun and the planet. If this gravitational force were to suddenly disappear for some reason, then the planet would continue to move in a straight line tangential to the original circular path.

Be sure that you keep *centripetal* force and *centrifugal* force separate in your mind. The centripetal force is the real force acting on an object in uniform circular motion and is directed inward toward the center of the circle. Many people will incorrectly say that there is an outward centrifugal force acting on anything moving in a circular path. This is incorrect. The real force is the inward centripetal force. The centrifugal effect is an apparent (not real) force acting on an object in circular motion because it is accelerated. As an analogy, think about being pushed back into the seat of a car that is accelerating forward rapidly. The real force acting on you is the car seat pushing you forward, but because you are in the accelerating car, there is an apparent (not real) force pushing you back into the seat that is caused by your inertial tendency not to accelerate. Again, centripetal force is the real force; centrifugal force is the apparent (not real) force.

Twirling a bucket of water overhead in a vertical circle provides a commonly misunderstood example here. People will often incorrectly say that the water does not fall out at the top of the circle because it has a centrifugal force directed outward from the circle acting on it. This idea is, once again, wrong. The *real* forces acting on the water are the downward forces of gravity and the normal force from the bottom of the bucket. Here, much of the confusion comes because people naturally ask the wrong question: "Why doesn't the water fall?" Instead, we should ask, "Why doesn't the water follow its inertial tendency to continue to travel in a straight line?" The answer is that the downward forces of gravity and the normal forces of the bucket combine to provide a centripetal force that causes the water to travel in a circular path rather than in a straight line.

ANGULAR MOMENTUM AND ITS CONSERVATION (C EXAM)

The **law of conservation of angular momentum** is the rotational analogue of the law of conservation of momentum. Thus, it will help to recall the law of conservation of momentum here: In the absence of external forces, the total momentum of a system remains constant. Analogously, in the absence of external torques, the total angular momentum of a system remains constant.

Ice skaters provide a familiar example. When a figure skater wants to go into a spin, he starts with his arms outstretched. To spin faster, he brings his arms in close to the axis of rotation. Bringing his arms inward reduces his moment of inertia. His angular velocity must, therefore, increase so that his total angular momentum remains constant.

Like the law of conservation of momentum, the law of conservation of angular momentum is a fundamental law that is also an extremely powerful tool for solving physics problems. Again, use the analogy to remind yourself how to proceed. If the momentum is given by the mass times the velocity, then the angular momentum, L, will be the moment of inertia, I, times the angular velocity, ω.

$$L = I\omega$$

In the case of an isolated system with no external torques, the angular momentum must remain constant. If the rotational axis does not change, then the vectors point in the same direction, and we can drop the vector notation. Hence,

$$L_i = L_f$$

$$I_i \, \omega_i = I_f \, \omega_f$$

The i and f susbscripts refer to the initial and final values of the quantity. You must be able to find the relevant moments of inertia and solve this equation for the unknown.

POINT PARTICLES

Finding the moment of inertia for point particles is easier than finding the moment of inertia for extended bodies. If a point particle of mass, m, is a distance, r, from the axis of rotation, then the moment of inertia for a point particle is given by

$$I = mr^2$$

The angular momentum for a point particle will then be

$$L = I\omega = mr^2\omega$$

In the case of a system of point particles, simply add the moments of inertia of each of the point particles to get the total moment of inertia.

EXTENDED BODIES, INCLUDING ROTATIONAL INERTIA

Finding the moment of inertia of an extended body is more complex. Rather than simply adding the moments for discrete point masses, one needs to integrate over the mass distribution. The moment of inertia of an extended object is

$$I = \int r^2 \, dm$$

This integral will typically be a three-dimensional volume integral. Actually being able to do these integrals is, however, much less important than understanding how to use the moment of inertia to solve problems. The resulting moment of inertia will typically be a fraction, between 0 and 1, multiplied by the total mass times the maximum radius squared. This fraction accounts for the fact that the mass is not all at the maximum radius of the object. For example, the fraction is $\frac{2}{5}$ for a sphere, so $I_{sphere} = \frac{2}{5} MR^2$; and $\frac{1}{2}$ for a disk, so $I_{disk} = \frac{1}{2} MR^2$. If all of the mass is at the maximum radius, as it is for a hoop, then the fraction will be 1, and $I_{hoop} = MR^2$. Most physics textbooks give a table of moments of inertia for common object shapes. Don't spend your time trying to memorize this table, because required moment of inertia equations should be given with the exam problem.

To solve angular momentum problems, start by drawing a picture that will help you to visualize the initial and final situations. Find the initial and final moments of inertia in terms of the known and unknown quantities. Do the same for the angular velocities. Finally, solve the conservation of angular momentum equation for the unknown quantity.

TORQUE AND ROTATIONAL STATICS

Torque is the rotational analogue to force. Just as the net force on an object in translational equilibrium must be zero, the net torque on an object in rotational equilibrium must also be zero. Students often misunderstand the concept of equilibrium. Saying that an object is in equilibrium does *not* mean that the object isn't moving. It means that the object is *not accelerating*. Hence, when applied to rotational motion, an object or system with zero angular acceleration (i.e., an object that is either not rotating or rotating at a constant angular velocity) must have a net torque of zero acting on it.

To understand how torque works, think about children playing on a seesaw. The seesaw is a system that can rotate about the axis formed by its central supporting point. Two children of the same weight will balance the seesaw when they are the same distance from the rotational axis (supporting point) because they produce equal but opposite torques on the seesaw. A heavier individual on a seesaw with a child must sit closer to the center than the child for the torques and the seesaw to balance. The shorter distance from the rotational axis compensates for the heavier person's greater weight. This example tells us that the torque must involve multiplying the force by the distance.

There is, however, a complicating factor. Exactly how is the distance measured? Imagine applying a force on a wheel to produce a torque and to turn the wheel, as shown in the figure below.

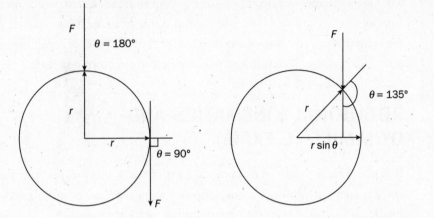

If you push tangential to the wheel, then the force vector is perpendicular to the distance vector, and the distance is simply the radius of the wheel. If, however, you push straight down on the top center of the wheel, then the line extending the force vector goes straight through the axis at the center of the wheel. Even though you are actually doing the pushing at the edge of the wheel, the distance in this case is zero rather than the radius of the wheel. Why? To compute the torque, you need to use the shortest, or perpendicular, distance between the axis and the extended line of the force vector. Do *not* use the distance vector from the axis to the point where the force is applied. To calculate this perpendicular distance, multiply the distance from the axis to the point where the force is applied by the sine of the angle between the distance and force vectors. The torque, τ, is then

$$\tau = rF\sin\theta$$

When computing the torque, be certain that you get the right angle for θ. Draw a picture to make sure that you are getting the perpendicular distance between the force vector line and the axis of rotation. Extend and draw *both* the distance and force vectors so that the lines actually cross at the vertex point. The angle between the vectors is measured between the two arrows pointing away from the vertex.

The torque can be used to solve rotational equilibrium problems. When the angular acceleration is zero, the net torque is also zero

$$\alpha = 0 \Leftrightarrow \Sigma\,\tau = 0$$

If the system is in both rotational and translational equilibrium, the forces also sum to zero (see Chapter 3). To solve rotational statics problems, use this equation along with the x and y translational equilibrium equations, if appropriate. Always start by drawing a free-body diagram. Include torques as well as forces. To compute torques, you will need a rotational axis. If the system is static and therefore not rotating, the choice of rotational axis is somewhat arbitrary.

AP EXPERT TIP

When drawing the weight vector of an extended object, like a plank, draw the weight vector straight down from the center of mass.

Don't spend a lot of time worrying about it. Just select an axis that will reduce the computations required by picking a point where one or more of the force vectors will have a distance of zero. The resulting torques are then zero, which means fewer computations. Compute the x and y components of all the forces, as well. Now set the sums of the torques as well as the sums of the x and y force components equal to zero. Finally, solve the resulting equations for the unknown quantities.

ROTATIONAL KINEMATICS AND DYNAMICS (C EXAM)

Rotational kinematics and dynamics are analogous to translational kinematics and dynamics. Replacing the quantities in the translational equations with their rotational analogues gives the corresponding rotational equations. If you have mastered translational kinematics and dynamics, you should be able to master rotational kinematics and dynamics.

Kinematics is the analysis of motion without paying attention to the forces that cause these motions. For rotational kinematics, the angle θ is analogous to the displacement, s (or possibly x or y), in translational motion. However, the angle must be measured in radians rather than the more familiar degrees. If you encounter a rotational kinematics problem with angles given in degrees, convert them to radians. For a segment of a circle, the angle in radians is defined as the arc length of the segment divided by the radius: $\theta = \dfrac{s}{r}$. For an entire circle, the arc length is simply the circumference of the circle. Therefore, an angle of 360° corresponds to $\dfrac{2\pi r}{r} = 2\pi$ radians. If you understand this definition of an angle in radians, you should be able to remember the conversion factor $360° = 2\pi$ rad.

Angular velocity and **angular acceleration** are defined in a way that is exactly analogous to the translational definitions of velocity and acceleration. In the case where the acceleration is constant, there are three kinematic equations that are useful for solving translational kinematic problems. Similarly, corresponding rotational kinematic equations are useful for solving problems when the angular acceleration is constant. One advantage of rotational kinematics is that you will not be asked to solve two-dimensional rotational kinematic problems, whereas you may be asked to solve two-dimensional (projectile motion) translational problems. The relevant equations can be found below. By comparing the rotational equations to the corresponding translational equations, you can better understand how to apply the principles for solving translational motion problems to rotational motion problems.

Translational Motion	**Rotational Motion**
$v_{avg} = \dfrac{\Delta s}{\Delta t}$	$\omega_{avg} = \dfrac{\Delta \theta}{\Delta t}$
$v_{inst} = \lim\limits_{\Delta t \to 0} \dfrac{\Delta s}{\Delta t}$	$\omega_{inst} = \lim\limits_{\Delta t \to 0} \dfrac{\Delta \theta}{\Delta t}$
$a_{avg} = \dfrac{\Delta v}{\Delta t}$	$\alpha_{avg} = \dfrac{\Delta \omega}{\Delta t}$
$a_{inst} = \lim\limits_{\Delta t \to 0} \dfrac{\Delta v}{\Delta t}$	$\alpha_{inst} = \lim\limits_{\Delta t \to 0} \dfrac{\Delta \omega}{\Delta t}$

For the case of constant translational or angular acceleration, we also have

$a = \text{constant}$	$\alpha = \text{constant}$
$v = v_0 + at$	$\omega = \omega_0 + \alpha t$
$s = s_0 + v_0 t + \dfrac{1}{2} at^2$	$\theta = \theta_0 + \omega_0 t + \dfrac{1}{2} \alpha t^2$
$v^2 = v_0^2 + 2a(s - s_0)$	$\omega^2 = \omega_0^2 + 2\alpha(\theta - \theta_0)$

You will also occasionally need to go back and forth between translational and rotational velocity and acceleration. To do so, use

$$\omega = \frac{v}{r}$$

and

$$\alpha = \frac{a_t}{r}$$

where a_t represents the component of the acceleration tangential to the circle. Notice the similarity between these equations and the definition of an angle in radians. Recalling this similarity can help you to remember the equations.

To solve rotational kinematic problems, apply the equations above in the same way that you would to solve translational kinematic problems.

In kinematics, there is an acceleration or angular acceleration, but the force or torque causing it is ignored. However, dynamics problems *do* consider the force or torque that causes the acceleration or angular acceleration. Newton's second law is relevant here. In the one-dimensional translational case, you should know that this law is $\Sigma F = ma$. By analogy, the rotational equivalent to Newton's second law is $\Sigma \tau = I \alpha$.

Rotational dynamics problems will typically involve both solving a kinematic component and applying the equation to find either the torque or the angular acceleration. To solve these problems, always start with a free-body diagram. Include both the forces and the torques. If the translational forces are relevant to the problem, divide them into x and y components. Next, sum the torques about a convenient point and (if appropriate) the translational forces. Apply Newton's second law. Equate this sum of the torques to the moment of inertia times the angular acceleration. If appropriate to the particular problem, you should also set the sums of the x or y forces equal to the mass times the x or y acceleration. Finally, solve the resulting equations for the unknowns. You may have to solve a system of two or three equations for two or three unknowns.

SIMPLE HARMONIC MOTION (DYNAMICS AND ENERGY RELATIONSHIPS)

Simple harmonic motion is related to circular motion in a way that might not be immediately obvious. When viewed from an edge-on angle, an object in uniform circular motion will appear as though it is oscillating in simple harmonic motion.

Simple harmonic motion is oscillatory motion, but not all oscillatory motion is simple harmonic motion. **Simple harmonic motion** is the oscillating motion that occurs when the force restoring the system to its equilibrium position is directly proportional to its displacement from the equilibrium position. Like a spring, it follows Hooke's law:

$$F = -kx$$

The constant k is called the spring constant. Be careful about x; it is *not* the length of the spring. Instead, x represents the amount that the spring departs from its natural length (via stretching or compression). Don't let the negative sign bother you. It simply refers to the fact that the restoring force is in the opposite direction of the change in length of the spring, denoted by x. A spring is the best-known system that follows Hooke's law, but it is not the only such system. This law also provides a good approximation of a pendulum oscillating at small angles. Thus, springs and pendulums are both examples of simple harmonic motion. Oscillatory motions that are not governed by this law are not simple harmonic motion.

You can often use the law of conservation of energy to solve problems involving the speed or position of simple harmonic oscillators. The total kinetic energy plus potential energy remains constant at all points in the oscillating path. At the equilibrium position, the potential energy is zero if you select this position as the zero reference level for the gravitational potential energy. At this position, the kinetic energy—and hence velocity—will have its maximum value. At the maximum deviation from the equilibrium position, the oscillatory motion reverses direction; at this instant the kinetic energy and velocity will be zero, and the energy is entirely potential energy. If you know the maximum displacement from equilibrium, you can use the law of conservation of energy to find the speed at other positions. Conversely, if you know the speed at the equilibrium position, energy considerations allow you to find the maximum displacement from equilibrium.

To apply energy conservation to simple harmonic motion, you will need to know that for a spring, Hooke's law gives the potential energy formula:

$$U_s = \frac{1}{2} kx^2$$

MASS ON A SPRING

Because springs obey Hooke's law, $F = -kx$, a mass oscillating on the end of a spring is a prime example of simple harmonic motion. To solve problems involving other oscillatory motions that also follow this law, you can use the simple harmonic motion equations developed for an oscillating spring. Substitute the analogous expression for k from Hooke's law into the simple harmonic motion equations.

We apply Newton's second law ($F = ma$) to the force equation for a spring ($F = -kx$). For acceleration, we use $a = \dfrac{dv}{dt} = \dfrac{d^2x}{dt^2}$. Combining both equations, we find

$$-kx = m\frac{d^2x}{dt^2} \quad \text{or} \quad \frac{d^2x}{dt^2} = -\left(\frac{k}{m}\right)x$$

Note that the second derivative of x is proportional to the negative of x. You should recall from calculus that sine and cosine functions have this property. In fact, only those functions have this property while being bounded like the harmonic motion we are analyzing.

To find a solution to the differential equation, assume a solution $x = X_0 \cos(\omega t + \theta)$.

We find

$$\frac{dx}{dt} = -\omega X_0 \sin(\omega t + \theta)$$

$$\frac{d^2x}{dt^2} = -\omega^2 X_0 \cos(\omega t + \theta) = -\omega^2 x$$

Therefore, $\omega^2 = \dfrac{k}{m}$ or $\omega = \sqrt{\dfrac{k}{m}}$.

Since a cosine function repeats itself every 2π radians, it is interesting to know the time for the harmonic oscillator to complete a cycle. The time it takes for a spring, or any other simple harmonic oscillator, to complete one cycle is called the **period**, T. We can derive T by noting that the solution for x will repeat at a time $t = T$ when $\omega T = 2\pi$ or $T = \dfrac{2\pi}{\omega}$. Therefore,

$$T = 2\pi\sqrt{\frac{m}{k}}$$

The **frequency**, f, is the number of complete oscillations in one second, and the angular frequency, ω, is given by $\omega = 2\pi f$.

This gives the equations for the frequency, f:

$$f = \frac{\omega}{2\pi} = \frac{1}{2\pi}\sqrt{\frac{k}{m}}$$

For a spring, k is called the spring constant. It is a property of an individual spring and must be determined experimentally for each individual spring. This constant is our way of measuring the stiffness of the spring. If k is large, then the spring is very stiff, whereas a spring that compresses or stretches easily has a small k value. Hooke's law is not a fundamental law in the sense that it is always obeyed, like the law of conservation of energy. Rather, Hooke's law is an empirical law that is true to a good approximation most of the time. However, if a spring is stretched so much that it is no longer elastic, then it will no longer follow Hooke's law.

These equations allow you to solve problems involving the period or frequency of an oscillating spring. In the case of a horizontal spring, gravitational forces have no effect on the horizontal oscillations. In the case of a vertical spring, it might seem that the complicating effect of the gravitational force would affect how the spring oscillates. However, that is not the case. The period and frequency of the oscillation do not change. What does change is the point around which the mass on the end of the spring oscillates. It oscillates around the equilibrium point of the spring (i.e., the equilibrium point of the spring when the mass is hanging on the end of the spring *without* oscillating). It is also important to note that the period and frequency of the oscillation are independent of the amplitude of the oscillations.

PENDULUM AND OTHER OSCILLATIONS

A swinging pendulum is not strictly simple harmonic motion. The restoring force is not exactly proportional to the displacement; in other words, a pendulum does not exactly follow Hooke's law. However, if the angle through which the pendulum oscillates is small, then the pendulum follows Hooke's law to a very good approximation. Therefore, a pendulum oscillating through small angles is a good approximation of simple harmonic motion.

In this small-angle approximation, the period and frequency of a pendulum are given by

$$T_\text{p} = 2\pi\sqrt{\frac{\ell}{g}} = 2\pi\left(\frac{\ell}{g}\right)^{\frac{1}{2}}$$

and

$$f_\text{p} = \frac{1}{T_\text{p}} = \frac{1}{(2\pi)\sqrt{\dfrac{g}{\ell}}} = \frac{1}{(2\pi)\left(\dfrac{g}{\ell}\right)^{\frac{1}{2}}}$$

where ℓ is the length of the pendulum. Notice that the period of a pendulum does not depend on the mass in the way that the period of a spring does. However, the period of a pendulum depends on the local acceleration of gravity, which is *not* the case for a spring. Also notice that—as for a spring—the period and frequency are independent of the amplitude of the oscillation, which in this case is the maximum angle from the vertical as long as the amplitude is "small."

Other simple harmonic oscillations occur when a system obeys Hooke's law. If the restoring force has the form

$$F = -Ax$$

where x is the displacement from an equilibrium, take the equation for the period of the spring and substitute A for k to get the period of that particular simple harmonic oscillator.

$$T_s = 2\pi\sqrt{\frac{m}{k}}$$

gives

$$T_s = 2\pi\sqrt{\frac{m}{A}}$$

NEWTON'S LAW OF UNIVERSAL GRAVITATION

Gravity is one of the four fundamental forces of nature, along with the electromagnetic force, the strong nuclear force, and the weak nuclear force. Newton's law of gravity, which describes gravitational force, is a very important law, one that you will need to understand thoroughly.

Gravity is the force that holds the universe together on a large scale, because there is an attractive gravitational force between any two massive objects in the universe. Earth is the most massive object close to us (we measure the distance from the center of the Earth), so the gravitational force from the Earth is the strongest gravitational force we feel. The sun is much more massive than the Earth, but its gravitational force does not draw people into the sun because its distance from us makes the gravitational attraction it exerts on us much weaker than the force exerted by the Earth. There is also a gravitational force acting between you and the person sitting next to you in your physics class. The Earth's center is much farther away from you than that person is, but he or she is also much less massive than the Earth. Thus, if you feel an attraction to that person, it is not caused by the gravitational force he or she is exerting on you.

The **gravitational force** between any two objects depends on both the masses of the objects and the distance between the objects. The equation describing Newton's law of gravity is

$$F_G = \frac{-Gm_1m_2}{r^2}$$

G is called the **universal gravitational constant**. It is a fundamental constant of nature that had to be measured experimentally. Its value is $G = 6.67 \times 10^{-11}$ N · m^2/kg^2. Do not confuse G and g; they are not the same thing. The minus sign indicates that this gravitational force, F_G, is always an attractive force; any two objects are always pulled toward each other—*not* pushed apart—by gravity. Be careful how you measure the distance, r, between the two objects; correctly understood, it is the distance between the *centers* of the objects. For example, you might think that the distance between someone lying on the ground and the Earth is zero, but the distance between them is actually the radius of the Earth.

The masses of the two objects are m_1 and m_2. Notice the way this equation is written: the force that object 2 exerts on object 1 is exactly the same magnitude (but in the opposite direction) as the force that object 1 exerts on object 2. That means that if you weigh 75 N, Earth is pulling down on you with a force of 75 N and you are pulling up on the Earth with exactly the same 75 N of force. This symmetry in the law of gravity is required by Newton's third law (action-reaction). The gravitational forces between the two objects must be equal and opposite because they are an action-reaction pair.

ORBITS OF PLANETS AND SATELLITES

Most people know that gravity is the force that causes the moon to orbit the Earth and the planets to orbit the sun. However, for the purposes of this exam, you will need to know something about *how* gravity works in that role. Gravity does not push the moon or a planet along in its orbit. Rather, the gravitational force on a moon or planet acts toward the center of the attracting planet or sun. The force changes the direction and speed of the motion, causing the moon or planet to deviate from its natural tendency to continue moving in a straight line and follow its orbital path instead.

CIRCULAR

Circular orbits are the simplest case. To solve problems with circular orbits, you will need to remember two things. First, gravity is the centripetal force. Second, the orbital speed is the circumference of the orbit divided by the time for the orbit. These two principles, illustrated in the following figure, and a bit of algebra can get you through most circular orbit problems.

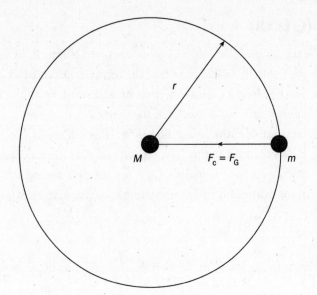

First equate the gravitational force with the centripetal force:

$$F_G = F_c$$

so

$$\frac{GMm}{r^2} = \frac{mv^2}{r}$$

Here, M represents the mass of the more massive body being orbited, and m represents the mass of the less massive body that is doing the orbiting. For example, M might be the mass of the sun and m the mass of a planet, or M might be the mass of the Earth and m the mass of the moon. G is, of course, the universal gravitational constant, and r is the orbital radius. Notice that m cancels out of both sides of the equation, so the mass of the orbiting object does not affect the orbit. The time that it takes for the object to make a complete orbit is called the period, T. The orbital speed is the circumference divided by the period; it is given by

$$v = \frac{2\pi r}{T}$$

Substituting this expression for v and performing some algebraic manipulation gives

$$T^2 = \frac{4\pi^2}{GM} r^3$$

This equation is Kepler's third law. It is very useful for solving circular orbital problems. You should not, however, rely on using just this formula because this equation is not given with the equations for the AP Physics exam. Instead of counting on memorized equations, you should understand how to equate the gravitational force with the centripetal force to derive it when you need it.

GENERAL ORBITS (C EXAM)

Circular orbits may be the simpler case, but real orbits are often elliptical. In the more general case of elliptical orbits, Kepler's law still holds. However, the form of the equation is modified in two ways. First, an ellipse does not have a radius as a circle does; instead, an ellipse has both a major (long-way) axis and a minor (short-way) axis. Half the major axis or the semimajor axis, a, can be thought of as the longest possible radius of an ellipse. For elliptical orbits, Kepler's third law uses the semimajor axis, a, rather than the radius. The second way that Kepler's law is modified for this general case is that it uses the sum of the masses, $(M + m)$, for the orbiting and orbited bodies in place of the mass, M, of the orbited body. Incorporating these changes gives

$$T^2 = \frac{4\pi^2}{G(M + m)} a^3$$

In an elliptical orbit, the orbiting body (planet or moon) changes its distance from the orbited body (sun or planet) as it moves through the orbit. As this distance changes, so does the orbital speed. You can find the orbital speed at different orbital distances by using the law of conservation of energy. The total kinetic energy plus potential energy remains constant. However, in this case, do not use mgh for the potential energy; instead, use

$$U_G = \frac{-Gm_1m_2}{r} = \frac{-GMm}{r}$$

where m_1 and m_2 are replaced by the masses M and m, of the two bodies in the orbital problem. The total energy is conserved, so

$$E = \frac{1}{2}mv^2 - \frac{GMm}{r} = \text{constant}$$

Use the known values of v and r at a given orbital position to solve for the constant ($= E$). Then the speed at any other distance r is given by

$$v = \sqrt{\frac{2E}{m} + \frac{2GM}{r}}$$

REVIEW QUESTIONS

1. A car is traveling at a speed of 10 m/s. A 0.5 kg clump of mud is stuck to the outside edge of the tire, which has a radius of 0.2 m. What force is acting on the mud, keeping it from flying off?

   (A) 250 N outward

   (B) 250 N inward

   (C) 25 N outward

   (D) 25 N inward

   (E) 40 N inward

2. Which of the following is a correct expression giving the value of the centripetal acceleration on the moon as it orbits the Earth, where M_E is the mass of the Earth, R_E is the radius of the Earth, M_M is the mass of the moon, and D_M is the center-to-center distance between the Earth and the moon? Assume that the orbit is approximately circular.

   (A) $\dfrac{GM_E}{R_E^2}$

   (B) $\dfrac{GM_M}{R_E^2}$

   (C) $\dfrac{GM_M}{D_M^2}$

   (D) $\dfrac{GM_M}{R_E^2}$

   (E) $\dfrac{GM_E}{D_M^2}$

Use the following figure to answer question 3.

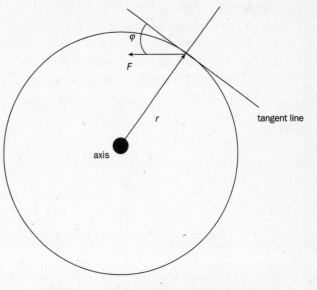

3. A force, F, is applied to a wheel of radius r, as shown in the above figure. Which of the following gives the magnitude of the torque, τ, acting on the wheel?

   (A) $\tau = rF \sin \varphi$

   (B) $\tau = rF \cos \varphi$

   (C) $\tau = rF\varphi$

   (D) $\tau = rF \tan \varphi$

   (E) $\tau = rF$

4. Starting from rest, a 2-meter-long pendulum swings from an angle of 12.8°. Of the choices below, which is closest to its speed at the bottom position, where the angle is 0°? (Use sin 12.8° = 0.22, cos 12.8° = 0.975, and $g = 10$m/s^2.)

   (A) 100 m/s

   (B) 0.1 m/s

   (C) 10 m/s

   (D) 1 m/s

   (E) It is not possible to answer this question without knowing the mass of the pendulum bob.

5. In a laboratory experiment, you are asked to measure the spring constant of a spring. The length of the spring when no mass is attached to it is 0.2 m. With a 0.5 kg mass attached to the spring, its length is 0.3 m. Which of the values below is closest to the value of the spring constant?

   (A) 1.7 N/m
   (B) 5 N/m
   (C) 17 N/m
   (D) 25 N/m
   (E) 50 N/m

6. A mass oscillating on a spring is transported—still oscillating—to the moon, which has about $\frac{1}{6}$ the gravitational acceleration of the Earth's surface. What is the period on the moon, T_M, compared to the period on the Earth, T_E?

   (A) $T_M = \left(\frac{1}{6}\right)T_E$

   (B) $T_M = 6T_E$

   (C) $T_M = \left(\frac{1}{\sqrt{6}}\right)T_E$

   (D) $T_M = \left(\sqrt{6}\right)T_E$

   (E) $T_M = T_E$

7. A swing consists of a 5 kg seat tied to the end of a 10 m-long rope, which is suspended from a tall tree. If you push a friend on the swing, which of the following is closest to the time it will take for your friend to make one complete oscillation so you can push her again?

   (A) $\sqrt{2\pi}$ s

   (B) $\frac{1}{\sqrt{2\pi}}$ s

   (C) 2π s

   (D) $\frac{1}{2\pi}$ s

   (E) It is not possible to answer this question without knowing the friend's mass.

8. Far from any other masses, two masses, m_1 and m_2, are interacting gravitationally. The value for the mass of m_1 suddenly doubles. What happens to the value of the gravitational force that mass m_2 exerts on mass m_1?

   (A) It doubles.
   (B) It decreases by a factor of 2.
   (C) It quadruples.
   (D) It decreases by a factor of 4.
   (E) Nothing, because the mass m_2 did not change.

9. Which of the following expressions gives the value of the acceleration due to gravity for an object of mass m near the surface of the Earth, if M_E and R_E are the mass and radius of the Earth?

   (A) $g = \dfrac{GM_E m}{R_E^2}$

   (B) $g = \dfrac{Gm}{R_E^2}$

   (C) $g = \dfrac{GM_E}{R_E^2}$

   (D) $g = \dfrac{GM_E}{R_E}$

   (E) $g = \dfrac{GM_E m}{R_E}$

10. Two planets are orbiting the same star in circular orbits. Planet B is 3 times as far from the star as planet A. Which of the following is closest to the time it takes planet B to orbit the star?

    (A) About the same time as planet A

    (B) About 3 times as long as planet A

    (C) About 5 times as long as planet A

    (D) About 9 times as long as planet A

    (E) About 27 times as long as planet A

C Exam

11. A spherical star with mass M and radius R_0 is in the process of expanding to its red giant stage. It is spinning with an angular velocity of ω_0. If the radius, R, doubles while the star keeps the same mass, what is the new angular velocity, ω, of the star? For a sphere, use $I = \dfrac{2}{5} MR^2$.

    (A) $\omega = 2\omega_0$

    (B) $\omega = 4\omega_0$

    (C) $\omega = \dfrac{1}{2}\omega_0$

    (D) $\omega = \dfrac{1}{4}\omega_0$

    (E) $\omega = 0$

12. A wheel is spinning with an angular velocity of 2 rad/s. It accelerates at a constant angular acceleration of 3 rad/s² for 5 s. What is its final angular velocity?

    (A) −13 rad/s

    (B) 6 rad/s

    (C) 15 rad/s

    (D) 17 rad/s

    (E) 30 rad/s

13. Children are playing on a merry-go-round in the city park. After the children take their places sitting on the merry-go-round, an adult who is not riding the merry-go-round starts pushing it to get it spinning faster. Is the angular momentum of the merry-go-round, by itself, conserved?

   (A) No, because the adult pushing the merry-go-round is an external torque.

   (B) Yes, because angular momentum is always conserved.

   (C) No, because the children sitting on the merry-go-round affect its rotation.

   (D) Yes, because there is nothing affecting the rotation of the merry-go-round.

   (E) Yes, because there is no external torque on the merry-go-round.

14. A disk-shaped rotating flywheel has a radius of 2 m and a total mass of 5 kg. A torque of 20 N · m is applied tangential to the wheel. What is its angular acceleration? For a disk, $I = \frac{1}{2}MR^2$.

   (A) 0 rad/s^2

   (B) 0.2 rad/s^2

   (C) 2 rad/s^2

   (D) 30 rad/s^2

   (E) 200 rad/s^2

15. As the Earth orbits the sun in its elliptical orbit, its distance from the sun varies. Its closest approach to the sun is called perihelion. Which of the statements below is true?

   (A) Earth's speed is fastest at perihelion because energy and angular momentum must be conserved.

   (B) Earth's speed is slowest at perihelion because energy and angular momentum must be conserved.

   (C) Earth's speed is fastest at perihelion because torque and angular velocity must be conserved.

   (D) Earth's speed is slowest at perihelion because torque and angular velocity must be conserved.

   (E) Earth's speed does not change during its orbit because energy and angular momentum must be conserved.

FREE-RESPONSE QUESTION

1. In a laboratory situation, you have a spring and a half dozen or so known masses that are sufficient to stretch the spring but not enough to exceed its elastic limit. You have a bracket that is suitable to hang the spring on. You also have a meter stick or a similar device for measuring lengths and an accurate timing device. You have no other equipment.

   (a) Devise and describe an experimental procedure to measure the spring constant of the spring by direct application of Hooke's law. Describe each step in sufficient detail so that virtually anyone reading your procedure could do the experiment correctly.

   (b) Devise and describe a second experimental procedure for measuring the spring constant of the spring. This time, apply the principles of simple harmonic motion. Again, describe each step in sufficient detail so that no one reading your procedure could do the experiment incorrectly.

   (c) After doing the experiments, you find that the measured value of your spring constant is 30 N/m. What would be the period and frequency of oscillation of a 4 kg mass hanging on the end of this spring?

   (d) You are now given a spool of cord strong enough to support this mass and asked to design a pendulum that has the same period as the oscillating spring. What is the length of this pendulum?

ANSWERS AND EXPLANATIONS

1. B

This question requires you to compute the centripetal force *and* to recognize that it acts in an inward direction. For the mud to stick to the tire and thus to continue moving in uniform circular motion, it must have an inward centripetal force acting on it. Hence, the choices containing outward forces can immediately be eliminated. Now compute the amount of the entripetal force, F_c, using

$$F_c = \frac{mv^2}{r}$$

where m is the mass of the clump of mud, v is the linear speed of the mud, and r is the radius of the circular motion. In this case, the linear speed of the rotating tire is the same as the speed of the car because the mud is on the outside edge of the tire. The radius of rotation is the same as the radius of the tire for the same reason. Substituting these numbers into the equation gives

$$F_c = \frac{(0.5 \text{ kg}) (10 \text{ m/s})^2}{0.2 \text{ m}}$$
$$F_c = 250 \text{ N}$$

So the answer is 250 N inward.

2. E

This question requires you to recognize that the centripetal force acting on an orbiting object is the gravitational force exerted on the orbiting object by the object being orbited. This is one of the key concepts involved in working circular orbit problems. You must also know the equations for the law of gravity and for centripetal force. After putting these key concepts together, you must remember Newton's second law. You can also use this law to quickly eliminate some of the answer choices. The centripetal force acting on an object (moon) is the

mass of the object times the centripetal acceleration of the object, and the centripetal force, which in this case is gravity, also depends on the mass of the object (moon). This means that the mass of the moon must cancel out. Therefore, the answer choices that include the mass of the moon cannot be correct. You can also deduce this by remembering that falling objects, which are acting only under the influence of gravity, accelerate at a rate independent of the mass of the object. An object orbiting another object is acting only under the influence of gravity; thus, it is importantly like a falling object. Again you can conclude that the mass of the moon cancels out and eliminate the choices containing the moon's mass.

To solve this problem, you will need to recognize that the centripetal force is supplied by gravity and equate the centripetal and gravitational forces. The centripetal and gravitational forces are

$$F_c = \frac{mv^2}{r}$$

and

$$F_G = \frac{GMm}{r^2}$$

In the law of gravity, M and m represent the masses of the two objects, which in this case are the Earth and moon. Because the mass, m, in the centripetal force equation represents the mass of the object moving in a circle, it is the mass of the moon. In the centripetal force equation, v is the orbital speed. In both equations, r physically represents the distance between the two objects, which is the distance between the Earth and the moon. Applied to this problem, the equations therefore become

$$F_c = \frac{M_M v^2}{D_M}$$

and

$$F_G = \frac{GM_E M_M}{D_M^2}$$

Equating them gives

$$\frac{M_M v^2}{D_M} = \frac{GM_E M_M}{D_M^2}$$

Notice that the mass of the moon, M_M, cancels out. You also now need to recall that the centripetal acceleration is the centripetal force divided by the mass, so

$$a_c = \frac{v^2}{r}$$

Thus,

$$\frac{v^2}{D_M} = \frac{GM_E}{D_M^2} = a_c$$

So the correct answer for the centripetal acceleration is

$$a_c = \frac{GM_E}{D_M^2}$$

3. B

Be very careful here; this problem is deceptive. By design, this problem will lead you astray if you memorize equations without thoroughly understanding what the symbols in the equations represent. You need to make sure you completely understand the physical meaning of all the symbols in the equations that you learn. Simple memorization is not enough.

Torque is given by this formula:

$$\tau = rF \sin \theta$$

which may lead you to select (A). If, however, you think about what θ represents in the formula, you won't fall into the trap. As shown in the figure below, θ in this formula for torque represents the angle between the extended lines of the force vector, F, and the radius vector, r, and the angle between two vectors should always be drawn between the two arrows. Hence, the angle that should be used for θ in the torque formula is φ plus 90 degrees: $\theta = \varphi + 90°$.

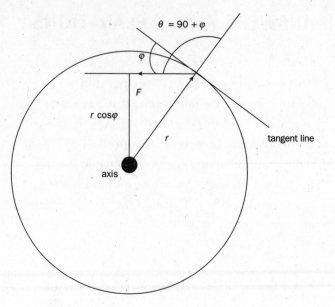

Using the correct angle between the vectors gives

$$\tau = rF \sin (\varphi + 90°) = rF \cos \varphi$$

An alternate way to arrive at the correct answer is to understand why the torque formula contains the sine function in the first place. It is there because the distance must be the shortest, or perpendicular, distance from the rotation axis to the extended force vector line. As the diagram shows, this perpendicular distance is $r \cos \varphi$; it is not $r \sin \varphi$. So multiplying the force by the perpendicular distance also gives the correct answer:

$$\tau = rF \cos \varphi$$

The lesson here is that you should not simply memorize formulas. You must completely understand what all the symbols and equations mean. Otherwise you run the risk of applying the equations incorrectly.

4. D

To answer this question, you must apply conservation of energy to a swinging pendulum. When the pendulum is released from rest at the top, its kinetic energy is zero, and its potential energy is at its maximum. If you select the bottom position to be the zero-level reference for potential energy, then the potential energy at this position is zero, and all the energy is kinetic energy. So the potential energy at the top is equal to the kinetic energy at the bottom. To solve for the potential energy at the top, you need to know the vertical height. From the diagram below, you can see that the vertical height is given by

$$h = \ell - \ell \cos\theta = \ell\,(1 - \cos\theta)$$

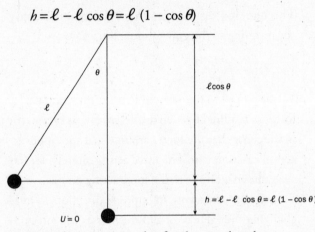

Now use the energy to solve for the speed at the bottom:

$$U_{top} = K_{bottom}$$
$$mgh = E = \frac{1}{2}mv^2 = constant$$

Notice that the mass cancels out, eliminating (E). Substitution gives

$$g\ell(1 - \cos\theta) = \frac{1}{2}v^2$$
$$v^2 = 2\,(10\text{ m/s}^2)\,(2\text{ m})\,(1 - 0.975)$$
$$v^2 = 2\,(10\text{ m/s}^2)\,(2\text{ m})\,(0.025)$$
$$v^2 = 1\text{ m}^2/\text{s}^2$$
$$v = 1\text{ m/s}$$

The correct answer is 1 m/s. The numbers in this problem were deliberately chosen so that it is possible to do the numerical calculations without a calculator.

5. E

This question applies Hooke's law as one might apply it in a laboratory to measure the spring constant of a spring. Hooke's law is

$$F = -kx$$

In this situation, ignore the negative sign, because the laboratory procedure measures the force applied to the spring rather than the restoring force indicated by the negative sign. The applied force is the weight hanging on the spring. Be sure that you don't confuse mass and weight. This weight is given by

$$w = mg$$
$$w = (0.5\text{ kg})(10\text{ m/s}^2)$$
$$w = 5\text{ N}$$

If you do not know what x represents physically in Hooke's law, you can easily use the wrong value for x. In Hooke's law, x represents the amount the spring is stretched (or compressed) from its equilibrium length—it does not represent the length of the spring. So in this case

$$x = 0.3\text{ m} - 0.2\text{ m}$$
$$x = 0.1\text{ m}$$

Substituting these numbers into Hooke's law gives

$$F = kx$$
$$5\text{ N} = k\,(0.1\text{ m})$$
$$k = 50\text{ N/m}$$

So the correct answer is 50 N/m.

6. E

In addition to requiring you to apply the formula for the period of a spring, this question requires you to understand the distinction between mass and weight clearly. An object's mass is a fundamental property of the object that does not change when the object's location changes. Mass is therefore exactly the same on the moon as it is on Earth. Weight, on the other hand, is the force of gravity acting on an object, which does change with the object's location. Any object will weigh about $\frac{1}{6}$ as much on the moon as on Earth, even though its mass remains the same.

Now apply the equation for the period of an oscillating spring:

$$T_s = 2\pi\sqrt{\frac{m}{k}}$$

Notice that in this equation the period depends on mass but not on weight or acceleration due to gravity. Transporting the spring to the moon will also in no way affect the spring constant. There is nothing in the equation that is different on the moon than on the Earth, so the correct answer is that the period of an oscillating spring does not change when it is transported from the Earth to the moon: $T_M = T_E$.

You should beware, however. Not all simple harmonic oscillators will keep the same period if they are transported to the moon. You cannot just mechanically apply this result to any simple harmonic oscillator; you must look at the equation. For example, the period of a pendulum does not depend on mass; however, it does depend on the acceleration of gravity, g. So a pendulum on the moon will have a period that is about $\sqrt{6}$ times longer than the same pendulum on the Earth.

7. C

Unlike an oscillating spring, the period of a pendulum does not depend on its mass, so you should immediately eliminate (E). Now look at the equation for the period of a pendulum:

$$T_p = 2\pi\sqrt{\frac{\ell}{g}}$$

If the length, ℓ, is 10 m, then $\frac{\ell}{g}$ equals 10 m/(10 m/s^2) = 1 s^2, and $\sqrt{\frac{\ell}{g}} = 1$ s. Therefore, the correct answer for the period is

$$T_p = 2\pi\ \text{s}$$

8. A

This question requires you to recall and understand the meaning of the gravitational force law:

$$F_G = \frac{Gm_1m_2}{r^2}$$

You also need to understand that the gravitational force, as required by Newton's third law, is symmetric in the sense that the forces are an equal and opposite action reaction pair. The force acting on mass m_1 has the same magnitude as the force acting on mass m_2, though the forces work in opposite directions. Thus, changing the value of one of the masses will affect the force acting on either mass. You can immediately eliminate (E). Now look at the form of the force equation. If the other variables do not change, then the force is directly proportional to either of the masses. So doubling the value of either one of the masses, while not changing anything else, will double the value of the force. Therefore, the correct answer is that the force doubles, choice (A).

9. C

To answer this question, you should recall that there are two expressions for the gravitational force acting on an object near the surface of the Earth. The weight, w, of an object of mass m is given by

$$w = mg$$

Applying the law of gravity to an object near the Earth's surface gives

$$F = \frac{GM_E m}{R_E^2}$$

Because an object's weight is the force of Earth's gravity acting on the object, $F = w$, and we have

$$mg = \frac{GM_E m}{R_E^2}$$

Canceling out the mass, m, of the object leaves us with the answer:

$$g = \frac{GM_E}{R_E^2}$$

Dimensional analysis provides a nice shortcut for this problem. You should readily recall that $g = 9.8 \text{ m/s}^2$. Notice that the units are m/s^2. If you check the units of the other choices, you will notice that only (B) and (C) have the correct units. If you also recall that g is independent of the mass of the falling object, you can also eliminate (B), which contains that mass. This elimination only leaves (C), which is the correct answer.

10. C

This question requires applying Kepler's third law:

$$T^2 = \frac{4\pi^2}{GM} r^3$$

where T is the orbital period of the planet, r is the orbital radius, and M is the mass of the star being orbited. If you don't remember this equation, you should be able to derive it by equating the centripetal force to the gravitational force. However, you don't need to remember the entire equation to solve this problem. It is sufficient to remember that

T^2 is proportional to r^3

or

$T^2 = (\text{some constants}) \, r^3$

Because you are taking the ratio for the two planets, the constants will cancel out. Taking the ratio of planet B over planet A gives

$$\frac{T_B^2}{T_A^2} = \frac{(\text{constants}) \, r_B^3}{(\text{constants}) \, r_A^3}$$

The constants will cancel. Now the question states that $r_B = 3r_A$, so:

$$\frac{T_B^2}{T_A^2} = \frac{r_B^3}{r_A^3}$$

$$\frac{T_B^2}{T_A^2} = \frac{(3 \, r_A)^3}{r_A^3}$$

$$\frac{T_B^2}{T_A^2} = 3^3 = 27$$

Taking the square root of both sides gives:

$$\frac{T_B}{T_A} = 5.2$$

or

$$T_B = 5.2 \, T_A$$

This tells us that the orbital period for planet B is 5.2 times the orbital period for planet A. So the correct answer is about 5 times the orbital period of planet A.

Note: If you know that $5^2 = 25$, you don't need a calculator to compute the square root of 27. It is sufficient to know that it is a little more than 5. You should also notice that taking ratios will cause constant values to cancel out and save a considerable amount of arithmetic in many problems.

C Exam

11. D

This question uses the principle of conservation of angular momentum. There are no external torques, so the angular momentum, L, is conserved.

$$L_0 = L$$

$$L_0 \omega_0 = L\omega$$

The moment of inertia of a sphere is $\frac{2}{5} MR^2$.

However, if you don't remember this formula, just remember that moments of inertia take the form of a fraction times MR^2. Because the shape remains spherical, the same fraction will appear in both sides of the equation and cancel out. In this case, the mass does not change, so it, too, will cancel out.

$$\frac{2}{5} MR_0^2 \omega_0 = \frac{2}{5} MR^2 \omega$$

Letting $R = 2R_0$ and canceling $\frac{2}{5} M$ gives the correct answer:

$$\omega = \frac{1}{4} \omega_0$$

12. D

To solve this problem, you need to apply the angular kinematic equations. To remember these equations, it helps to use the analogy between translational and rotational motion. The relevant equation in this case is translational motion:

$$v = v_0 + at$$

So the analogous rotational form is

$$\omega = \omega_0 + \alpha t$$

where ω and ω_0 are the final and initial angular velocities, α is the angular acceleration, and t is the time. Notice the analogous form between the translational and rotational equations. Use this analogy to aid you in recalling the angular kinematic equations when you need them. Now substitute the values from the problem into the above equation: $\alpha = 3$ rad/s^2, $t = 5$ s, and $\omega_0 = 2$ rad/s. The substitution gives

$$\omega = 2 \text{ rad/s} + (3 \text{ rad/s}^2)(5 \text{ s})$$

$$\omega = 2 \text{ rad/s} + 15 \text{ rad/s}$$

$$\omega = 17 \text{ rad/s}$$

So the correct answer is 17 rad/s.

13. A

To answer this question, you must recall the law of conservation of angular momentum in its entirety. Many people will incompletely recall this law as stating, "Angular momentum is always conserved," which is not correct. The correct statement of the law is "When the net external torque is zero, the angular momentum of an isolated system is conserved." You should notice that the same thing is true for the conservation of momentum in the translational case. Momentum is not always conserved; rather the momentum of an isolated system with a net external force of zero is always conserved. Conservation of angular momentum is analogous to conservation of momentum.

Now in this situation, the adult is pushing the merry-go-round. The adult is not on the merry-go-round, so he or she is supplying an external torque to it. The merry-go-round is, therefore, not an isolated system. The correct answer is that the angular momentum is not conserved because the adult is supplying an external torque to the merry-go-round.

14. C

Solving this problem requires two steps. First, you must find the moment of inertia of the disk using the formula given. Second, you must apply Newton's second law in its rotational analogue to calculate the angular acceleration.

The moment of inertia formula for a disk of total mass M and radius R is

$$I = \frac{1}{2} MR^2$$

Inserting the numbers provided in the problem gives

$$I = \frac{1}{2} (5 \text{ kg}) (2 \text{ m})^2$$

$$I = 10 \text{ kg} \cdot \text{m}^2$$

Now apply the rotational analogue of Newton's second law. You can use the analogy with the more familiar translational form to assist your memory. The translational form for one-dimensional motion is

$$F = ma$$

Torque, τ, is the rotational analogue to force, F. Moment of inertia, I, is the analogue to mass, m, and angular acceleration, α, is the analogue to acceleration, a. So the rotational form of Newton's second law is

$$\tau = I\alpha$$

We know that $I = 10 \text{ kg} \cdot \text{m}^2$ and $t = 20 \text{ N} \cdot \text{m}$, so

$$\alpha = \frac{20 \text{ N} \cdot \text{m}}{10 \text{ kg} \cdot \text{m}^2}$$

$$\alpha = 2 \text{ rad/s}^2$$

The correct answer is therefore 2 rad/s^2.

15. A

This question applies conservation of energy and angular momentum to an orbiting object. First you should recognize that energy and angular momentum are quantities that are conserved. You should similarly recognize that torque and angular velocity are quantities that are not conserved. You have heard the phrases "conservation of energy" and "conservation of angular momentum." You have not (or should not have) heard the phrases "conservation of torque" and "conservation of angular velocity". There is a reason you haven't heard those phrases: these principles do not exist. So you should immediately eliminate (C) and (D).

Now you should think about what conservation of energy and conservation of angular momentum tell us. Consider energy first. When you drop an object near the Earth's surface, it moves faster as it falls.

Conservation of energy tells us that potential energy is converted to kinetic energy, so the object speeds up. The same principle applies to an orbiting object. As Earth moves closer to the sun, potential energy is converted to kinetic energy. Hence, by conservation of energy, Earth has its fastest orbital speed when it is closest to the sun.

Conservation of angular momentum leads to the same conclusion. Think about a figure skater going into a spin. She starts with her arms outstretched. As she brings her arms closer to her body, her moment of inertia decreases, and her angular velocity must increase to conserve angular momentum. The same thing is true for a planet orbiting the sun. As Earth moves closer to the sun, its moment of inertia decreases. Hence its angular velocity, and therefore linear speed, must increase to conserve angular momentum. Therefore, by conservation of angular momentum, Earth has its fastest orbital speed when it is closest to the sun.

If you have one orbital speed and distance, you can apply the equations of conservation of angular momentum or conservation of energy to calculate the orbital speeds at other distances. However, that is not the form of this question. The correct answer to this question is that by both conservation of energy and conservation of angular momentum, the Earth has its fastest speed when it is at perihelion.

FREE-RESPONSE QUESTION

You should expect to see questions on the AP Physics exam that cover laboratory situations. To answer this question, you will need to apply your understanding of simple harmonic motion to a laboratory context. You must design two experiments to measure the spring constant of a spring, one by a direct application of Hooke's law and the other by application of simple harmonic motion. When designing these experiments, you should remind yourself of the good experimental technique that is required for accurate results as well as the principles of physics that you are applying. After having you design experiments, this question asks you to make some standard simple harmonic motion calculations.

(a) Hooke's law states that the restoring force is proportional to the distance the spring is stretched from its equilibrium position: $F = -kx$. F is the force applied by the spring, k is the spring constant, and x is the distance the spring has moved from its equilibrium length when force is acting on it. To measure the spring constant applying Hooke's law, first measure the length of the spring when no force is stretching or compressing the spring (e.g., when it is lying on a table). Now hang the spring from the hook. Hang various masses on the spring and measure the length of the spring with each mass hanging on the spring. In principle, one mass is sufficient to measure the spring constant, but using several different masses and taking the average of the resulting experimental values for the spring constant will increase the accuracy. To find the value of x for each mass, subtract the unstretched length of the spring from the length of the spring with the mass hanging on it. This value is the amount the spring is stretched from its equilibrium length. Now you must find the force corresponding to the mass. The force the mass applies to the spring is its weight, so use $w = mg$ to compute the weight of each mass. Now use $F = kx$. Solving for k gives $k = \dfrac{F}{x} = \dfrac{mg}{x}$. It is possible to compute the value of k from the values of F and x for each mass and then take the average value of k. Alternatively, one can plot a graph of F versus x. The slope of the best-fitting straight line is then the spring constant.

(b) To measure the spring constant using simple harmonic motion, suspend the spring from the hook. Hang a mass on the end of the spring. Allow the mass to oscillate and time the period of oscillation. Because of human reaction time, timing a single oscillation will not be very accurate. Measure the time for several complete cycles (perhaps 10, but it might depend on the time a cycle takes) and divide the total time by the number of cycles to compute the period. Now apply simple harmonic motion equations to find the spring constant. The period of an oscillating spring is $T = 2\pi \left(\dfrac{m}{k} \right)^{\frac{1}{2}}$. Solving for k gives $k = \dfrac{4\pi^2 m}{T^2}$, where T is the measured period and m is the mass on the spring. Using the mass and measured period, compute the spring constant. For greater accuracy, repeat this procedure for several masses and take the average value of k.

(c) To find the period of the oscillating spring, use

$$T_s = 2\pi \left(\frac{m}{k} \right)^{\frac{1}{2}}$$

$$T_s = 2\pi \left(\frac{4 \text{ kg}}{30 \text{ N/m}} \right)^{\frac{1}{2}}$$

$$T_s = 2.3 \text{ s}$$

The frequency is just 1 over the period:

$$f_s = \frac{1}{T_s}$$

$$f_s = \frac{1}{2.3 \text{ s}}$$

$$f_s = 0.43 \text{ Hz } (1 \text{ Hz} = 1 \text{ s}^{-1})$$

(d) To find the length of a pendulum with a period of 2.3 s, use the equation for the period of a pendulum, which is

$$T_P = 2\pi \sqrt{\frac{\ell}{g}}$$

where ℓ is the length of the pendulum. Solving for ℓ gives

$$\ell = \frac{T_p^2 \, g}{4\pi^2}$$

$$\ell = (2.3 \text{ s})^2 \frac{(9.8 \text{ m/s}^2)}{4\pi^2}$$

$$\ell = 1.3 \text{ m}$$

CHAPTER 6:   FLUID MECHANICS

NOTE: This topic appears *only* on the B exam.

INTRODUCTION

Fluid mechanics is a very old and well-established area of study. Some of the topics in this chapter were known in one form or another to the ancient Greeks; some of these topics may have been known to the ancient Egyptians as well, for whom the irrigation of fields meant the difference between life and death.

A **fluid** is anything that is capable of flowing (both liquids and gases are included under this definition). Water is the most common **working fluid** (i.e., fluid used to absorb and transfer heat energy) on the planet. Because it is so common, physicists have a good understanding of both its properties and its behavior. They understand water so well, in fact, that much of what is contained in this chapter has been relegated to the domain of engineering rather than being seen as a respectable area of study for a serious physics student. However, there is a great deal to gain from studying the behavior of a seemingly simple phenomenon such as water flowing through pipes.

The object of this chapter is to provide a solid review of the things you need to know for any fluid mechanics questions you may see on the AP Physics B exam. This topic accounts for about 6 percent of the exam; even though this might seem like a small percentage, you cannot afford to ignore this topic. It is common that one of the first parts of a free-response question will deal with fluid mechanics, and the answer to that part will be necessary if you want to receive full credit for the question.

The good news is that only certain types of fluid mechanics questions can be asked. After a review of each of the major divisions of this topic, we will try to give you a heads-up about these common question types as well as any predictable "trick" questions that are likely to pop up on the test.

HYDROSTATICS

Hydrostatics is the study of static fluids (i.e., fluids that are not moving). The behavior of static fluids is important for a wide range of applications, ranging from dam building to stopping your car.

HYDROSTATIC PRESSURE

In the mechanics chapters of this book, you dealt with quantities such as mass, velocity, energy, and force, among others. In this chapter, we will typically consider these quantities only indirectly. For example, instead of mass, a more convenient quantity for fluid mechanics is **density**, ρ, defined as

$$\rho = \frac{mass}{volume} = \frac{m}{V}$$

The units for density are kg/m^3. A related quantity that you will see less frequently is **specific gravity**, which is the ratio of a substance's density to the density of water (assumed to be $1,000 \ kg/m^3$). This ratio is dimensionless and must usually be converted back into the required density before you do any calculations.

Another quantity that is used is **pressure**, P, defined as

$$P = \frac{force}{area} = \frac{F}{A}$$

The units for pressure are N/m^2, which are given the derived unit name of *pascals* (Pa). You should note that pressure is a scalar quantity, though the force involved is a vector and is always assumed to be directed perpendicular to the area in question. Suppose you are standing on the floor. The force you exert on the floor is equal to your weight and is directed straight downward, perpendicular to the floor. The pressure that you are exerting on the floor, on the other hand, is the magnitude of that force divided by the contact area between you and the floor (the outlines of your feet). But because you are not a fluid, the pressure is felt only by the part of the floor directly beneath your feet.

When a fluid is the cause of a force, it can flow until it reaches an obstruction or boundary. This means that the pressure generated is felt by all of the surfaces with which the fluid comes into contact. For example, atmospheric pressure is a result of the weight of all of the air in the atmosphere in a column above a given area, and this pressure is felt on every surface of your body. The value for atmospheric pressure is typically taken to be approximately 100 kPa. This pressure is also defined to be 1 atm (using an old, but still common, unit for pressure, the *atmosphere*). This special pressure is given the symbol P_0, or sometimes P_{atm}.

We can now use these ideas to describe the pressure in a stationary fluid. On the AP exam we will always be dealing with fluids that are **incompressible**, meaning that they have a constant density. Even though this is not true of air and other gases, it is usually taken to be true for liquids. To find the density inside some stationary fluid that is open to the atmosphere, we simply start from the air interface, or some other known pressure, and start tacking on the weight of the incompressible fluid down to whatever depth we are interested in. For example, say we are interested in knowing the pressure at a depth of h meters below the surface of some incompressible fluid having density ρ kg/m^3. If we consider the figure below, we can see that the total force on some arbitrary area, A, is going to be the weight of the air plus the weight of the fluid directly above the area.

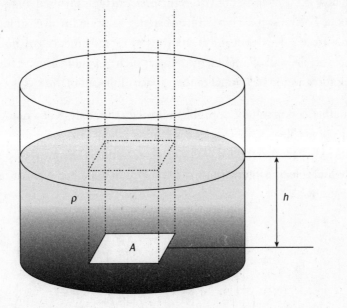

To find the pressure, we note that the total weight is

$$w_{total} = w_{air} + w_{fluid}$$
$$= P_0 A + \rho g A h$$

where g is the acceleration due to gravity, 9.8 m/s². If we then divide both sides by the arbitrary area, A, we get the formula for the hydrostatic pressure:

$$P = P_0 + \rho g h$$

It is important to note that it does not matter where we are in a given fluid or what the shape of the container holding the fluid happens to be. As long as the fluid has no breaks in it, all points at the same horizontal elevation are at the same pressure. Another implication is that the pressure 100 m below the surface of the ocean is the same as the pressure at the bottom of a hypothetical 100-meter-long soda straw filled with seawater.

When we calculate pressure with the hydrostatic pressure equation, we are calculating something called the **absolute pressure**. This means that it includes the contribution of the atmospheric pressure, and this absolute pressure is *always* a positive quantity. Sometimes, however, it is convenient to ignore the contribution of the atmospheric pressure and define a quantity called **gauge pressure**. Gauge pressure is simply equal to absolute pressure minus atmospheric pressure; this quantity is often negative. For example, when you breathe, you create a negative gauge pressure in your lungs that allows the atmosphere to push air into them.

So how is the concept of hydrostatic pressure typically used on the AP Physics B exam? The most basic type of hydrostatics question involves simply finding the pressure at some depth and then using it in some other calculation, say to find the force on the wall of a submerged object such as a submarine. As described earlier, this force would be normal to the surface of the submarine.

Another common type of problem involves the analysis of a **manometer**. In this type of problem, a hollow piece of tubing is bent into some shape (typically one or more U-shapes) and filled with one or more fluids separated by distinct interfaces. For example, the figure below shows a typical manometer situation as seen on the AP exam.

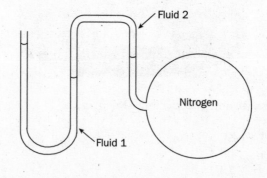

In the figure, we see a manometer containing two fluids that is connected to a holding tank containing gas at some pressure. The quantities involved in the problem are the densities of the fluids, the pressures at various points in the manometer, the elevations of the fluid and gas interfaces, and the pressure in the gas. It is typically assumed that the density of the gas in such a problem is small enough that the difference in pressure from one elevation in the gas to another is negligible. A typical problem may ask you to calculate any of the relevant quantities, which at first glance may seem insurmountable. However, all of these problems involve only the formulas we've already discussed.

To attack such problems, the best plan is to start by labeling all of the interfaces. In the following figure, these are labeled as A, B, and C. We then label any known pressures. For example, since the manometer is open to the atmosphere, the pressure at A is known to be $P_0 = 1.0 \times 10^5$ Pa. Next, label each fluid with its density. We then check for any other points in the same continuous stretch of a fluid that happen to be at the same elevation (and therefore the same pressure) as either some known pressure or one of the interfaces. Finally, we label any elevation differences in the fluids that arise due to the identification of these interfaces.

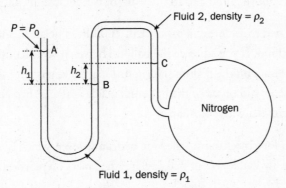

Now we simply use the hydrostatic pressure equation starting from any known pressures and working toward our goal. Say the problem tells us that the density of the first fluid is $\rho_1 = 1,000$ kg/m^3, the density of the second fluid is $\rho_2 = 700$ kg/m^3, the distance $h_1 = 300$ cm, and the distance $h_2 = 200$ cm. Then the problem would be to find the pressure at each interface and, ultimately, the pressure in the nitrogen gas.

We start by noting again that the pressure at point A is the standard atmospheric pressure of 1.0×10^5 Pa. We also know that the pressure at interface B will be the same as the pressure at the elevation $h_1 = 3.0$ m below A. Remember to pay careful attention to units in this kind of problem.

$$P_B = P_A + \rho_1 g h_1$$
$$= 1.0 \times 10^5 \text{ Pa} + 1,000 \text{ kg/m}^3 \times 9.8 \text{ m/s}^2 \times 3.0 \text{ m}$$
$$= 1.29 \times 10^5 \text{ Pa}$$

Then we use the pressure at B to find the pressure at C. Because point B is at a lower elevation than C, we must start with

$$P_B = P_C + \rho_2 g h_2$$

Solving for P_C and substituting values gives

$$
\begin{aligned}
P_C &= P_B - \rho_2 g h_2 \\
&= 1.29 \times 10^5 - 700 \text{ kg/m}^3 \times 9.8 \text{ m/s}^2 \times 2.0 \text{ m} \\
&= 1.15 \times 10^5 \text{ Pa}
\end{aligned}
$$

Now, because the gas is assumed to be at the same pressure at all elevations, we can say that the nitrogen is at a pressure of 1.15×10^5 Pa.

PASCAL'S PRINCIPLE

Another important aspect of hydrostatics is known as **Pascal's principle**. This principle states that when pressure is applied to an *enclosed* static fluid, it will be instantaneously transmitted—undiminished—throughout the fluid and to the walls of the enclosing container.

Consider a balloon inflated to an internal pressure of 200 kPa. The air inside the closed balloon constitutes an enclosed fluid. If you squeeze the balloon between your hands and exert an additional 50 kPa of pressure on the balloon, the additional 50 kPa is felt instantaneously in *all* portions of the enclosed air and on *all* of the balloon's inside surfaces, not just on the portions that happen to be in direct contact with your hands.

This principle is the basis for a mechanism called the **hydraulic lever**, which is most commonly used in hydraulic jacks and in the hydraulic system that operates the brakes on practically every automobile.

In a hydraulic lever, we enclose a fluid between two pistons that have different cross-sectional areas (see the figure below). When an outside force is applied to the piston that has the smaller area, the resulting pressure increase is transmitted throughout the fluid, including (and most importantly) to the larger piston.

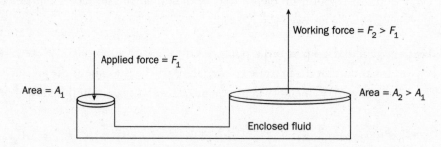

The pressure increase on the fluid due to the applied force on the smaller piston is

$$P_{\text{increase}} = \frac{F_1}{A_1}$$

Setting this equal to the pressure increase on the underside of the larger piston, we get

$$P_{piston2} = \frac{F_2}{A_2} = P_{increase} = \frac{F_1}{A_1}$$

Solving this for F_2, we see the net multiplication of the applied force is given by

$$F_2 = \frac{A_2}{A_1} F_1$$

The larger the ratio of the piston areas, the larger the force multiplication becomes.

BUOYANCY AND ARCHIMEDES' PRINCIPLE

In a fluid, pressure varies with depth. This fact gives rise to a very important effect. We are all familiar with the fact that objects seem to weigh less when immersed in water. Some objects even float on the water's surface, apparently weightless. This effect is due to a net imbalance in the hydrostatic pressure between the top and bottom of an object immersed in a fluid. The next figure shows how this imbalance arises.

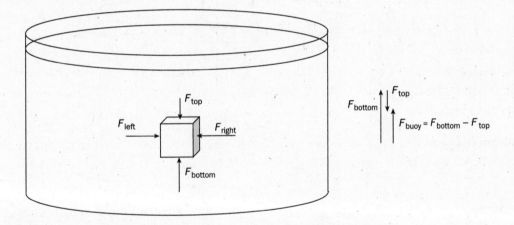

Because the force associated with the pressure at a given depth is always directed normal to the surface of the object, the force on the bottom surface is directed upward and the force on the top surface is directed downward. The magnitude of the forces is equal to the product of the pressure at that depth and the surface area. If we consider a cubic body of area A on each face, we can find the net upward force, called the **buoyant force**:

$$
\begin{aligned}
F_{buoy} &= F_{bottom} - F_{top} \\
&= P_{bottom}A - P_{top}A \\
&= (P_0 + \rho_{fluid}gh_{bottom})A - (P_0 + \rho_{fluid}gh_{top})A \\
&= \rho_{fluid}g(h_{bottom} - h_{top})A \\
F_{buoy} &= \rho_{fluid}Vg
\end{aligned}
$$

Note that in this formula, the force depends on the density of the surrounding fluid and the volume of the immersed body. If we consider how this buoyant force compares to the weight of the body, $w = mg = \rho_{body} Vg$, we can see that if the weight is greater than the buoyant force, the object will sink. This situation occurs only when the density of the body is greater than that of the fluid. The **apparent weight** of the body is then the difference between the actual weight and the buoyant force and is always less than the actual weight.

If the density of the body is less than that of the fluid, then buoyant force is greater than the weight, and the body feels a net upward force. However, we note that the buoyant force is proportional to the *immersed* volume of the body. Therefore, the net upward force works only until a sufficient portion of the body's volume is no longer immersed in the fluid. When enough of the body is above the surface of the fluid, the buoyant force is exactly equal to the weight, and the body floats.

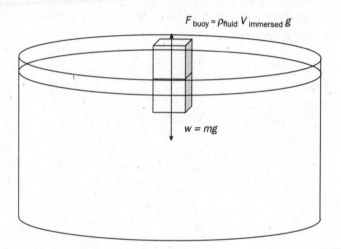

$$F_{buoy} = \rho_{fluid} V_{immersed} g$$

$$w = mg$$

Considering the formula for the buoyant force, we say that the buoyant force is equal to the weight of the fluid displaced by the body. This idea was first enunciated long ago and is known as **Archimedes' principle**, an idea captured in the buoyant force equations discussed earlier.

HYDRODYNAMICS

The previous section dealt with the behavior and characteristics of static fluids. In the following section, we consider fluid in motion. While these concepts can be made completely general for all fluids, the AP Physics B exam typically considers only the hydrodynamics of an **ideal fluid**. An ideal fluid satisfies the following set of four conditions:

1. The fluid is nonviscous. This means that the fluid is able to flow freely with no friction.

2. The fluid is incompressible. The density of the fluid is constant.

3. The flow rate does not change with time. We say that the flow is steady.

4. The flow is irrotational. This is another way of saying that the flow is smooth without any swirling or turbulence. This doesn't mean that the flow goes only in straight lines but simply that there are no vortices along the flow.

The following sections describe the formulas that are used to describe the behavior of moving fluids. As was true in the mechanics chapters, it is often convenient to resort to conservation equations when dealing with moving fluids. The equations we will focus on are versions of the conservation of mass and the conservation of mechanical energy for fluids.

THE CONTINUITY EQUATION

For an ideal fluid flowing in a pipe, the conservation of mass says that whatever amount of fluid flowing past one point in the pipe in a given amount of time must flow past any other point in the same amount of time. Consider the fluid flowing in the pipe depicted below.

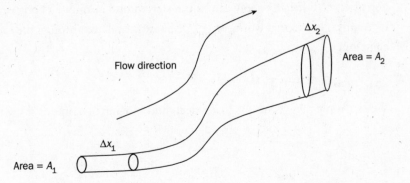

The mass, m_1, moving past A_1 in a given time, Δt, has a volume of V_1 given by

$$V_1 = A_1 \Delta x_1$$

This mass can then be given in terms of the density of the fluid as

$$m_1 = \rho_1 V_1 = \rho_1 A_1 \Delta x_1$$

Similarly, the mass, m_2, moving past A_2 in the same time has a mass given by

$$m_2 = \rho_2 A_2 \Delta x_2$$

Equating these two masses, and dividing each by the common time taken for each to move, we get

$$\frac{\rho_1 A_1 \Delta x_1}{\Delta t} = \frac{\rho_2 A_2 \Delta x_2}{\Delta t}$$

But $\dfrac{\Delta x}{\Delta t}$ is just the speed, v, of the fluid at a given position in the pipe. Further, if we assume an ideal fluid, the density at both points is the same. Canceling this common density gives

$$A_1 v_1 = A_2 v_2$$

As we have seen, this formula, known as the **continuity equation**, is merely a statement of the conservation of mass for an ideal fluid flowing in a pipe of varying cross-sectional area. It says that if the fluid meets a constriction in a given pipe, it must speed up. This is the physics behind the common backyard trick of placing one's thumb over the end of a garden hose to make the water shoot farther from the end.

> **AP EXPERT TIP**
>
> The quantity Av is often referred to in AP questions as the "volume flow rate."

BERNOULLI'S EQUATION

The final topic we will consider is a form of the conservation of mechanical energy for moving fluids known as **Bernoulli's equation**. Recall from the mechanics chapters that there are two categories of mechanical energy: kinetic, K, and potential, U. Further, energy can be transferred into or out of a body through the mechanism of external work, $W_{external}$. These are related through the following equation:

$$W_{external} = \Delta K + \Delta U$$

We can apply these concepts to an ideal fluid moving in a pipe as shown here.

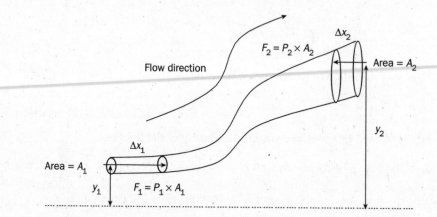

At both locations, the kinetic and gravitational potential energies are given by their common formulas:

$$K_i = \frac{1}{2} m_i v_i^2$$

$$U_i = m_i g y_i$$

Then, if we recall that the external work is given by

$$W_{external} = \sum F_i \Delta x_i \cos \theta_i$$

we can write:

$$F_1 \Delta x_1 \cos 0° + F_2 \Delta x_2 \cos 180° = \frac{1}{2}(m_2 v_2^2 - m_1 v_1^2) + (m_2 g y_2 - m_1 g y_1)$$

Because the preferred quantity when dealing with fluids is pressure rather than force, we substitute to get this:

$$P_1 A_1 \Delta x_1 - P_2 A_2 \Delta x_2 = \frac{1}{2} m_2 v_2^2 + m_2 g y_2 - \left(\frac{1}{2} m_1 v_1^2 + m_1 g y_1\right)$$

Finally, notice that the product of the cross-sectional area and the length is the volume of fluid. The conservation of mass and ideal fluid assumption ensure that the volumes, masses, and density at both ends of the pipe are the same. We can therefore divide by the volume and combine like terms to write:

$$P_1 + \rho g y_1 + \frac{1}{2} \rho v_1^2 = P_2 + \rho g y_2 + \frac{1}{2} \rho v_2^2 = constant$$

This is the final form of Bernoulli's equation, which is, as we saw, merely a form of the conservation of mechanical energy for the fluid flowing through the pipe.

We should take a moment to observe some of the useful information that follows from Bernoulli's equation. First, in a situation where there is no difference in elevation between the two locations in the pipe, we see that the pressure decreases as the flow rate increases. This fact is responsible for a wide range of useful phenomena, from throwing a curveball (in which the air pressure is greater on one side of the ball than the other) to providing lift for an aircraft.

Second, in the case where the flow is not moving ($v = 0$), Bernoulli's equation simplifies to the equation we saw above for the hydrostatic pressure at some depth below the surface. Because the hydrostatic pressure equation was the starting point for deriving the formula for buoyant force and Archimedes' principle, we could say that all of the information that we discussed in the hydrostatics section is contained as a special case of Bernoulli's equation. In fact, we could say that thoughtful applications of the conservation of energy and the conservation of mass are all that are necessary to handle any problem we may see in fluid mechanics.

We will conclude this chapter with a final example of the use of Bernoulli's equation in a third type of special case. Consider a tank of water with a leak in its side somewhere below the water line.

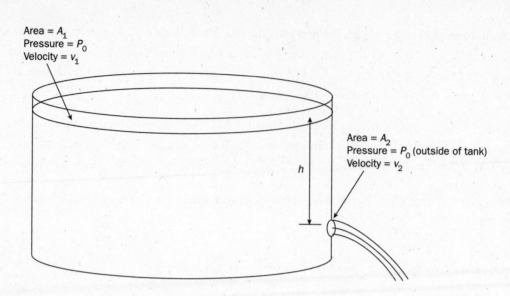

The problem in this case is to determine the flow rate of the liquid escaping from the leak. To analyze this situation, we should first note that we are concerned with a moving fluid, which means that the continuity equation and Bernoulli's equation are our most useful tools. The trick in this situation is to realize that the pressure at the leak just outside the tank is actually atmospheric pressure, P_0, exactly as it is at the surface of the water in the tank. Of course, just inside the tank, the pressure could be approximated by the hydrostatic pressure at a depth h below the surface, but we are interested in the speed of the flow out of the leak, so we must use the pressure acting on the fluid outside of the leak. In this situation, Bernoulli's equation becomes

$$\frac{1}{2}\rho v_1^2 + \rho g y_1 = \frac{1}{2}\rho v_2^2 + \rho g y_2$$

If we cancel the density from both sides and rearrange just a bit, we get this:

$$v_2^2 - v_1^2 = 2g(y_1 - y_2) = 2gh$$

Note the similarity to the kinematic equation given in the mechanics section. We see that in a situation in which the initial and final pressures are equal, Bernoulli's equation again reverts to something familiar. This equation says that in such a situation, the water coming out of the leak behaves as if it has "fallen" from a height h under the influence of gravity. Further, the stream behaves exactly like a projectile with initial velocity given by v_2.

But there are two velocities to be determined, and we have only one equation. To find the complete solution, we would use the continuity equation solved for one of the velocities:

$$v_1 = \frac{A_2}{A_1}v_2$$

This can then be substituted into the previous equation and solved for v_2. It is important to note that as the level of the liquid falls, the value of h changes. So this analysis is applicable for only a single instant in time. The analysis of this problem for finite periods of time requires the use of calculus and is beyond the scope of this exam.

However, in situations like this, it is often assumed that the area of the liquid's surface is much greater than the area of the hole. In this case, the continuity equation can be used to justify the assumption that the velocity of the falling surface is essentially zero, and the analysis can be greatly simplified.

REVIEW QUESTIONS

Use the following figure to answer question 1.

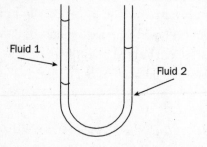

1. A manometer contains two fluids at rest and open to the atmosphere. The density of fluid 1 is ρ_1, and the density of fluid 2 is ρ_2. Based on the figure, which of the following is true about the two densities?

    (A) $\rho_1 > \rho_2$
    (B) $\rho_2 > \rho_1$
    (C) $\rho_1 = \rho_2$
    (D) $\rho_1 + \rho_2 = g$
    (E) The problem does not supply sufficient information to make a judgment.

2. A cup contains water up to a level of 15 cm. A 29 cm straw is placed straight down into it so that the end of the straw is 1 cm from the bottom of the cup. If a person sucks on the straw, what gauge pressure is needed at the top of the straw to allow the water to run out of the top and into his mouth?

    (A) 9.85×10^4 Pa
    (B) 1.5×10^3 Pa
    (C) -1.5×10^3 Pa
    (D) 1.015×10^5 Pa
    (E) -2.9×10^3 Pa

3. Approximately how deep below the surface of the ocean must a scuba diver go before the absolute pressure on her is 2 atm? Assume the density of seawater is 1,000 kg/m^3.

    (A) 1 m
    (B) 2 m
    (C) 10 m
    (D) 20 m
    (E) 100 m

4. Water flows from a 4 cm diameter pipe into a 2 cm diameter pipe. By what ratio does the velocity of the flow change?

    (A) 2
    (B) 4
    (C) 8
    (D) $\dfrac{1}{2}$
    (E) $\dfrac{1}{4}$

Use the following figure to answer question 5.

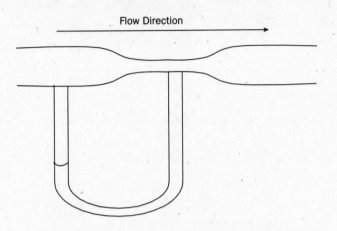

Flow Direction

5. The figure above shows a device known as a venturi. It consists of a manometer connected to a narrowed section of pipe as shown and is used to determine the flow rate in a pipe. If the pipe is horizontal, the fluid level in the right side of the manometer will be

(A) higher than the level in the left side.

(B) lower than the level in the left side.

(C) the same as the level in the left side.

(D) dependent on the density of the liquid in the manometer.

(E) dependent on the length of the venturi.

6. A cube of an unknown material is immersed in a container of water. When the top of the unknown material is placed at a depth of h meters below the surface and released, the cube remains in place, neither sinking farther nor rising toward the surface. If the cube is now moved so that the top is at a depth of $2h$ and released, what will happen to the cube?

(A) It will rise toward the top and stop at a depth of h.

(B) It will sink to the bottom of the container.

(C) It will rise and oscillate about a position at a depth of h.

(D) It will remain at a depth of $2h$.

(E) It will rise to the top and float.

7. A body having a density of 300 kg/m^3 is floating in a container of liquid of unknown density. If one-third of the volume of the body remains above the surface of the liquid, what is the density of the liquid?

(A) 200 kg/m^3

(B) 100 kg/m^3

(C) 400 kg/m^3

(D) 900 kg/m^3

(E) 450 kg/m^3

8. A balloon is filled with a gas having a density that is half that of the surrounding air. When the balloon is released from rest, what will be its acceleration? (Ignore the mass of the balloon material.)

(A) $\frac{g}{2}$ upward

(B) $\frac{g}{2}$ downward

(C) $2g$ upward

(D) $2g$ downward

(E) g upward

Use the following figure to answer question 9.

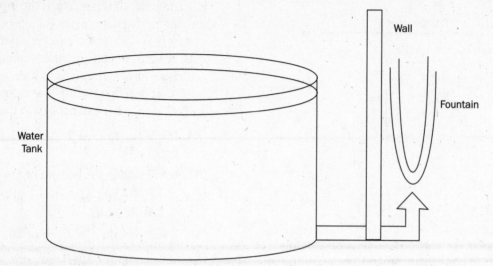

9. A water fountain is created from a water tank connected to a pipe arranged to point straight upward, as shown in the figure above. If the surface of the water in the tank is 15 m above the pipe, how high does the fountain shoot into the air? (Ignore friction effects.)

(A) Higher than the surface in the tank
(B) Lower than the surface in the tank
(C) The same height as the surface in the tank
(D) It depends on the area of the pipe.
(E) The answer cannot be determined from the information given.

Use the following figure to answer question 10.

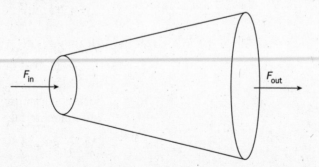

10. A conical container, shown above, is filled with water. The diameter of the small end is 1 m; the diameter of the larger end is 2 m. If a 100 N force is applied to the small end, F_{in}, what is the size of the force, F_{out}, felt at the large end?

(A) 100 N
(B) 50 N
(C) 200 N
(D) 25 N
(E) 400 N

FREE-RESPONSE QUESTION

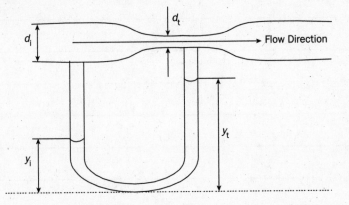

A venturi is a device that is used to measure the flow rate of fluid in pipes. It consists of an inlet that is the same diameter as the main pipe, a throat that has a smaller diameter, and an outlet that is again the same diameter as the main pipe. A manometer is connected between the inlet and the throat, but the flow in the pipe does not enter the manometer.

In the venturi above, the diameter of the inlet section is d_i, and the diameter of the throat is d_t. The elevation of the fluid in the left side of the manometer from the bottom is y_i, and the elevation on the right side is y_t. The density of the fluid in the manometer is ρ_m, and the density of the flowing liquid is ρ_f. Answer all questions in terms of given quantities and fundamental constants.

(a) Derive an expression for the difference in pressure between the inlet and the throat of the venturi $(P_i - P_t)$.

(b) Derive an expression for ratio of the inlet velocity v_i to the throat velocity v_t.

(c) Derive an expression for the inlet velocity alone.

(d) If the diameter of the throat is made smaller, explain what will happen to the height of the fluid in the throat side of the manometer, if anything.

ANSWERS AND EXPLANATIONS

1. B

Manometer questions almost always involve the hydrostatic pressure equation. In this case, you know that the pressure at the elevation of the interface between the two fluids is the same as the pressure at the same elevation in the other vertical tube in fluid 2. If both tubes are open to the atmosphere, the surfaces are both at P_0; therefore, the height of the column above that elevation would be creating the same pressure. Because the pressure increases like the term ρgh, a greater value of h implies a smaller value of ρ.

2. C

Note that it is common to approximate the acceleration of gravity as $g = 10$ m/s^2. Recall that gauge pressure is the absolute pressure *minus* the value of atmospheric pressure, P_0, which can be approximated as 10^5 Pa. The geometry in this problem works out that there is 15 cm (= 0.15 m) of the straw above the surface of the water in the cup. Because the pressure in the straw at the elevation of the water surface in the cup is equal to atmospheric pressure, the pressure at the top of the straw needs to be such that the column of water in the straw creates that much pressure. The absolute pressure at the top of the straw is therefore derived from

$$P_0 = P_{top} + \rho gh$$

Recalling the density of water to be 10^3 kg/m^3, we see that the absolute pressure is

$$P_{top} = 10^5\,\text{Pa} - 10^3\,\text{kg/m}^3 \times 10\ \text{m/s}^2 \times 0.15\ \text{m}$$
$$= 9.85 \times 10^4\,\text{Pa}$$

This is one of the given choices, but it is *absolute* pressure, which is *not* the answer we want. To obtain the *gauge* pressure, the value of $P_0 = 1 \times 10^5$ Pa must be subtracted, leaving a value of -1.5×10^3 Pa. The other choices are also results of common errors in this calculation.

3. C

Although the units of pressure are officially pascals, atmospheres are still often used. The information sheet you will have during this portion of the exam will include the conversion from atm to Pa (1 atm = 10^5 Pa). Using this conversion and the hydrostatic pressure formula, you will arrive at the correct answer, choice (C).

4. B

The continuity equation says that the velocity in the second section of the pipe is given by

$$\frac{v_2}{v_1} = \frac{A_1}{A_2}$$

Therefore, the flow velocity increases as the area decreases. Since the area is proportional to the square of the diameter, the ratio is $\left(\dfrac{4}{2}\right)^2 = 4$.

5. A

This question tests your understanding of the relationship between pressure and velocity in a flowing ideal fluid, the relationship between the velocity and flow area in a flowing fluid, and even the behavior of manometers. The continuity equation tells you that the velocity in the constricted area (known as the *throat* of the venturi) will be greater than the velocity in the wider inlet. Bernoulli's equation tells you that if the elevation of a flowing fluid doesn't change, the pressure will be lower where the velocity is greater. Finally, since the manometer must have equal pressures at the same elevations in the fluid, the column on the right must be higher than the one on the left.

6. D

Archimedes' principle states that the buoyant force is equal to the weight of the fluid displaced by an object. Another way of saying this is that an object sinks when its density is greater than that of the surrounding fluid and floats when its density is less. When the object did not rise or sink at a depth of h, we knew that the density of the object is exactly the same as that of water. This is true no matter how deep it is placed in the water. At all depths, the weight of the displaced fluid is exactly equal to the weight of the object, so the object does not move.

7. E

As noted in the answer to the previous problem, Archimedes' principle states that when an object is floating, the weight of the object is exactly equal to the weight of the displaced liquid. In this case, the amount of displaced liquid is two-thirds of the total volume of the floating body, since one-third is above the surface. If you equate the buoyant force with the weight of the body, you get

$$\rho_{liquid} V_{displaced} g = \rho_{body} V_{body} g$$

Cancel the gravitational accelerations and solve for ρ_{liquid}:

$$\rho_{liquid} = \frac{\rho_{body} V_{body}}{V_{displaced}} = \frac{\rho_{body} V_{body}}{\frac{2}{3} V_{body}} = \frac{3}{2} \rho_{body}$$

Plugging in values gives us

$$\rho_{liquid} = \frac{3}{2}(300 \text{ kg/m}^3) = 450 \text{ kg/m}^3$$

8. E

Because the density of the gas is less than that of the surrounding fluid (air), the balloon will certainly

rise. Therefore responses (B) and (D) are not correct. To find the correct response, we must perform a force balance on the balloon. The two forces are the buoyant force and the weight. The net force will then yield the acceleration from Newton's second law.

$$F_{buoyant} = \rho_{air} V_{balloon} g$$
$$F_{weight} = \rho_{gas} V_{balloon} g$$

Using the fact that $\rho_{air} = 2 \times \rho_{gas}$, we can write:

$$F_{net} = F_{buoyant} - F_{weight} = m_{gas} a_{balloon}$$
$$2\rho_{gas} V_{balloon} g - \rho_{gas} V_{balloon} g = \rho_{gas} V_{balloon} a_{balloon}$$

If we cancel the term $\rho_{gas} V_{balloon}$ from both sides, we see that the correct answer is $a = g$ upward.

9. C

Recalling that Bernoulli's equation is actually a form of the conservation of energy for a fluid, we noted earlier that water escaping to atmospheric pressure from the bottom of a tank behaved as if it had "fallen" from the height of the surface. If this is true, then the kinetic energy of the fluid at the exit is exactly equal to the change in gravitational potential energy due to the "fall." If the fluid is directed straight upward with that kinetic energy, it will rise, gaining gravitational potential energy until the kinetic energy is exhausted. In the absence of friction, this will result in the water returning to the original elevation.

10. E

This is an example of a hydraulic lever, as discussed in the chapter. The solution involves Pascal's principle, which says that the pressure exerted by F_{in} is transmitted throughout the fluid. The force on the other end, therefore, is multiplied by the ratio of the areas. This ratio is the same as the ratio of the squares of the diameters, or $\frac{2^2}{1^2} = 4$. Therefore, the force will be $F_{out} = 4F_{in} = 400 \text{ N}$.

FREE-RESPONSE QUESTION

In this problem, the results from steps (a) and (b) are used in the solution to step (c). *Never* give up on a problem completely. For example, if you do not know how to find the answer to step (a) but do know how to use it in step (b), it may be in your best interests simply to guess an answer for (a). Then use that guess in step (b) and all succeeding steps. A variation of this strategy would be to write a *short* note to the grader that you realize that the answer to (a) is necessary to solve (b) and that you don't know how to solve (a). Then use a dummy value for the solution to step (a). If you go on to solve the later steps correctly, albeit using an incorrect assumption, you may very well get a respectable score on the problem. Don't forget, overall score of 3 out of 5 on the exam can typically be obtained by getting as little as 35 percent of the total points available. A score of 5 out of 5 is often to be had by getting no more than 65 percent of the total points.

(a) The pressure difference can be found from the differences in height in the two sides of the manometer. The pressure in the throat (right) side of the manometer must be the same as that in the inlet (left) side at the same elevation. Therefore, the hydrostatic pressure equation for the right side is

$$P_i = P_t + \rho_m g(y_t - y_i)$$

Solving for the required difference yields

$$P_i - P_t = \rho_m g(y_t - y_i)$$

(b) The required ratio can be found from the continuity equation:

$$A_i v_i = A_t v_t$$

The required ratio is this:

$$\frac{v_i}{v_t} = \frac{A_t}{A_i} = \left(\frac{d_t}{d_i}\right)^2$$

(c) Having used the continuity equation to find the ratio of the two velocities, use Bernoulli's equation to find the inlet velocity. For a horizontal pipe,

$$P_i + \frac{1}{2}\rho_f v_i^2 = P_t + \frac{1}{2}\rho_f v_t^2$$

Solving this for the square of the inlet velocity yields the following:

$$\frac{1}{2}\rho_f v_i^2 = P_t - P_i + \frac{1}{2}\rho_f v_t^2$$

$$v_i^2 = \frac{2(P_t - P_i)}{\rho_f} + v_t^2$$

Substituting from the first parts (a) and (b) yields the following:

$$v_i^2 = \frac{-2\rho_m g(y_t - y_i)}{\rho_f} + \left(\frac{d_i}{d_t}\right)^4 v_i^2$$

$$v_i^2 \left[1 - \left(\frac{d_i}{d_t}\right)^4\right] = \frac{-2\rho_m g(y_t - y_i)}{\rho_f}$$

$$v_i = \sqrt{\frac{-2\rho_m(y_t - y_i)}{\rho_f\left[1 - \left(\frac{d_i}{d_t}\right)^4\right]}}$$

(d) If the throat diameter becomes smaller, the velocity will increase, as indicated in the continuity equation. This greater velocity will lead to a lower pressure, as shown by Bernoulli's equation. This lower pressure will then require a higher column of fluid in the throat side of the manometer.

CHAPTER 7:   THERMAL PHYSICS

IF YOU MEMORIZE ONLY NINE EQUATIONS IN THIS CHAPTER . . .

$$\Delta \ell = \alpha \ell_0 \Delta T \qquad\qquad W = -P\Delta V$$

$$PV = nRT = Nk_B T \qquad\qquad e = \left| \frac{W}{Q_H} \right|$$

$$K_{avg} = \frac{3}{2} k_B T \qquad\qquad e_C = \frac{T_H - T_C}{T_H}$$

$$v_{rms} = \sqrt{\frac{3k_B T}{\mu}} = \sqrt{\frac{3RT}{M}} \qquad\qquad H = \frac{kA\Delta T}{L}$$

$$\Delta U = Q + W$$

NOTE: This topic appears *only* on the B exam.

INTRODUCTION

Thermal physics is probably the single area of physics most directly responsible for the 19th-century spurt of technological advancement known as the Industrial Revolution. It is the branch of physics that deals with converting heat energy into energy in the form of mechanical motion. This is the ultimate basis for virtually every machine that uses fuel to produce usable work (e.g., automobiles and airplanes) as well as virtually every nuclear or fossil fuel plant that provides electricity to power the world's appliances.

The study of heat and temperature and their effects on the behavior of enclosed systems of gas led to many scientific insights. Two particularly important insights were the recognition that energy was of central importance in the physical world and the realization that it was the single quantity connecting the macroscopic and microscopic worlds. In fact, as far as we know, all of the energy that has ever existed still exists and will exist forever in one form or another. This idea is the concept of energy conservation, and it was first articulated in a universal way as part of the study of thermal physics.

Thermal physics questions will make up approximately 9 percent of the AP Physics B exam. After a review of this topic's major subject divisions, we will try to give you some insight into the more common types of thermal physics questions as well as any "trick" questions that are likely to pop up on the test.

TEMPERATURE, HEAT, AND THERMAL EXPANSION

In this section, we will be dealing with the fundamental definitions and material properties that are important to thermal physics. These concepts will be useful throughout the remainder of this chapter.

ZERO-TH LAW OF THERMODYNAMICS AND TEMPERATURE

When you think of the word *thermal*, the ideas of temperature and heat probably come to mind. However, these concepts are not as straightforward as one might assume. The terms *hot* and *cold* are actually quite subjective and can lead you to make inaccurate judgments about the temperature of things. For example, when you walk from a hardwood floor onto a carpeted floor in your bare feet, the hardwood floor *feels* colder than the carpeted floor, but they are usually the same temperature.

To allow for a more objective assessment of temperature, the concept of **thermal equilibrium** was developed. Two objects are said to be in thermal equilibrium if no change in their temperatures occurs when they are brought into contact. This concept is fundamental to the principle known as the **zero-th law of thermodynamics**, which states that if two objects are both in thermal equilibrium with some third object, the two objects will then be in thermal equilibrium with one another. That is to say, the two objects are at the same **temperature**.

While this principle's claims may seem perfectly obvious, the example about how our bodies subjectively perceive temperature indicates otherwise. For example, the zero-th law allows us to create an object that can be used to test whether two independent, unconnected objects are in thermal equilibrium (i.e., are at the same temperature). This "test object" is what we typically refer to as a **thermometer**.

Thermometers can be constructed in various ways, but all of them must be calibrated in some manner so that the temperatures of different things can be compared quantitatively. Over the years, several different calibration schemes have been used, each with its own units for temperature. The Fahrenheit scale, which is still used in the United States for weather forecasts, derived its zero-degree standard from the coldest temperature to which a saturated salt-water solution could be brought before freezing. It was then found that distilled water froze at about 32°F and boiled at 212°F.

This rather inconvenient scale eventually gave way to the centigrade, or Celsius, scale. This scale was more systematically calibrated, with 0°C being defined as the temperature at which distilled water freezes and 100°C being the temperature at which it boils (at a pressure of 1 atm). Using this scale, it was found that if one cools a sufficiently rarefied constant volume of gas, a linear relationship can be seen between the gas pressure and the gas temperature. For different types of gas, different slopes were observed for this relationship, but all of the slopes appeared to intersect at an extrapolated point where the pressure was assumed to be zero.

The temperature at which these pressure-temperature lines intersect is used to define a new scale called the absolute temperature scale, or the Kelvin scale. The intersection temperature is taken to be 0 K, called **absolute zero**. The "degree" is taken to be the same size as that on the Celsius scale. Using these conventions, it is found that the temperature of freezing water is 273.15 K.

You may have noted the absence of the degree symbol on the temperatures above. The Kelvin scale is the SI temperature scale; temperatures on this scale are measured in *kelvins*, not "degrees Kelvin." Notice, too, that temperature differences are the same whether one uses the Celsius scale or Kelvin scale, but in situations that call for a single temperature, only the Kelvin (or absolute) scale is appropriate. The absolute temperature can be obtained from the Celsius temperature using this formula:

$$T(\text{K}) = T(°\text{C}) + 273.15$$

Heat

We are all familiar with the fact that hot things cool down (their temperatures drop) and cold things warm up (their temperatures rise). A hot object loses something as it cools down, while a cold object gains something in warming up.

The "stuff" that is lost or gained was once believed to be a special type of fluid called "caloric fluid." The unit used to track this fluid was called the **calorie**, which was defined to be the amount of fluid needed to change the temperature of 1 gram of water by 1°C (more specifically, to move its temperature from 14.5°C to 15.5°C).

It has since been determined that temperature is actually correlated with the amount of energy that is tied up in a substance in the various types of motion and bonds of its atoms and molecules. This energy inside the substance is known as its **internal energy**. We'll discuss this in greater detail later in this chapter, but for now, suffice it to say that when a substance's temperature drops, it is losing some of this internal energy.

This is directly analogous to what you saw in the mechanics section of this book, when energy was transferred into or out of a system through *work*. However, when the energy is transferred into or out of something due to differences in temperature, the process is called not work but **heat transfer**, and the energy that is transferred is called **heat**. Because heat is a form of energy, the proper units for it are joules (J). It has been found that the comparison between the outmoded concept of caloric fluid and the more accurate concept of internal energy leads to the **mechanical equivalent of heat** (a constant factor relating the calorie to the joule), which is 1 calorie = 4.186 J.

HEAT TRANSFER

Temperature is a measure of the average kinetic energy of the individual molecules in a substance. Energy tends to flow from objects at higher temperature to objects of lower temperature because the activity of the molecules equalizes over time. **Heat** is the term we use for the amount of energy that flows from one body to another because of a difference in temperature. As with all types of energy, the unit for heat is the joule. Heat may transfer through conduction, convection, and radiation.

Energy must be added to matter in order to increase its temperature. The amount of heat energy (Q) transferred to a certain substance depends on three quantities:

1. The change in temperature (ΔT) the substance experiences
2. The mass (m) of the substance
3. The specific heat capacity (c) of the substance, with units of J/(kg°C)

The equation looks like this:

$$Q = cm\Delta T$$

The specific heat capacity of a material can be determined using an instrument called a **calorimeter.** To use a calorimeter, you begin by heating a sample of the material to a known temperature (like the boiling point of water). Then, the sample is placed in an insulated container that is filled with a known mass of water. Eventually, both the water and the sample reach a common temperature (equilibrium). Because the system is isolated in an insulated container, a certain amount of the heat energy of the sample transfers to the water and the container. According to the law of conservation of energy, the heat energy lost by the sample equals the heat energy gained by the water and the container. Using this law along with the equation for heat above, the specific heat capacity of the material may be calculated.

If there is a difference in temperature from one end of a substance to the other, heat will conduct. The rate at which heat conducts depends:

- directly on the temperature difference between the ends (ΔT).
- directly on the cross-sectional area the heat is flowing through (A).
- directly on the thermal conductivity of the substance (k).
- inversely on the distance between the two ends (L).

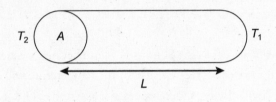

Therefore, the equation for the heat transfer rate (H) is

$$H = \frac{\Delta Q}{\Delta t} = \frac{kA\,(T_2 - T_1)}{L}$$

$$H = \frac{kA\Delta T}{L}$$

The thermal conductivity, k, is a property of the material and is measured in units of $J/(s \cdot m \cdot {}^\circ C)$. Substances with large k values (such as metals) conduct heat rapidly and are considered to be good conductors. Substances with low values for k (such as wood and air) are considered to be good insulators. Imagine walking barefoot on ceramic tiles on a cold day. The tiles are good conductors and transfer heat rapidly from your feet, making them feel cold. Now imagine walking on a carpet instead. Because of all the pockets of air in the carpet, the carpet conducts heat less rapidly from your feet, and they feel warmer.

THERMAL EXPANSION

It has long been observed that most substances change their physical size when subjected to a change in temperature. When temperature increases, most substances get larger, and when temperature decreases, most substances get smaller. Between the temperatures of 273 K and 277 K, water is a notable and important exception to this rule. We can understand this change in size by recalling that temperature is actually a measure of how much internal energy a substance has.

EXPANSION IN ONE DIMENSION

When a rod of length ℓ experiences an increase in temperature or, equivalently, an increase in internal energy, it means that the kinetic energy of the atoms and molecules in the rod has increased. The faster motion of its atoms and molecules requires more room, and the rod expands. For a relatively narrow rod, this expansion takes the form of a lengthening of the rod, to a good approximation. The amount of expansion is experimentally observed to be proportional to the change in the temperature and to the original length of the rod, ℓ_0. The constant of proportionality is called the **coefficient of linear expansion**, α. These quantities are related through the equation

$$\Delta\ell = \alpha\ell_0\Delta T$$

The coefficient of linear expansion has the SI units K^{-1}, but because we are dealing with changes in temperature, the use of degrees Celsius for the temperature will also produce a correct result. For this reason, you will often see these values given in units of $(°C)^{-1}$. The coefficient of linear expansion is a property of a given substance and, therefore, differs in value among different substances. Typical values for solids' coefficients of linear expansion are in the range of approximately 1×10^{-6} $(°C)^{-1}$ to 50×10^{-6} $(°C)^{-1}$.

EXPANSION IN TWO AND THREE DIMENSIONS

When a material that cannot be approximated as a thin rod experiences a change in temperature, it also changes its size. For example, if a flat plate of metal is heated, it gets both longer and wider. And if a cube of metal is heated, it gets longer, wider, and taller.

To account for these effects, we typically assume that each linear dimension expands according to the equation given above. Then we simply define new parameters called the **coefficient of area expansion**, γ, and the **coefficient of volume expansion**, β. These are used to calculate changes in area and volume, respectively, similar to the linear expansion formula above.

$$\Delta A = \gamma A_0\Delta T$$
$$\Delta V = \beta V_0\Delta T$$

To a good approximation, we can say:

$$\gamma \cong 2\alpha$$
$$\beta \cong 3\alpha$$

A common type of "trick" question that arises when dealing with thermal expansion involves holes in material. For example, if the flat plate of metal shown in the following illustration has a hole in the center, what happens to the hole if we heat the metal?

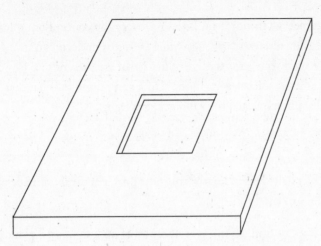

One common misconception is that as the plate heats up, the sides of the hole expand *away* from the solid portion of the plate and the hole actually gets smaller. This is, however, *not* the case. In all cases of thermal expansion, holes change shape as if they were filled with the same material as the rest of the plate or block. So the hole depicted in the figure above will actually get *larger* if the plate undergoes an increase in temperature.

MACROSCOPIC DESCRIPTION OF GASES (IDEAL GAS LAW)

Historically, the first investigations into the behavior of gases were carried out on relatively large samples. Various relationships were developed between quantities of interest (e.g., pressure, volume, temperature, quantity of gas, etc). These relationships are all categorized under the term **equations of state**. Many such relationships exist, all of which are basically semi-empirical curve fits to sets of data, and each of which is more or less accurate under different sets of circumstances.

Of these equations of state, the only one relevant to the AP Physics B exam is the familiar **ideal gas law:**

$$PV = nRT$$

where P is the absolute pressure of the gas, V is the volume of the gas, n is the amount of gas in **moles**, and T is the temperature of the gas. Moles (mol) are the SI unit of amount of a substance. This equation of state is applicable to a wide range of situations in which the gas has a relatively low pressure or density. A gas in such a state is known as an **ideal gas**.

The parameter R is the universal gas constant. It has various values, depending on the units used for the other parameters in the ideal gas law. In the SI system, the units of pressure are Pa, the units of volume are m^3, and the units of temperature are K (kelvins). In this system, the value of R is 8.31 J/(mol · K). Notice that the temperature in the ideal gas law must *always* be given in absolute units.

The most confusing aspect of the ideal gas law is probably the concept of moles. The number of moles of a gas is a measure of how many individual atoms or molecules of the gas are present. The mole is defined as the number of atoms of carbon-12 that are present in a 12 g sample. It is also, then, the number of atoms/molecules in a sample of a compound that has a mass in grams equal to its molar mass, M (also often referred to as molecular weight).

$$n = \frac{mass(g)}{M}$$

The molecular mass is usually approximated for these problems as the sum of the atomic masses of the molecule's constituent parts.

For example, say we want to find the number of moles of helium in a 12 g sample and the number of moles of CO_2 in an 88 g sample. To find these values, we first note that we need to know the molecular mass of each gas. For the monatomic helium, the molecular mass is approximately 4 g/mol, and for CO_2, it is approximately 44 g/mol (= 12 for carbon + (2 × 16) for each oxygen). Plugging this information into the molecular mass equation, we get

$$n_{He} = \frac{12 \text{ g}}{4 \text{ g/mol}} = 3 \text{ mol He}$$

$$n_{CO_2} = \frac{88 \text{ g}}{44 \text{ g/mol}} = 2 \text{ mol } CO_2$$

The number of atoms/molecules that make up one mole has been found to be ~6.02×10^{23} $(\text{mol})^{-1}$. This number of *things*/mol is known as **Avogadro's number**, N_0 (also often denoted by the symbol N_A). We should think of moles in exactly the same way that we think of dozens or scores. Where there are 12 *things* in a dozen, or 20 *things* in a score, there are ~6.02×10^{23} *things* in a mole. If we take the number of *things* to be N, the number of moles is given by

$$n = \frac{N}{N_0}$$

When dealing with gases, we must be careful how we count up these *things*. If the gas we are considering is monatomic, such as helium, the number of *things* refers to the number of individual *atoms* of the helium gas. If the gas we're considering is carbon dioxide (CO_2), the number of *things* refers to the number of carbon dioxide *molecules*, each containing one carbon atom and two oxygen atoms.

When working with the ideal gas law, it is sometimes convenient to work with the actual number of gas atoms/molecules rather than the number of moles. In this case, we can substitute and collect constants to obtain an equivalent version:

$$PV = \frac{N}{N_0} RT$$

$$PV = N k_B T$$

where the new parameter k_B is known as Boltzmann's constant and is equal to

$$k_B = \frac{R}{N_0} = 1.38 \times 10^{-23} \, \text{J/K}$$

Once the number of moles or atoms/particles is known, application of the ideal gas law is fairly straightforward. For a given isolated sample of gas ($n = $ constant),

$$\frac{PV}{T} = \text{constant}$$

For example, if you hold the volume of an isolated sample of gas fixed and double the temperature (in absolute units, K), the pressure would also double. Or if you hold such a sample at a constant temperature and double the volume, the pressure would be cut in half.

We'll stop this discussion of the ideal gas law for now, but we will see many more applications when we discuss the first law of thermodynamics later in this chapter.

MICROSCOPIC DESCRIPTION OF GASES (KINETIC THEORY OF GASES)

Although the success of the ideal gas law was very satisfying to scientists of the period, they recognized that it was ultimately merely a "curve fit" to empirical data. The kinetic theory of gases was developed in an attempt to provide a firm theoretical underpinning for the ideal gas law. This gas model involves several basic assumptions and applications of the well-known laws of mechanics. The assumptions include the following:

1. The gas consists of a number of identical particles.

2. The number of particles is very large, but at the same time, the density is low enough that there is a large amount of separation between individual particles.

3. The particles interact only through elastic collisions with other molecules.

4. The particles make elastic collisions with the walls of the container.

5. The particles have random directions and distribution of speeds, but they obey Newton's laws.

The first assumption implies that the gas particles all have the same mass, denoted by μ on your AP Physics B exam equation sheet. This simplifies the calculation of momentum and kinetic energy conservation in the elastic collisions between molecules (from assumption #3).

Assumption #2, together with assumption #5, allows the gas to be treated via statistical means while maintaining the range of validity of the ideal gas law for "low" pressures and densities.

Finally, assumption #4 allows us to determine the interaction of the gas with the "outside world" through the impulse it imparts to the walls of its container.

With these assumptions in hand, we can analyze a sample of gas consisting of N particles confined in a cubical container with sides of length d. In such an analysis, the particles colliding off of the rigid walls of the container impart an impulse due to the change in their momentum during the collision. It can be shown that this impulse is equivalent to a total force over each side during a given time:

$$F = \frac{N\mu}{d}\overline{v_x^2}$$

where $\overline{v_x^2}$ is the average squared x component of the speed of all the particles bouncing from a given wall in a given time. Here, the wall happens to be along the x-axis, so we consider the x component of the total velocity of the particle.

If we then divide this force by the area of the wall, $A = d^2$, we get

$$P = \frac{F}{A} = \frac{N}{Ad}\mu\overline{v_x^2}$$

$$= \frac{N}{d^3}\mu\overline{v_x^2} = \frac{N}{V}\mu\overline{v_x^2}$$

If we then recognize that assumption #5 implies that the average squared speed in any given direction is one-third the average squared total velocity, or

$$\overline{v_x^2} = \frac{\overline{v^2}}{3}$$

we can substitute this into the previous equation to get

$$P = \frac{N}{V}\left(\frac{1}{3}\mu\overline{v^2}\right)$$

Since $K_{avg} = \frac{1}{2}\mu\overline{v^2}$, $\mu\overline{v^2} = 2K_{avg}$. Substituting into the equation for P, we get

$$P = \frac{N}{V}\frac{2}{3}K_{avg}$$

or

$$PV = N\left(\frac{2}{3}K_{avg}\right)$$

If we compare this with the ideal gas law for N particles, we see that, eliminating like terms, we get

$$\frac{2}{3}K_{avg} = PV = k_{B}T$$

or

$$K_{avg} = \frac{3}{2}k_B T$$

So for an ideal gas, the temperature is actually a measure of the average kinetic energy of the gas particles. Pressure is actually caused by the collection of gas particles rebounding from the walls of the container. This discovery was one of the first direct confirmations that matter was indeed made up of atoms.

For a monatomic gas, there are no other types of energy available, so the average kinetic energy is directly related to the total internal energy of the gas. In fact, the total kinetic energy of the N particles is the total internal energy, or

$$U = NK_{avg} = \frac{3}{2}Nk_B T = \frac{3}{2}nRT$$

where we have used the relationship between number of particles and number of moles.

We can also use these facts to define a special type of average velocity for the gas particles. We saw that the important quantity is the average of the square of the velocities of the N particles. We then define the **root-mean-square (rms) velocity** of the gas as the square *root* of the *mean* of the *squares* of the velocities, or

$$v_{rms} = \sqrt{\frac{3k_B T}{\mu}} = \sqrt{\frac{3RT}{M}}$$

FIRST AND SECOND LAWS OF THERMODYNAMICS

If we include the zero-th law of thermodynamics, there are a total of four laws of thermodynamics. The third law of thermodynamics is outside the scope of the AP Physics B course. Having previously discussed the zero-th law, we will now turn to the two laws that are relevant to this discussion.

FIRST LAW OF THERMODYNAMICS

At its most fundamental, the **first law of thermodynamics** is a statement of the conservation of energy for a system that accounts not only for mechanical energy but also for thermal energy and internal energy. The first law deals with macroscopic quantities, so it should not be surprising that the ideal gas law comes in to play in its application on the AP Physics B exam.

The standard type of problem deals with a confined sample of an ideal gas. For such a system, the conservation of energy yields the first law as

$$\Delta U = Q + W$$

where U is the internal energy of the system, related to the temperature; Q is the energy transferred in or out of the system as heat; and W is the energy transferred in or out of the system as work. Of course; the unit on all of these quantities is joules.

A word of caution is necessary here. In many—if not most—physics texts, the first law equation is written differently than the equation above, owing to a different convention on the sign of the work term. Both are correct and conventions are, after all, arbitrary. Here we will follow the convention as applied in the AP Physics B exam.

As used here and on the exam, both heat, Q, and work, W, are considered to be positive if they result in energy being transferred *into* the system and lead, therefore, to an *increase* in the internal energy. Specifically, if we do 20 J of work on the system, the W term in the first law is taken as +20 J. If the system does work on the environment, then the W term is taken as −20 J.

Just what does it mean when we say we do work on an enclosed gas? In the mechanics chapters, we saw that work was the result of force acting over some distance, or $W = \mathbf{F}\Delta s = F\Delta s \cos \theta$.

Here, we typically consider a sample of gas confined in a cylinder with a movable, tight-fitting piston, as shown in the figure below.

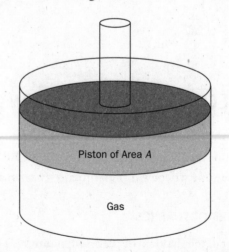

Piston of Area A

Gas

If we push down on the piston, compressing the gas, we apply a force through some small distance. If we assume the force to be normal to the piston, the angle between the applied force and the displacement is 0°. We can use the area of the piston to convert the force to the more convenient quantity of pressure as follows:

$$W = F\Delta r = \frac{F}{A} A\Delta r = P\Delta V$$

where ΔV is the change in the volume of the gas due to the compression. The sign convention for work is such that W must be positive if the gas is compressed ($\Delta V < 0$). Therefore, we must add a minus sign to the equation above to make it consistent with this convention. Thus, the equation for work as used during the AP Physics B exam is actually

$$W = -P\Delta V$$

Now for a reasonable change in volume, the force (and therefore the pressure) would not necessarily remain constant over the course of the change in volume. Therefore, we need to add up all the little bits of work for the different incremental changes in volume (each with an assumed constant pressure). The work can be determined graphically by finding the area underneath the P-V diagram.

THERMODYNAMIC STATES, PROCESSES, AND CYCLES

One of the primary uses of the ideal gas law and the first law of thermodynamics on the AP Physics B exam is in the analysis of how gases subjected to a series of processes behave. A process is something that is done to cause a sample of gas to change from one thermodynamic state to another. A thermodynamic state is simply the condition of the sample of gas in terms of the quantities that are important in a thermodynamic description, the state variables. These state variables are in turn related to one another through the relevant equation of state. Any set of sequential processes that eventually leads the gas back to its initial state so that the sequence may be repeated is called a thermodynamic cycle.

For example, if we are dealing with an ideal gas, our equation of state is the ideal gas law, and the state variables are those related by the ideal gas law; namely; the amount of gas in moles, n, the pressure, P, the volume, V, and the absolute temperature, T. For ideal gases, then, processes are things that can be done to the gas to change one or more of these quantities.

A careful examination of the ideal gas law, $PV = nRT$, shows that if we have a constant amount of gas, we can completely describe the state, or condition, of the gas by fixing any two of the remaining three state variables. For example, if we have 2 mol of an ideal gas and we fix the pressure at 2×10^5 Pa and the volume at 1.0×10^{-3} m^3, we can then deduce the absolute temperature to be

$$T = \frac{PV}{nR} = \frac{(2 \times 10^5 \text{ Pa})(1 \times 10^{-3} \text{ m}^3)}{(2 \text{ mol/J})(8.31 \text{ J/mol} \cdot \text{K})} \cong 12 \text{ K}$$

Changes in the state of a fixed sample (constant n) of an ideal gas can then be tracked by plotting the sequence of (P,V) ordered pairs for each step in the process as the gas undergoes the change from one state to another. Plots of these processes are called P-V diagrams.

Every point in the *P-V* plane corresponds to a well-defined state of the gas. There are any number of ways to get from one state to another (i.e., there are an infinite number of possible processes that could connect some initial state to a final state). The following figure shows three examples of processes between state *A* and state *B*. Choosing any of the three, continuing on to state *C*, and moving from there back to state *A* constitutes a cycle.

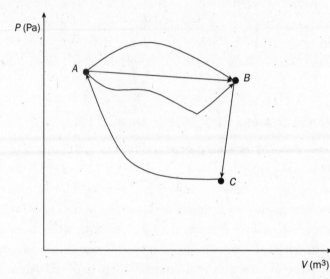

When dealing with *P-V* diagrams, states, processes, and cycles, remember that the ideal gas law deals with *states*, whereas the first law of thermodynamics ties information about states (through the ΔU) together with information about *processes* (through the *W* and *Q*).

It is important to note that when the cycle returns to a given state, all state variables have also returned to the same values where they were when the cycle started from that state. In particular, because the temperature returns to its original value at the end of a complete cycle, the internal energy does as well. Therefore, ΔU for a complete cycle is *always* equal to zero.

Let's return to the issue raised at the end of the last section concerning the work done on the gas. It can be shown that the magnitude of the work done between states *A* and *B* is equal to the area under the curve corresponding to a given process. Therefore, we see that different amounts of work will be done depending on the path we take from one state to another. This fact is crucial to correct analysis of thermodynamic processes.

Further, note that the sign on this work (positive or negative) depends on which direction the process actually takes on the *P-V* diagram. For example, if we consider the process in which the state of the gas is changed from B to C, it would result in *positive* work on the gas, since the volume decreases ($\Delta V < 0$). On the other hand, if we were to consider the opposite process that transforms the gas from C to B, it would result in *negative* work on the gas, since the volume increases ($\Delta V > 0$). In both directions, however, the amount of work would be the same.

Taken together, this information allows us to state that the net work done on a gas during one complete thermodynamic cycle is equal to the area enclosed by the process curves on the P-V diagram. If the direction of the processes in the cycle is clockwise, the net work is negative, and the cycle actually transforms heat into outside work. If the direction of the processes is counter-clockwise, the net work is positive, and the cycle transforms work into a transfer of heat (similar to what happens with a refrigerator).

The analysis of thermodynamic processes and cycles often must start with the determination of all aspects of the states on each end of all the processes. For example, say we have a sample of n moles of a monatomic ideal gas at state A (pressure given as P_A and volume as V_A). To find the temperature of the gas, we could simply apply the ideal gas law (the appropriate equation of state for an ideal gas) to state A to get

$$T_A = \frac{P_A V_A}{nR}$$

Next, we could find the internal energy for state A, since for a monatomic ideal gas

$$U_A = \frac{3}{2} nRT_A$$

We now have all the information we need to analyze any change in this initial state.

While change can be represented by any arbitrary process, certain processes are typically used to simplify the calculations. These special processes are of four types:

1. **Isobaric processes**, in which pressure is constant. An isobaric process is therefore represented by a horizontal line on a P-V diagram, and the work can be calculated directly from its definition ($W = -P\Delta V$).

2. **Isothermal processes**, in which the temperature is held constant. If we consider the ideal gas law, we see that such a process is represented by a "$1/V$" type of curve on a P-V diagram. Calculus can be used to calculate the amount of work performed during such a process to be $W = nRT \ln\left(\dfrac{V_{\text{initial}}}{V_{\text{final}}}\right)$. We also note that the internal energy remains constant during such a process ($\Delta U = 0$ or $W = -Q$).

3. **Adiabatic processes**, in which there is no heat transfer between the gas and the outside world. This can happen due either to insulation of the gas or to the process happening so

quickly that heat does not have an opportunity to cross the boundary of the gas before the process completes itself ($Q = 0$ or $\Delta U = W$).

4. **Isovolumetric processes**, in which the volume remains constant (these processes are sometimes also called **isochoric processes**). They are represented as vertical lines on a P-V diagram. Because there is no change in volume, no work is done during this type of process ($W = 0$ or $\Delta U = Q$).

On the AP Physics B exam, these types of processes are typically strung together to form cycles that must then be analyzed. In fact, the analysis of such cycles is the most common type of free-response question dealing with thermal physics.

Consider the following example of a gas undergoing a thermodynamic cycle. Suppose a sample of n moles of a monatomic ideal gas, at room temperature T_0, is enclosed in a cylinder of cross-sectional area A by a weighted piston, as shown in the following figure.

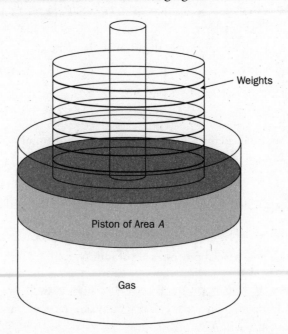

This cylinder is then placed in a furnace at temperature $3T_0$ until it comes into thermal equilibrium with the furnace. It is then removed from the furnace, and the weight on the piston is adjusted as the gas cools back to room temperature in such a way that the piston neither rises nor falls from its position when removed from the oven. Finally, the weight on the piston is again slowly brought back to its original value on the piston.

On the surface, this may not appear to be a cycle, but consider what is going on. The weight on the piston is setting the initial pressure. That, together with the initial temperature, allows us to determine the initial volume for a given amount of gas. We do not need to know the amount of gas for this analysis; we need only know that the amount is constant through the cycle. As long as the weight stays constant on the piston, the pressure remains constant as well; thus, when the cylinder is in the oven, the gas is undergoing an *isobaric* process in which the temperature increases from T_0 to $3T_0$. Obviously, heat is being added to the gas. During this process, we will show that the volume of the gas must also change (i.e., that work is done either on or by the gas).

Next, when the cylinder is removed from the oven and the weight (i.e., pressure) is adjusted so that the height of the piston remains the same, this means that the gas is undergoing an *isovolumetric* process while cooling back to room temperature. We will show that during this process, heat must be transferred into or out of the gas but no work is done.

Finally, when the weight is brought back to its original value while the temperature remains at room temperature, the gas returns its original state via an *isothermal* process. During this process, again, both work and heat transfer must occur, as we shall see.

The first step in analyzing a cycle is to create a *P-V* diagram. On a *P-V* diagram, the cycle described on the previous page would look like this:

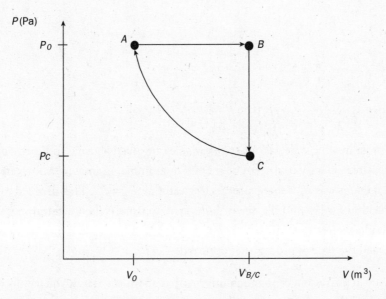

To analyze this cycle, we start with the initial state and define the relevant state variables in terms of the given quantities. For the initial state, called A, we find that

$$P_A = P_0 = \frac{weight_{piston}}{A}$$

$$T_A = T_0 \text{ (assumed in kelvins)}$$

$$V_A = \frac{nRT_A}{P_A} = \frac{nRT_0}{P_0}$$

$$U_A = \frac{3}{2}nRT_A = \frac{3}{2}nRT_0$$

Note that we have used the fact that the gas was described as monatomic to determine the internal energy.

We can then use what we now know together with the ideal gas law (it deals with states, remember?) to find out a great deal about the other two states: B (at the end of the heating in the oven) and C (at the end of the cooldown).

$$P_B = P_0$$

$$T_B = 3\,T_0$$

$$V_B = \frac{P_A V_A}{T_A} \times \frac{T_B}{P_B} = 3V_A$$

$$U_B = \frac{3}{2}\,nRT_B = 3\,U_A$$

and

$$V_C = V_B = 3V_A$$

$$T_C = T_0 = T_A$$

$$P_C = \frac{P_B V_B}{T_B} \times \frac{T_C}{V_C} = \frac{P_B}{3} = \frac{P_0}{3}$$

$$U_C = \frac{3}{2}\,nRT_C = \frac{3}{2}\,nRT_A = U_A$$

Note that the order in which the various state variables are given above is indicative of the calculations that are necessary to determine them. For example, in moving from state A to state B, we used the fact that it was an isobaric process to glean that $P_B = P_A$. This then led to the value of V_B by using the ideal gas law and the newly discovered pressure. When determining state variables, it is usually advantageous to start with what you know, either from given information or from previous states and the type of process that occurred.

Now that you are in possession of so much information about the states, you can move on to finding information about the processes. This involves determining the amount of work and heat transfer either into or out of the gas for all three processes. To do this, use the first law of thermodynamics, which deals with processes, as noted earlier.

For the isobaric process $A{\rightarrow}B$, this gives us

$$\Delta U = U_B - U_A = 3U_A - U_A = 2\left(\frac{3}{2}\,nRT_0\right) = 3nRT_0$$

$$W = -P\Delta V = -P_0(V_B - V_A) = -P_0(3V_A - V_A) = -P_0 \times 2\left(\frac{nRT_0}{P_0}\right) = -2nRT_0$$

$$Q = \Delta U - W = 3nRT_0 - (-2nRT_0) = 5nRT_0$$

During this process, the work done on the gas was negative (i.e., the gas expanded and did work on the environment). The heat transferred into the gas was positive, as expected when something is placed into an oven.

Next, for the isovolumetric process $B \rightarrow C$, we find

$$\Delta U = U_C - U_B = U_A - 3U_A = -2U_A = -2\left(\frac{3}{2}nRT_0\right) = -3nRT_0$$

$$W = -P\Delta V = 0$$

$$Q = \Delta U - W = -3nRT_0$$

During this process, the temperature decreases, so the change in internal energy is negative. This loss of energy is shown to result entirely from heat leaving the gas because no work is done in an isovolumetric process.

Finally, for the isothermal process $C \rightarrow A$, we obtain

$$\Delta U = U_C - U_A = 0$$

$$W = nRT_0 \ln\left(\frac{V_C}{V_A}\right) = nRT_0 \ln \frac{3V_A}{V_A} = nRT_0 \ln 3$$

$$Q = \Delta U - W = -nRT_0 \ln 3$$

As expected, positive work was performed on the gas because the volume decreased, and this work was exactly offset by heat transferred out of the gas. The following figure captures the spirit of these results.

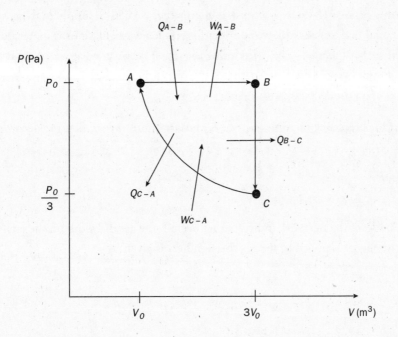

Second Law of Thermodynamics

In its most general sense, the second law of thermodynamics is very confusing. This is because it can be stated in a number of different ways, each of which deals with a different physical quantity. It is one of the consequences of the second law of thermodynamics that we can *never* get more energy out of a cycle as work than we provide as heat. This is the ultimate principle standing in the way of any type of perpetual motion device.

However, on the AP Physics B exam, the second law of thermodynamics deals almost exclusively with the concept of **thermodynamic efficiency**, *e,* a measure of how much useful energy (in the form of work) is derived from a thermodynamic cycle divided by how much energy we have to supply in the form of heat. This concept is sometimes understood in terms of a device that absorbs heat from the environment and transforms some portion of this heat into mechanical work. Such a device is known as a **heat engine**.

Again allowing for the convention on the sign of work on the AP exam, and noting that efficiencies are always taken to be greater than zero, we use an absolute value in the definition as follows:

$$e = \left| \frac{W}{Q_H} \right|$$

The energy that we supply as heat is always taken from a reservoir that is at a higher temperature than the gas, so we call this heat Q_H. The useful output energy is in the form of *work,* but we must be careful to separate output work from any input work we need to supply. We can see this in the example from the last section, in which work was involved during two of the three processes. However, during process $C \rightarrow A$, this work was energy that we had to supply to the gas for it to complete its cycle. The usable work we consider for efficiency calculations is therefore net work, which is taken as the algebraic sum (considering the signs) between what we take out of the gas and what we put into the gas. As discussed earlier, net work is always given as the area enclosed by the cycle of processes on the P-V diagram.

If we consider the preceding example, especially the last figure, we see that the net work is given by

$$W = W_{A-B} + W_{C-A}$$
$$= -2nRT_0 + nRT_0 \ln 3$$
$$= (\ln 3 - 2)nRT_0 = -0.901 \, nRT_0$$

The negative value of the net work indicates that overall, the gas did work on the environment. The amount of energy supplied to the gas through heat is given by

$$Q_H = Q_{A-B} = 5nRT_0$$

You can use these results to calculate the efficiency of the thermodynamic cycle:

$$e = \left| \frac{-0.901nRT_0}{5nRT_0} \right| \approx 0.18$$

or about 18 percent efficiency. This means that of all the energy we supplied to the gas in the form of heat, we only recovered 18 percent of it as useable mechanical work. The rest was lost as heat during processes $B{\to}C$ and $C{\to}A$, as shown in the figure. This lost heat represents energy that was wasted.

It may occur to you that it is economically beneficial to increase the efficiency of a cycle as much as possible. Keeping in mind that a cycle must end up at the same temperature from which it starts, you will see that some heat must be absorbed to be converted into work (we hope) and then we must exhaust anything that is left over. However, the zero-th law of thermodynamics implies that a thermodynamic cycle can absorb energy as heat only from some higher-temperature reservoir and can transfer energy as heat only to some lower-temperature reservoir.

Under these restrictions, the most efficient heat engine cycle is a special set of processes known as a **Carnot cycle**, which can be represented on a P-V diagram as follows:

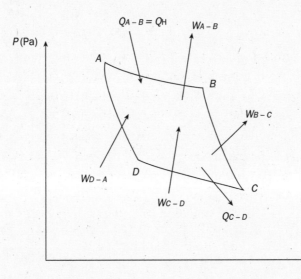

This cycle consists of the following steps:

1. **Isothermal expansion** ($A{\to}B$)
2. **Adiabatic expansion** ($B{\to}C$)
3. **Isothermal compression** ($C{\to}D$)
4. **Adiabatic compression** ($D{\to}A$)

When a gas undergoes a Carnot cycle, it is absorbing heat from a hot reservoir at temperature T_H during step 1 and exhausting it to a cold reservoir at temperature T_C during step 3. For this cycle, the efficiency has been shown to be

$$e_C = \frac{T_H - T_C}{T_H} = 1 - \frac{T_C}{T_H}$$

As we can see, anything we can do to either raise T_H or lower T_C will increase our efficiency. However, the as-yet-undiscussed third law of thermodynamics states that there is an absolute lower limit on T_C that is *not* zero, so we can never get 100 percent efficiency from any cycle.

REVIEW QUESTIONS

Use the following figure to answer questions 1 and 2.

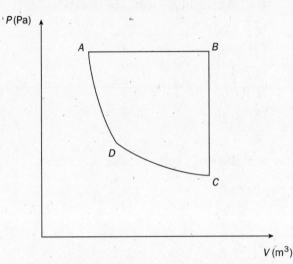

1. In which process does no work take place?

    (A) Process $A \rightarrow B$
    (B) Process $B \rightarrow C$
    (C) Process $C \rightarrow D$
    (D) Process $D \rightarrow A$
    (E) None of the above

2. Which process is isobaric?

    (A) Process $A \rightarrow B$
    (B) Process $B \rightarrow C$
    (C) Process $C \rightarrow D$
    (D) Process $D \rightarrow A$
    (E) None of the above

3. A sample of 10 moles of an ideal gas is originally at a pressure of 300 kPa. It undergoes an isothermal compression from an original volume of 1.2 m^3 to a final volume of 0.4 m^3. The final pressure is most nearly

    (A) 100 kPa
    (B) 600 kPa
    (C) 90 kPa
    (D) 9,000 kPa
    (E) 900 kPa

4. A 1 mole sample of a monatomic ideal gas (^{4}He) has a temperature of 40 K. The rms velocity of the helium atoms is most nearly

    (A) 10 m/s
    (B) 17 m/s
    (C) 500 m/s
    (D) 2,500 m/s
    (E) The answer cannot be determined from the information given.

5. A metal bar has a coefficient of linear expansion of $5 \times 10^{-6} (°C)^{-1}$. If the bar is originally 1.0 m long at 20°C, how much longer is it at 120°C?

    (A) 1.0005 m
    (B) 19×10^{-4} m
    (C) 5×10^{-6} m
    (D) 0.5 mm
    (E) 2.0 mm

Use the following figure to answer question 6.

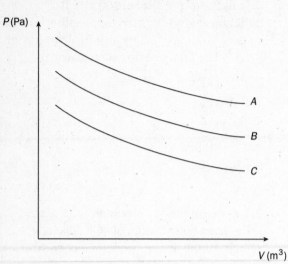

6. The figure above shows three isothermal processes for an ideal gas. Which one takes place at the highest temperature?

(A) Process A

(B) Process B

(C) Process C

(D) They all take place at the same temperature.

(E) The answer cannot be determined from the information given.

7. A monatomic ideal gas undergoes an adiabatic expansion. During this process, the temperature of the gas

(A) remains constant.

(B) increases.

(C) decreases.

(D) increases, then decreases.

(E) decreases, then increases.

8. A gas undergoes a thermodynamic cycle during which it absorbs 50 J of heat and exhausts 30 J of heat. What is the efficiency of the heat engine described by this cycle?

(A) 0.60

(B) 0.875

(C) 0.25

(D) 0.67

(E) 0.40

Use the following figure to answer question 9.

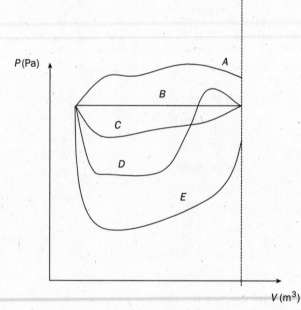

9. Which process in the figure is related to the greatest amount of work being performed?

(A) Process A

(B) Process B

(C) Process C

(D) Process D

(E) Process E

10. Two samples of a monatomic ideal gas are being kept at the same temperature and pressure in different containers. Container A has twice the volume of container B. What can you say about the internal energy of container A compared to container B?

(A) The internal energy of both containers is identical.

(B) The internal energy of container A is half the internal energy of container B.

(C) The internal energy of container A is twice the internal energy of container B.

(D) The internal energy of container A is more than twice the internal energy of container B.

(E) The internal energy of container A is less than half the internal energy of container B.

FREE-RESPONSE QUESTION

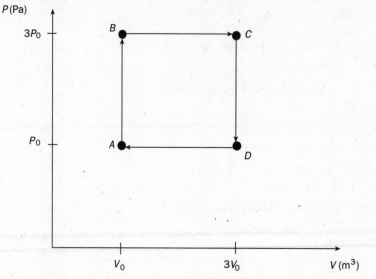

A sample of n moles of a monatomic ideal gas undergoes the cycle shown in the figure above. Processes $A{\to}B$ and $C{\to}D$ are isovolumetric, and processes $B{\to}C$ and $D{\to}A$ are isobaric. At state A, the pressure is P_0, and the volume is V_0. Give all answers in terms of the given quantities and fundamental constants.

(a) Determine the temperature at states A, B, C, and D.

(b) Determine the heat transfer during process $A{\to}B$ and state whether the heat is transferred into or out of the gas.

(c) Determine the heat transfer during process $B{\to}C$ and state whether the heat is transferred into or out of the gas.

(d) Determine the net work performed during the cycle and state whether it is done on the gas or by the gas.

(e) Determine the efficiency of this cycle and the maximum possible efficiency for a cycle operating between the highest and lowest temperatures in the cycle.

ANSWERS AND EXPLANATIONS

1. B

The work done during a process is given by the formula $W = -P\Delta V$. Choice (B) is a vertical line, and therefore there is no change in volume during this process. Thus, no work takes place.

2. A

The term **isobaric** means constant pressure. So the correct answer must be a horizontal line on the P-V diagram.

3. E

The clue words "ideal gas" indicate that the ideal gas law applies. The term **isothermal** implies that the temperature does not change during the course of the process. For a fixed amount of gas, here 10 moles, this leads to the relationship

$$P_1 V_1 = P_2 V_2$$

or

$$P_2 = \frac{P_1 V_1}{V_2}$$

Because the final volume is $\frac{1}{3}$ of the original volume, the final pressure must be 3 times the original pressure.

4. C

This result can be obtained from the formula

$$v_{\text{rms}} = \sqrt{\frac{3RT}{M}} = \sqrt{\frac{3 \times 8.31(\text{J/mol} \cdot \text{K}) \times 40(\text{K})}{4(\text{g/mol}) \times 10^{-3}(\text{kg/g})}}$$

Note that the quantity $3R$ is approximately 25. Next, cancel and collect powers of 10 to get

$$v_{\text{rms}} = \sqrt{25 \times 10 \times 1000} = 500 \text{ m/s}$$

A common error in this problem is to forget to convert the molar mass into kilograms from grams.

5. D

Getting the right answer here requires you to pay attention to units. You must also read the question carefully. Choice (A) is the total length of the heated rod, but that is not what is asked. The problem asks for the difference in the length. This is given by

$$\Delta \ell = \alpha \ell_0 \Delta T = 5 \times 10^{-6} (°\text{C})^{-1} \times 1\text{m}$$
$$\times (120°\text{C} - 20°\text{C})$$
$$= 5 \times 10^{-4} \text{ m} = 0.5 \text{ mm}$$

6. A

When temperature remains constant, the ideal gas law demands that an increase in pressure correspond to an increase in volume. Therefore, if we follow an imaginary line of constant volume through the curves, the one at the highest pressure must be at the highest temperature.

7. C

During an adiabatic process, the heat transfer, Q, is zero. The first law of thermodynamics then states $\Delta U = W$. Because this process is an expansion, the volume is increasing, which corresponds to work being done by the gas on the environment, so the sign of W is negative. Therefore ΔU is negative, meaning that the internal energy decreases, which implies a decrease in temperature.

8. E

If a gas undergoes a cycle, then the overall change in internal energy must be zero. That means the net work is equal to whatever heat is absorbed minus whatever is exhausted. Don't forget: at its most fundamental, the first law is merely a statement of the conservation of energy. So the work output is $50 - 30 = 20$ J. The efficiency is then this amount divided by the total heat absorbed, or $\frac{20}{50} = 0.40$.

9. **A**

The work involved in a given process is equal to the area under the path the process takes on the *P-V* diagram. Obviously, path *A* encloses the most area and so the most work is performed during that process.

10. **C**

The internal energy of a monatomic ideal gas is given by

$$U = \frac{3}{2}nRT$$

The only difference between containers A and B is volume. If we consider the ideal gas law, twice the volume must contain twice the amount of gas under identical conditions. Twice the gas means that the number of moles in A is twice the number of moles in B.

FREE-RESPONSE QUESTION

Thermal physics free-response questions are almost always based on the analysis of a P-V diagram. Some rules of thumb are useful when analyzing these diagrams. First, be aware of the special types of processes described in this chapter and the simplifications that they allow. Second, if a process with a given initial state is not adiabatic, then $Q > 0$ for any process from the same initial state that lies above the adiabatic line, and $Q < 0$ for any process below the adiabatic line. Isovolumetric processes are vertical paths on the diagram and have $W = 0$. If a process is not isovolumetric, then paths toward the right (V increasing) have $W < 0$, and paths toward the left have $W > 0$. Finally, adiabatic expansions from a given initial state lie below the isothermal path from the same initial state, while adiabatic compressions lie above the isothermal path from the same initial state.

(a) The problem is dealing with an ideal gas. Any questions about individual states, therefore, should employ the equation of state (i.e., the ideal gas law). Starting with state A, we get

$$P_A V_A = nRT_A$$

$$T_A = \frac{P_A V_A}{nR} = \frac{P_0 V_0}{nR}$$

Because the quantity of gas is a known constant, we can get the temperature at each of the other states in the same way:

$$T_i = \frac{P_i V_i}{nR}$$

$$T_B = \frac{3P_0 V_0}{nR} \, (= 3T_A)$$

$$T_C = \frac{3P_0 \times 3V_0}{nR} = \frac{9P_0 V_0}{nR} \, (= 9T_A)$$

$$T_D = \frac{P_0 \times 3V_0}{nR} = \frac{3P_0 V_0}{nR} \, (= 3T_A = T_B)$$

(b) Dealing with processes generally calls for the first law of thermodynamics. Process $A \rightarrow B$ is isovolumetric, so $W = 0$. Therefore,

$$\Delta U = Q$$

But for a monatomic ideal gas,

$$U = \frac{3}{2} nRT$$

Thus,

$$Q = \Delta U = \frac{3}{2}nR(T_B - T_A) = \frac{3}{2}nR(3T_A - T_A) = 3nRT_A = 3P_0V_0$$

Because the value of Q is positive, the heat is being transferred *into* the gas.

(c) Again, when dealing with a process, we must use the first law. In this case, the work is not zero, so we must account for it. However, because we know the temperature at both ends of the process, we can calculate the change in internal energy as we did above in part (b). This all leads to

$$\Delta U = Q + W$$

or

$$Q = \Delta U - W$$
$$= \frac{3}{2}nR(T_C - T_B) - (-P\Delta V)$$
$$= \frac{3}{2}nR\left(\frac{9P_0V_0}{nR} - \frac{3P_0V_0}{nR}\right) + 3P_0(3V_0 - V_0)$$
$$= 15P_0V_0$$

Again, since the value is positive, the heat transfer is into the gas.

(d) The magnitude of the net work for a cycle is equal to the area enclosed by the cycle on the *P-V* diagram. The sign of the net work is negative for cycles that run clockwise on the diagram. Therefore,

$$W_{net} = -(3P_0 - P_0) \times (3V_0 - V_0)$$
$$= -4P_0V_0$$

Because the sign of the work is negative, the gas does net work on the environment.

(e) The efficiency of this cycle is given by the fraction of the total added heat that is taken out as work. In the above cycle, heat is added during two processes: $A \rightarrow B$ and $B \rightarrow C$. The total added heat is

$$Q_H = 3P_0V_0 + 15P_0V_0 = 18P_0V_0$$

We have also found that the net work done by the gas is $W = -4P_0V_0$. Plugging these into the formula for efficiency gives

$$e = \left|\frac{W}{Q_H}\right| = \left|\frac{-4P_0V_0}{18P_0V_0}\right| = \frac{2}{9} \approx 22\%$$

The maximum efficiency for any cycle operating between two given temperatures is that of a Carnot cycle, given by

$$e_C = \frac{T_H - T_C}{T_H}$$

For this cycle, the temperatures in step (a) give us the highest temperature at state C, which is nine times the lowest temperature at state A. So the maximum possible efficiency is

$$e_C = \frac{9T_A - T_A}{9T_A} = \frac{8}{9} \approx 89\%$$

So we see that the cycle we've analyzed is only about a quarter as efficient as the theoretical maximum.

CHAPTER 8: ELECTRICITY

INTRODUCTION

From a physicist's perspective, the study of electricity can be divided into two major categories: electrostatics and electric currents. Electrostatics deals with the interactions of stationary charges as well as their fields and potentials. The systematic study of electrostatics dates back to the early 1700s; scientists who did work in this area include Coulomb, Volta, and Franklin. The study of electric currents is concerned with the motion of charged particles through wires in a circuit.

The study of electric currents is more recent (it did not begin until the mid- to late 1700s); familiar names in this field include Ohm, Kirchhoff, and Ampere.

We will begin this chapter with electrostatics, covering the concepts of charge, field, and potential. Then we will discuss systems of discrete charges, followed by continuous charge distributions. Gauss's law will be introduced as a tool for solving distributions with various types of symmetry. From there, we will move on to the concept of capacitance, including the theory of dielectric materials.

After capacitance, we will tackle electric currents in the form of simple resistor-battery circuits. Conservation laws play a big part in this discussion, as they do in most physics topics. We will also explore simple resistor-capacitor, or RC, circuits in the steady state and with their transients.

Throughout the chapter, we will attempt to bring in the experimental techniques used to study static electric systems and electric currents. Whenever applicable, we will discuss some of the well-known experiments that any good physicist should understand. Because the AP Physics exams are in the process of increasing their emphasis on laboratory questions, this information should be helpful when you take the tests.

ELECTROSTATICS

Electrostatics is concerned with stationary charges and their fields and potentials. Many electrostatic concepts can be extended to slowly moving charges. Pay close attention to this material; it is the foundation for our study of electricity and magnetism.

CHARGE, FIELD, AND POTENTIAL

On the AP Physics B exam, this material accounts for 5 percent of the test; on the C exam, it makes up 15 percent of the test. Even though these may not seem like large percentages, the basic ideas discussed in electrostatics are crucial for successfully completing questions on electricity and magnetism. Therefore, it is worth your time and effort to try to understand the concepts of charge, field, and potential.

There are two types of **charge**, positive and negative. It is an experimental fact that like charges exert repulsive forces on one another, while unlike charges exert attractive forces on one another. This can be demonstrated by rubbing two balloons on your hair. The balloons tend to repel one another (because they pick up the same charge), while each balloon will attract your hair (because it now bears a charge that is the opposite of each balloon's charge). This example also shows another important property of charge, namely the fact that total charge in the universe is *conserved*. There is no way to charge one object without simultaneously charging another object in the opposite way.

An object is **charged** if it has an excess number of charge carriers on it. The nature of the excess charges determines the charge on the object. An object with an equal number of positive and negative charges is neutrally charged. A **test charge** is a small positive charge used to establish the nature of an electric field; it is so small that its presence does not affect the rest of the world.

In general, charge is quantized. The smallest charge one can isolate (the charge on the electron) is so small that for all intents and purposes, we can consider charge to be a continuous quantity. This fact proves particularly useful when it comes to Gauss's law, as we will see later in this chapter.

From an electric standpoint, all objects can be divided into two types. The first type is the **conductor**. A conductor is an object or substance that allows charges to move through its volume. Metals and salt solutions, for example, are generally good conductors. The second type of object is the **insulator**. An insulator prevents charges from moving through its volume. Glass and plastic are examples of common insulators.

The space around any charged body is filled with an electric field. The **electric field** (E) can be defined as the electric force per unit charge acting on a point charge. This field is a vector quantity, and the direction the field points at one location in space is the same direction as the force a small positive charge (i.e., a test charge) would feel if placed at that point. That said, the oft-quoted statement that field lines point from positive charges to negative charges makes sense. The figure below illustrates the field concept, where electric field lines around a single positive charge (a), a single negative charge (b), and an electric dipole (c) are pictured.

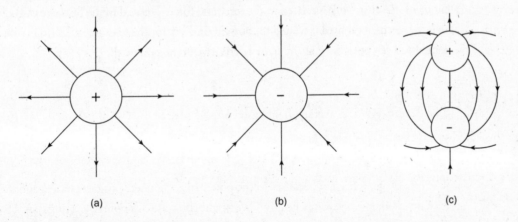

(a)   (b)   (c)

We can see from figure (c) that the field lines originate from a positive charge and terminate on a negative charge. In addition, the figure illustrates three basic properties about the field:

1. Two different field lines never cross.

2. In free space, the field lines are smooth curves.

3. The density, or spacing, of the field lines gives a qualitative estimate of the field's relative field strength at different points on the same diagram.

We'll come back to these points later. First, let's look at the more general issue of the field's behavior near an insulator or a conductor. The electric field penetrates the volume of an insulator. However, the field is generally weaker inside the insulator than it is in a vacuum. In the case of a conductor, the field is shielded from the interior region. This is because the charges in the conductor are free to move, so they will move in response to the field. After a short time (a few nanoseconds), the conductor's charges have moved so that they completely cancel the field inside.

You can also reach this conclusion by using the idea of energy conservation. Moving charges can do work; this is, for example, how a light bulb lights. If the electric field inside a conductor were not zero, charges would be moving continuously (because they will move until they no longer feel electric forces), and they would do work. If this were the case, using a conductor would allow you to build a device that puts out more energy than is put in, which violates the second law of thermodynamics. So the electric field inside a conductor is exactly zero, even if the conductor is hollow.

Electric potential is related to the concepts of field and potential energy. The easiest definition of the electric potential energy is the work ($Fs \cos \theta$) required to bring a small positive test charge from infinity to a particular point. There are some problems with this definition though, in that some charge distributions (such as the infinite line of charge) would create an infinite electric potential everywhere in space. However, certain techniques can be employed to get around these difficulties, so they are not generally a problem. Because work is related to the force and force is related to the field, we can see that the electric potential is related to the field. Indeed, for the C exam, the field in one dimension can be defined as the spatial derivative of the potential:

$$E = -\frac{dV}{dr}$$

The B exam gives an average field based on this idea as

$$E_{avg} = -\frac{V}{d}$$

The minus sign is present because electric potential is high near positive charges and the field must point away from a positive. We might also state this point by saying that the electric field points from higher electric potential to lower electric potential. In a way, this separates us from the charges so that we are dealing with the behavior of free space. This proves necessary when we are working with magnetism and want to discuss the way that Maxwell's equations lead to an understanding of light as an electromagnetic wave.

Both the electric field and the electric potential are defined at a given point in space. That is, these quantities are things that exist in space and can be measured at some location. If you measure them at a different location, you should generally expect different results. Thus, determining the field and potential at more than one point will probably require you to do more than one calculation. Remember this when you are doing problems.

AP EXPERT TIP

As a consequence of the electric field inside a conductor being zero, all excess charge possessed by a conductor must reside on its surface.

We now want to explore the specifics of the field and potential around some charges. To do this, we need Coulomb's law.

COULOMB'S LAW; FIELD AND POTENTIAL OF POINT CHARGES

Point charges are imaginary objects that carry charge (and possibly mass) but that have no physical size. As it turns out, the electrostatic forces that point charges exert on one another are very similar to the gravitational forces that point masses exert on one another. **Coulomb's law** gives a mathematical description of these forces between point charges. Suppose we have two point charges of magnitude q_1 and q_2 (q is the standard symbol for quantity of charge), and they are separated by a distance r. Coulomb's law tells us that the force between the charges is given by

$$F = \frac{1}{4\pi\varepsilon_0} \times \frac{q_1 q_2}{r^2}$$

In the figure below, charges q_1 and q_2 are separated by a distance r. The magnitude and direction of the force between them are given by Coulomb's law.

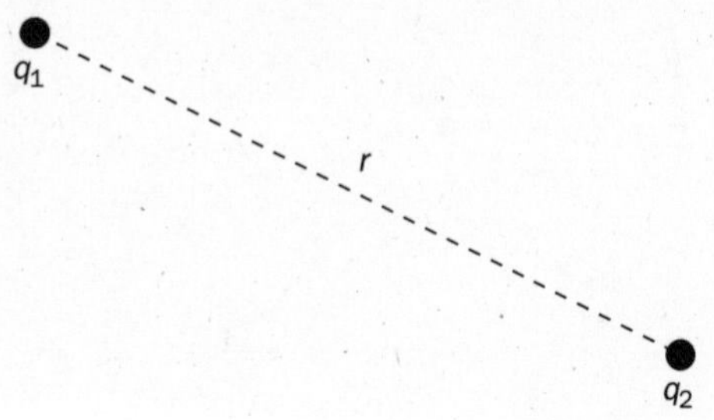

Sometimes the constant of proportionality $\frac{1}{4\pi\varepsilon_0}$ is written as k, but the AP exams do not use this notation, so you should get used to the long form. π is the usual $3.14\ldots$ and ε_0 (pronounced "epsilon zero" or "epsilon naught") is the vacuum permittivity. ε_0 has a value of $8.85 \times 10^{-12}\, C^2/(N \cdot m^2)$.

This value (along with other important constants) will be given to you on your AP Physics equation sheet.

Force is a vector, but the Coulomb's law equation above is not a vector equation. Where's the direction? Notice that if q_1 and q_2 are of the same type of charge (i.e., they are both positive or both negative) then the resulting force is positive, while if they are of different charges, the force is negative. Remember: if you do a problem using Coulomb's law and your answer comes out negative, the charges involved

AP EXPERT TIP

Notice that the AP equation sheet gives a value for the entire Coulomb's law constant. So you never need to plug "π" or "ε_0" into Coulomb's law.

should be attracting one another; if the force is positive, the charges should be repelling each other. At the level of the AP Physics exams, the direction of the force must be inferred using common sense. Still, you must ensure that the force lies on the straight line connecting the two charges.

You should compare Coulomb's law with Newton's law of gravitation. Note the minus sign associated with gravity, which is an attractive force for all mass. To study the relative strength of gravity and the electric force, it is convenient to use two electrons, as they are fundamental particles. If you calculate the ratio of the gravitational attraction to the electrostatic repulsion between two electrons, you will find a value of about 2×10^{-41}. Thus, electrical phenomena are on the order of 10^{41} times stronger than gravitational phenomena. And the electric force isn't even the strongest force there is!

One last point should be made regarding Coulomb's law. Electrostatic forces obey the property of superposition. That is, if you have multiple charges acting on a single charge, you should first determine the effect of each charge separately, then add them together as vectors. Superposition is a very handy property, because it creates a reasonably simple method for dealing with collections of charges and continuous charge distributions. In the figure below, charge Q is influenced by the other three charges. The total effect is the sum of the individual effects. Q and q_2 are both positive, while q_1 and q_3 are negative.

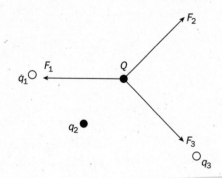

By relying on Coulomb's law, we can now determine the electric field around a point charge. Our basic working definition of the field as the force per charge is how we will get where we're going. The equation you'll receive on the equation sheet is

$$E = \frac{F}{q}$$

Here, F represents the force on the charge, and q is the magnitude of the charge feeling that force. This definition partially separates the concept of field from that of the charge producing the field. However, you should also know that around a point charge,

$$E = \frac{1}{4\pi\varepsilon_0} \times \frac{q}{r^2}$$

Here, q now represents the charge *producing* the field rather than a charge experiencing any force due to the field. Again, you must infer the direction of the field around a particular charge based on your knowledge of how a positive test charge would react. The expression for the field around a point charge is necessary because you are expected to know the fields around distributions of charges. Both the B exam and the C exam will require that you know how to handle a planar charge distribution; the C exam will present spherical and cylindrical distributions as well. We will discuss each of these distributions in detail later in this section.

For both exams, you will receive the following equation for electric potential:

$$V = \frac{1}{4\pi\varepsilon_0} \sum_i \frac{q_i}{r_i}$$

This expresses the potential at some point due to a collection of charges at various other points.

The figure below should help clarify this. As you can see, several point charges are in a region of space, and we want to measure V at the point X. Each q_i is a different distance from X, so we must use different r values for each q.

Make sure to not confuse electric field and electric potential, as the formulas are very similar. Also remember while the electric field results from the *vector* sum of the electric forces acting at a point, the electric potential is a *scalar* sum; this is an important distinction:

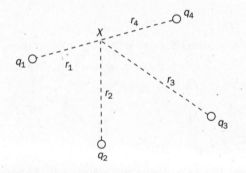

One handy thing about the electric potential equation above is that it is a helpful reminder of how to handle multiple charges and charge distributions.

If you move a charge in an external field, two things can happen:

1. The field does no work on the charge.
2. The field does work on the charge.

Condition #1 is met when the charge moves such that its path is always perpendicular to the field lines. This follows from the definition of work as force times displacement times the cosine of the angle between the force and the displacement. If the displacement is at right angles to the force, the cosine term is zero, so the particular force does no work. Motion along such a path in an electric field traces an equipotential curve in two dimensions or an equipotential surface in three dimensions.

Often these curves and surfaces are just called equipotentials, because the electric potential does not change along this sort of path.

Condition #2 is met when the charge moves such that *any* component of its displacement is along a field line. It is probably obvious that condition #1 is much more difficult to satisfy than condition #2. If the charge's path deviates from being perpendicular to the field lines even a little, the field will do work. This idea adds a small wrinkle to the relationship between potential and field. Recall, for the C exam,

$$E = -\frac{dV}{dr}$$

and for the B exam,

$$E_{\text{avg}} = -\frac{V}{d}$$

These equations, coupled with our "work" conditions, tell us that the field lines and equipotentials must always be perpendicular to one another. If you understand this concept, you should be able to sketch the equipotentials if you are given a picture of the field in some space, and you should be able to sketch the field if you are given the equipotentials. Remember the direction of the field is always from high potential to low potential.

As an example, consider the case of a positive point charge in the following illustration. The field lines are directed outward radially. The potential at some distance, *r*, from the charge is given by

$$V = \frac{1}{4\pi\varepsilon_0}\frac{q}{r}$$

Provided we are not directly on the charge ($r = 0$), there is no problem with this starting point. This expression has spherical symmetry (i.e., the potential does not depend on the direction from the charge, only the distance). Therefore, the equipotentials are going to be spherical shells—but not equally spaced shells—in three dimensions or circles in two dimensions. In both cases, the center of each shape is located on the charge. From geometry, we know that a radius of a circle is perpendicular to the tangent line on the circle where the radius terminates; this is the definition of *tangent*. Although more could be said about the mathematics here, you will not need to present a more complicated mathematical argument for this point on the AP Physics exams.

AP EXPERT TIP

You can think of a high-potential location as someplace that a free positive charge would most likely seek to avoid.

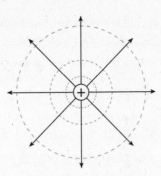

In the figure above, a positive point charge creates a radial electric field (solid arrows). The corresponding equipotentials (dashed curves) are circles centered on the charge, if we confine ourselves to a plane.

Suppose that a charge moves in a direction parallel to an electric field. In this situation, the field exerts a force on the charge as the charge moves through it. Exploring this idea further, consider a positive point charge Q located at the origin. A second positive charge q is infinitely far away, and you want to move it to some finite distance r from the origin. By moving the q closer, you moved it from an area of zero potential to one of a positive potential. In order to do this, you had to do work. How much work did you have to do? Now imagine you let q go. It is feeling a force from Q and would accelerate away, thus increasing its kinetic energy. To increase the kinetic energy of q, work must be done. How much work did Q now do?

In general, work is given by $W = \vec{F} \cdot \vec{s}$ or, using the geometrical definition of the dot product, $W = Fs \cos\theta$.

This approach works well if the force exerted is constant. However, in this case, the force exerted depends on the position, so we must resort to calculus to solve this problem. Students studying for the C exam should be able to follow this argument; those studying for the B exam should pay attention to the physics involved without getting caught up in the mathematics.

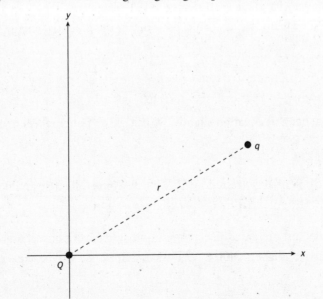

In the previous figure, a positive charge Q is situated at the origin. A second charge q is brought from infinity to a point a distance r away from Q. Due to the repulsion of the two charges, if q is released, it will accelerate away. This implies that there is some stored energy.

The force between the charges is given by Coulomb's law:

$$F = \frac{1}{4\pi\varepsilon_0} \frac{Qq}{r'^2}$$

The prime is used because we will be integrating over r' to end at r, so we need a "dummy" variable. If we move along a radial path, the angle between the Coulomb force and the displacement will be 180°. Thus, in moving some small distance $\Delta r'$, the Coulomb force will do a small increment of work given by

$$\Delta W = -F\Delta r' = -\frac{1}{4\pi\varepsilon_0} \frac{Qq}{r'^2} \Delta r'$$

The total work required to move from infinity to r is just the sum of all the increments of work:

$$W = \sum \Delta W = \sum -\frac{1}{4\pi\varepsilon_0} \frac{Qq}{r'^2} \Delta r'$$

If we let $\Delta r'$ go to zero and keep the sum finite, we have the physicist's version of an integral:

$$W = \int_\infty^r -\frac{1}{4\pi\varepsilon_0} \frac{Qq}{r'^2} \, dr'$$

You should verify that the solution to this integral is

$$W = \frac{1}{4\pi\varepsilon_0} \frac{Qq}{r}$$

By the work-kinetic energy theorem, this is also the change in the kinetic energy. If the charge q is moved at a constant velocity, there is no change in kinetic energy. Therefore, this work must go into potential energy, U. Finally, we have an expression for the potential energy between two point charges separated by a distance r:

$$U_E = \frac{1}{4\pi\varepsilon_0} \frac{Qq}{r}$$

Notice that if we rearrange this equation slightly, we can see q's potential energy:

$$U_E = qV$$

where V is the electric potential generated by Q. We have come full circle and now see that the electric potential is electric potential energy per charge.

We can use the material just presented to explore simple systems of point charges, specifically the dipole and "geometric" shapes. On the AP Physics exams, you will be expected to know how to solve for forces on the charges as well as for the fields and potentials in the space around these configurations. Let's take each in turn. All that will be presented here are the starting and finishing points. Filling in the steps that lead to the answer would be excellent practice.

THE DIPOLE

The electric dipole (see figure below) is composed of two charges (q_1 and q_2) that are equal in magnitude and opposite in sign, separated by a distance d. The line connecting the charges is often called the axis of the dipole. The field and potential are symmetric around the dipole axis. For our calculations, assume that the charges are situated at $+/-\dfrac{d}{2}$ on the x-axis.

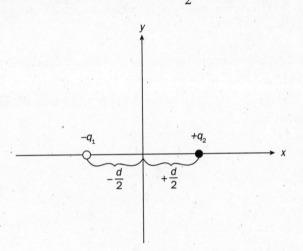

It is straightforward to show that the force each charge feels is

$$F = \frac{-1}{4\pi\varepsilon_0}\frac{q^2}{d^2}$$

This is just Coulomb's law. Further, you can see that the field at some point x on the axis between the charges is directed toward the negative charge and has the form

$$E = \frac{q}{4\pi\varepsilon_0}\left[\frac{1}{\left(x+\dfrac{d}{2}\right)^2} + \frac{1}{\left(x-\dfrac{d}{2}\right)^2}\right]$$

If you move to a point on the axis that is *not* between the charges, the field takes on the form

$$E = \frac{q}{4\pi\varepsilon_0}\left[-\frac{1}{\left(x+\dfrac{d}{2}\right)^2} + \frac{1}{\left(x-\dfrac{d}{2}\right)^2}\right]$$

In this case, the field is directed toward the negative charge if $x < 0$ or toward infinity if $x > 0$.

Finally, you can calculate the potential at any point on the axis as well. In this case, however, there is only one expression for the quantity of interest.

$$V = \frac{q}{4\pi\varepsilon_0}\left[\frac{1}{\left|x - \dfrac{d}{2}\right|} - \frac{1}{\left|x + \dfrac{d}{2}\right|}\right]$$

This is because electric field is a vector and must be added as such. Electric potential is a scalar, and it is added algebraically.

The figure below shows the field lines and equipotentials (dashed lines) around the dipole. Notice how the field lines "bend" and "flex" to satisfy the smoothness requirement. Notice, too, how the equipotentials are everywhere perpendicular to the field line being crossed. The dipole's associated field is indicated by a solid arrow.

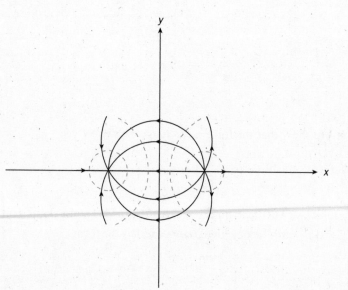

Students taking the C exam would be well advised to find expressions for the field and potential for any point near a dipole.

"GEOMETRIC" SHAPES

On the AP Physics exams, it is extremely common to see a collection of charges placed on the vertices of various well-known geometric shapes. You will be expected to know how to find distances, particularly across diagonals of squares and rectangles. When you see problems of this sort (like those shown in the following figures), just use the Pythagorean theorem. It's one of the most useful formulas to memorize for solving basic physics problems.

Once you have determined the distances, you just need to apply the principle of superposition to the basic formula for whichever quantity you are asked to find. The Pythagorean theorem has that nasty square root in the final answer for the distance between two points. Fortunately, the force and field equations are concerned with the square of the distance, which helps to eliminate some of the difficulty there.

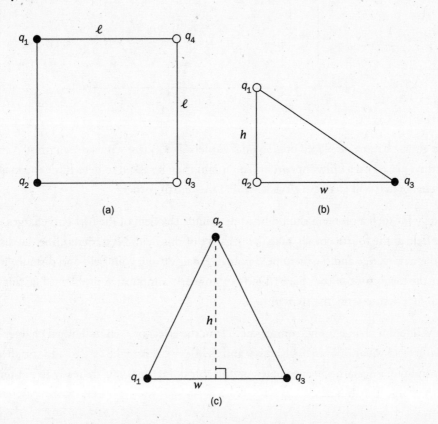

The figure above shows charges on the corners of a square (a), a right triangle (b), and an isosceles triangle (c). Note: Charges may not be the same, but they are generally multiples of one another.

A common testing technique asks students to compare two slightly different situations. For example, you might be given the situation in the following figure:

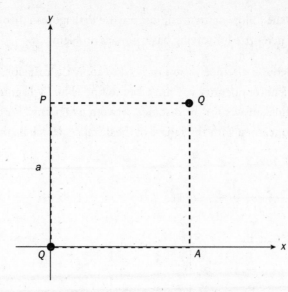

Here, two equal charges are placed on diagonal corners of a square with side length *a*. A third charge is then placed on a different corner. You might then be asked to determine the magnitude of the third charge that will give zero electric field at the fourth corner.

Your strategy for such a problem should be to determine the field of the first two charges. Since you want zero field at the fourth corner, take the negative of this field. Next, determine the distance between the new charge and the "zero field point." Finally, using your field and distance formulas, determine the magnitude of the charge. Don't try to work complicated problems like this one in your head—it's best to write the steps out.

Collections of point charges only require you to sum the effect of each individual charge. When dealing with vector quantities (namely, force and field), you must take care to add the effects as vectors and not as scalars. Once you are comfortable doing this, you are ready to master the continuous charge distributions.

FIELDS AND POTENTIALS OF OTHER CHARGE DISTRIBUTIONS

In this section, we will review the fields and potentials around planar, cylindrical, and spherical continuous charge distributions. **The AP Physics exams do *not* give you the equations we will discuss in this section.** If you need them to solve a problem, you will be expected to derive them. Because this is so, we will go over how this is done. You should have blank paper handy so that you can follow along. Most students can improve their understanding of a derivation by copying it out as they read an explanation of how it is done.

PLANAR DISTRIBUTIONS

This is the case of a sheet of charge either finite in extent (C exam) or infinitely large. At the B exam level, this is merely a qualitative problem. The important ideas here are that near a sheet of uniform surface charge density, the electric field is uniform in strength, perpendicular to the sheet, and directed according to the nature of the charge on the sheet. Any derivation of these properties requires the use of calculus or a simplified version of Gauss's law. Neither of these are B exam level topics.

Later we'll see how Gauss's law provides an elegant solution to this problem. But if you are working without Gauss's law, the most expedient way to find the field near a plane of charge is to use a collection of lines. Therefore, it is best that we first find the field of a line of charge. The following figure shows an infinitely long line of positive charge.

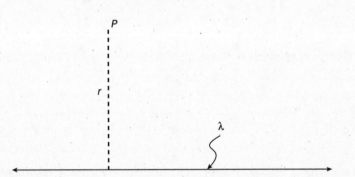

The charge is distributed such that in each meter of length, there is an equal amount of charge. Such a distribution is called a "uniform line of charge" or a "line of uniform charge density." In the case of the charge density of a line, the Greek letter λ (lambda) is used. So we have a line of uniform charge density, λ coulombs per meter. Our goal is to find the electric field strength at a point P that is a distance of r meters from the line, as shown in the figure above.

To accomplish this, first we must choose the line to lie on the x-axis. Also, let's take P to lie on the y-axis so that its x coordinate is zero. Notice that a small segment located at a point x on the line is a distance d from P, where d is given by

$$d = \sqrt{r^2 + x^2}$$

If the segment has length dx, then the total charge of the segment is given by

$$dq = \lambda dx$$

From Coulomb's law, we can determine that the segment produces a field of magnitude:

$$dE = \frac{1}{4\pi\varepsilon_0} \frac{dq}{r^2 + x^2} = \frac{1}{4\pi\varepsilon_0} \frac{\lambda dx}{r^2 + x^2}$$

For every segment at x, there is a corresponding segment at $-x$, so we expect the component parallel to the line to cancel as in the figure below.

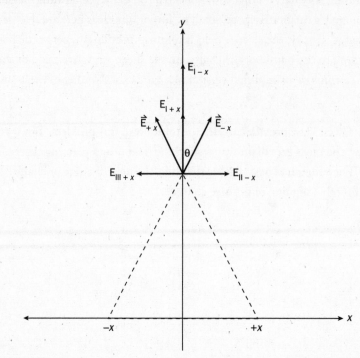

For each element of charge at $+x$, there is a corresponding element at $-x$. The components of the electric field parallel to the wire cancel each other, while the perpendicular components combine constructively.

Therefore, only the perpendicular component survives. Using trigonometry, the perpendicular component is found to be

$$dE_\perp = \frac{1}{4\pi\varepsilon_0} \frac{\lambda dx}{r^2 + x^2} \sin\theta$$

where θ is the angle shown in the diagram. Replacing $\sin\theta$ with its triangle identity (opposite over hypotenuse), we arrive at

$$dE_\perp = \frac{1}{4\pi\varepsilon_0} \frac{\lambda dx}{r^2 + x^2} \frac{r}{\sqrt{r^2 + x^2}}$$

If we integrate both sides over their respective differentials, we find that at a point P a distance r from the line of charge, the field is given by

$$E_\perp = \frac{\lambda}{2\pi\varepsilon_0 r}$$

To solve the right-hand side of this equation, you should make the trigonometric substitution $x = r\tan\theta$ and find the corresponding differentials. Notice that the value of E is the same at all points a distance r from the line, regardless of the direction. This is really a cylindrically symmetric system with the axis of symmetry on the line of charge, but it is needed to understand how the plane of charge works.

Now let's consider a disk of charge. The figure below shows a disk of radius R with a uniform surface charge density of σ coulombs per square meter. (The Greek letter σ, sigma, stands for surface charge density.) Here, we want to find the field at the point P on the axis of the disk at a distance d above the disk. You should approach the problem by setting your coordinate system as shown and dividing the disk into rings of charge.

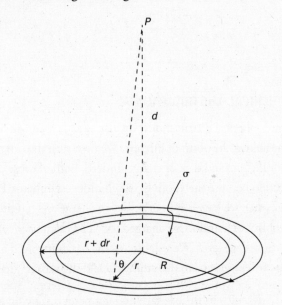

A ring with radius r and thickness dr contributes $2\pi\sigma r dr$ to the total charge of the disk. Any point on the ring is a distance $\sqrt{d^2 + r^2}$ from P. Thus, each ring contributes to the field an amount given by

$$dE_\perp = \frac{1}{4\pi\varepsilon_0}\frac{2\pi r\sigma\,dr}{d^2 + r^2}$$

Using the same trigonometric substitution $r = d\tan\theta$, we can integrate both sides. For a point on the axis of the disk, we find

$$E_\perp = \frac{\sigma}{2\varepsilon_0}\left[1 - \frac{d}{\sqrt{d^2 + R^2}}\right] = \frac{\sigma}{2\varepsilon_0}\left[1 - \frac{\dfrac{d}{R}}{\sqrt{\dfrac{d^2}{R^2} - 1}}\right]$$

Notice that the field is perpendicular to the disk, provided we are on the axis. If we take the limiting case in which d is significantly less than R, we are very close to the disk, and it looks like an infinite plane. A simple rewriting of the solution in which we factor R^2 out of each term in the second denominator leaves us with

$$E = \frac{\sigma}{2\varepsilon_0}$$

for the field. This is the field of an infinite plane of charge with the charge distributed on one surface—that is to say, this works for an infinite insulator. We arrived at this conclusion from the limiting case of a finite surface. If you take a finite surface, the field has a component along the surface. If that surface is a conductor, the charges will tend to congregate around the edges. For the time being, let's confine ourselves to insulating materials so that we avoid this kind of complication.

For the infinite sheet, E is uniform in strength and directed according to the sign of σ. E does not depend on the distance from the plane in this case, which will prove to be important when we discuss capacitors.

FOR C EXAM ONLY

CYLINDRICAL AND SPHERICAL DISTRIBUTIONS

To deal with cylindrical and spherical distributions, we just need to establish how a small charge element contributes to the field at the point of interest. We then sum (integrate) over all small elements, just as we did for the line and disk of charge. Indeed, both of those distributions could very well be put in this section because they both have cylindrical symmetry. However, here we are focusing on what happens when we have a spherical shell of charge and a uniform sphere of charge. We will also tackle the problem of a uniformly charged rod. The calculus involved in these problems generally exceeds a first-year college level. We will discuss the basics involved and explore the nature of the final result; Gauss's law will permit us to determine these distributions' fields in detail.

The rod of charge is a cylinder with infinite length and diameter d. The figure below shows a uniformly charged rod with volume charge density ρ. An imaginary cylinder (dotted line) with the point P on its surface contains all of the charge that contributes to the field at that point.

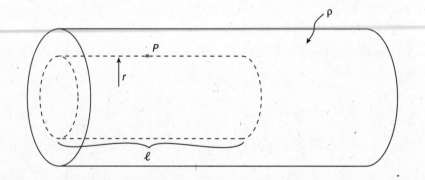

The charge is uniformly spread throughout the volume of the cylinder, giving a uniform volume charge density of ρ coulombs per cubic meter. If we want to find the field at a point P some distance r from the axis of the rod, we need only to consider the charge that lies inside an imaginary cylinder of radius r with arbitrary length. We can then treat the situation as if all of the charge inside that cylinder were located on the axis of the rod and treat it as a line of charge. This is very convenient, but the trick lies in determining how much charge is in the imaginary cylinder.

The spherical ball of charge (see figure below) also has a uniform volume charge density ρ. Just as was true for the previous rod diagram, when we want to determine the field at some point P a distance r from the center of the sphere, we are concerned only with the amount of charge inside an imaginary sphere of radius r concentric with the charge distribution. In this case, the charge acts as if it is concentrated at the center of the sphere, and Coulomb's law allows us to find the answer. Again, the trick lies in determining how much charge is enclosed by the imaginary sphere.

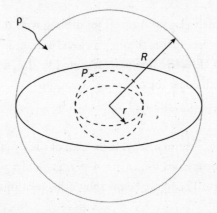

The above figure shows a uniformly charged sphere of radius R. The sphere has a charge density of ρ because the charge is uniformly spread through its volume. Again, an imaginary sphere with P on its surface contains all of the charge that contributes to the field at that point.

We can see from this qualitative argument that if the charge is distributed on a shell of radius R (see figure below), then there is no electric field inside the sphere.

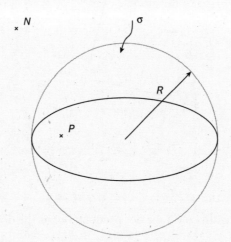

The spherical shell is uniformly charged. Because the charge is distributed over the surface of the sphere, there is a surface charge density, σ. A point inside the sphere has no electric field. A point outside the sphere has a field that is the same as if all of the charge were at the center of the sphere.

There is no electric field inside the imaginary sphere because no charge is enclosed by the sphere, which is quite handy when solving problems. To deal with these distributions in a more rigorous manner, we will now delve into Gauss's law.

GAUSS'S LAW (C EXAM)

Gauss's law provides an extremely elegant method for dealing with the fields of charge distributions. It states that the electric flux through a closed surface is proportional to the total charge enclosed by that surface. **Electric flux** is something like the "flow" of electric field through a surface. Electric flux, Φ_E, is defined by the following equation:

$$\Phi_E = EA\cos\theta$$

where E is the magnitude of the electric field; A is the area of the surface; and θ is the angle between the field vector and the area vector, an arrow perpendicular to the surface whose length represents the size of the surface. The figure below shows this relationship.

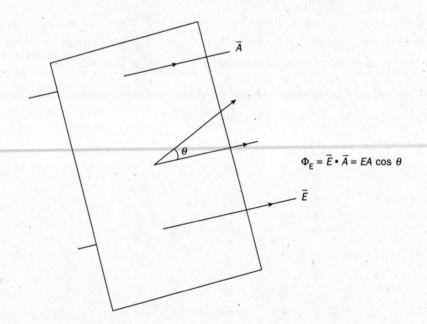

$$\Phi_E = \vec{E} \cdot \vec{A} = EA\cos\theta$$

A more succinct way of stating this is via the dot product: flux is the dot product of the field and the area.

$$\Phi = \mathbf{E} \cdot \mathbf{A}$$

When the field is always perpendicular to the surface, the flux is just the algebraic product of the field and the area. The astute reader may recall, however, that there is a catch to this flux business when it comes to Gauss's law.

Gauss's law deals with imaginary closed surfaces, often called **Gaussian surfaces**. When you are dealing with a closed surface, there are two regions of space: the region inside the surface and the region outside the surface. Thus, there are two different directions in which the area vector can point. By convention, the area vector of a closed surface points out. If a surface encloses positive charge, both the field and the area point outward, and the flux is positive. On the other hand, if the surface encloses negative charge, the field points in, the area points out, and the flux is negative.

With flux defined, we can examine Gauss's law. In integral form it is written as follows:

$$Q_{enc} = \varepsilon_0 \int_S E \bullet dA$$

This is a formidable formula. However, taken in bits it is quite simple. The left-hand side is just the total charge enclosed by the surface, S. The right-hand side has three parts: (i) ε_0 is the permittivity of free space, (ii) $E \bullet dA$ is a differential flux element, and (iii) the $\int_S$ just tells you to integrate over a closed surface. The next figure shows how to set up this type of integral. At this level (and beyond, for the most part), we are concerned only with very symmetrical problems (e.g., spherical, cylindrical, and planar). As a result, the integrals are usually fairly simple.

The following figure shows two differential flux elements for a cube around a point charge. Notice that the value of $E \bullet dA$ depends on the location of the element. To solve the integral in Gauss's law, you need a functional dependence for the dot product over the whole surface; this is a general property of Gauss's law. However, in highly symmetrical cases, this location dependence can be made to disappear.

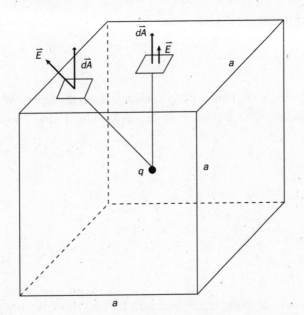

Let's review how Gauss's law makes field calculation simple by returning to the infinitely long line of uniform charge. The distribution has a constant linear charge density, λ coulombs per meter, and we need to calculate the field at a distance r from the line. This is a case of cylindrical symmetry, so we should think of the cylinder's axis as the line of charge.

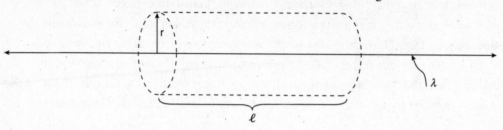

The cylinder is our Gaussian surface. Through the judicious choice of coaxial cylinder and line, we see that at the ends of the cylinder, the field and the area vectors are at right angles. Thus, the ends contribute nothing to our answer ($E \bullet dA_{end} = 0$). We also see that on the curved surface, the field and the area vectors are parallel everywhere. Therefore, the integrand in Gauss's law is just $E\,dA$, giving us

$$Q_{enc} = \varepsilon_0 \int_S E\,dA$$

The integral is just over the curved surface of the cylinder. Further, we expect the field to be independent of the area element. This is due to the fact that if we rotate the whole system around the line of charge, we should see no difference. As a result, E can come out of the integral:

$$Q_{enc} = \varepsilon_0 E \int_S dA$$

The integral is now just the area of the side, $2\pi rl$. We have this expression for E:

$$E = \frac{Q_{enc}}{\varepsilon_0 2\pi rl}$$

All that remains is to determine the total charge enclosed. From the figure, we see that the cylinder encloses l meters of the charge distribution. From the definition of λ, we know that

$$\lambda = \frac{Q}{l}$$

so

$$Q_{enc} = l\lambda$$

Substituting this gives us

$$E = \frac{\lambda}{2\pi\varepsilon_0 r}$$

for E, which matches what was found from Coulomb's law and that nasty integral over θ.

There are four key steps involved in using Gauss's law successfully:

1. Determine the symmetry of the problem: spherical, cylindrical, or planar.

2. Draw the appropriate Gaussian surface: sphere, cylinder, or pillbox.

3. Solve the simplified integral exploiting the symmetry.

4. Find the enclosed charge.

The figure below shows the Gaussian surfaces for (a) spherical, (b) cylindrical, and (c) planar symmetries.

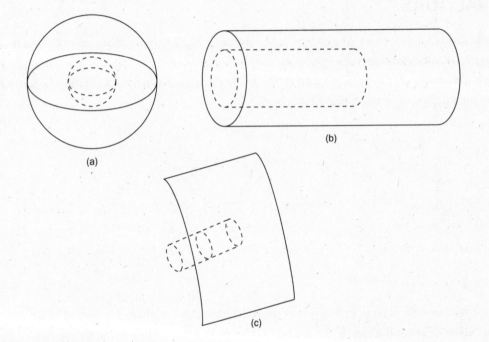

(a)

(b)

(c)

You should be comfortable with the methods used to determine total enclosed charge. You will typically be writing the charge density as a function of one variable only, which allows you to integrate over that variable to find the total charge enclosed. This is primarily important only when you have to deal with nonuniform charge densities. An example of such a distribution would be given by the piecewise function:

$\rho = ar, r \leq R$
$\rho = 0, r > R$

If you wanted to know the field at a point inside the sphere, you would have to write the proper integral for the total enclosed charge. In this case, if we are concerned with a point a distance d from the origin with $d < R$, then

$$Q_{enc} = 4\pi \int_{0}^{d} \rho r^2 \, dr = 4\pi \int_{0}^{d} (ar)r^2 \, dr$$

If you are not familiar with this expression, you should review your integral calculus text.

One final point about Gauss's law: we can use it to show that charge placed on a conductor resides on the outer surface of that conductor. Because we know that the field inside a conductor must be zero everywhere, we know that the integral in Gauss's law must be identically zero for any Gaussian surface entirely within a conductor. This implies that the total charge enclosed by that surface must be zero. The surface can be made so that it lies just inside the surface of the conductor. Therefore, there is no net charge inside an isolated conductor.

CAPACITORS

A **capacitor** is a device that stores energy in an electric field. Strictly speaking, any two conductors separated by some distance make up a capacitor. Typically we think of capacitors as having no net charge. One conductor has positive charge $+Q$, while the other has negative charge $-Q$ (see figure below for a depiction of an arbitrary capacitor).

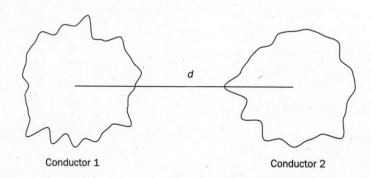

Conductor 1   Conductor 2

Taken together there is no net charge. Yet due to the separation of charges, a potential difference exists between the conductors. If this potential is known to be V volts, we can define the capacitance as follows:

$$C = \frac{Q}{V}$$

The **capacitance** (C) is measured in coulombs per volt, or farads (F). Q is the charge on the positively charged conductor, by convention. V is the potential difference between the conductors.

How does a capacitor store energy? When we charge a capacitor, it is typically assumed that we are moving electrons from one plate to the other. This leaves the donor plate with a net positive charge and the receiving plate with an equal but opposite charge. If we start with an uncharged capacitor, we can move the first charges for free. However, once those first charges are present, we must do work to move the next charges from one plate to the other. If we want to make the positive plate more positive, we have to move more electrons from that plate to the negative plate. Because the field around a charged object is related to the charge of the object, we are increasing the field between the plates of the capacitor. The work we do to move the charges and create this field is stored as energy in the field. Without going into a lengthy derivation, we can state that the energy stored in the field of a charged capacitor is

$$U = \frac{1}{2}CV^2$$

By substituting the definition for capacitance, we can obtain two other expressions for the energy stored:

$$U = \frac{1}{2}QV$$

$$U = \frac{1}{2}\frac{Q^2}{C}$$

You should feel comfortable switching between these expressions as needed. In Chapter 9, we will see how a capacitor plays the role of a spring in an oscillating inductor-capacitor circuit.

The capacitance is a constant for a given geometrical configuration (i.e., for conductors of a given size, shape, and separation). Typical capacitors are two parallel plates, concentric spherical shells, and coaxial cylinders. You should expect to see parallel plates on both the B and C exams; you will encounter spherical shells and coaxial cylinders on the C exam only.

PARALLEL PLATE CAPACITORS

A **parallel plate capacitor** consists of two conducting plates, each of area A, separated by a distance d. If the separation is small (d is significantly less than $\sqrt{A}$), then the capacitor is considered ideal. For the B exam, the space between the plates will be filled with a vacuum. For the C exam, the space may be filled with an insulator (dielectric), or it may be partially filled with a conductor. The following figure shows a typical **parallel plate capacitor**. The plates have the same shape, generally squares or circles. This is an ideal capacitor, so d is significantly less than $\sqrt{A}$, and the field between the plates is uniform.

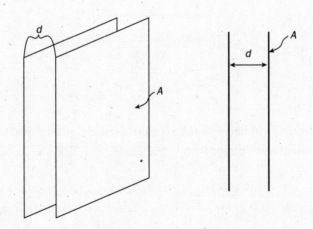

Because capacitance is related to geometry, we would like to have a formula that gives the capacitance for two parallel plates. Suppose we charge the plates so that one has charge $+Q$ and the other has charge $-Q$. There is a potential difference between the plates, so there is an electric field. Recall that the electric field near a charged plane is

$$E = \frac{\sigma}{2\varepsilon_0}$$

Each plate contributes this to the total field, so the field between the plates is

$$E = \frac{\sigma}{\varepsilon_0}$$

σ is the surface charge density, which is given by $\frac{Q}{A}$. The last bit we need is the relationship between field and potential difference when the field is uniform in space. Remember that this is

$$V = Ed$$

Now, based on the definition of capacitance $\left(C = \dfrac{Q}{V} \right)$, we can find the geometric formula for the capacitance of a parallel plate capacitor:

$$C = \frac{\varepsilon_0 A}{d}$$

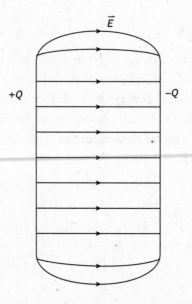

The above figure shows the field between the plates of a parallel plate capacitor. Near the edges of a real capacitor, the field bows out or fringes (we neglect this fringing in our geometrical calculation).

FOR C EXAM ONLY

SPHERICAL AND CYLINDRICAL CAPACITORS

The method used to determine the capacitance of a parallel plate capacitor from geometry may be employed to explore spherical and cylindrical capacitors as well. Regardless of shape, the conductors in a capacitor are often referred to as plates. All you have to do is find the field between the plates and relate it to the total charge and potential difference, which usually requires Gauss's law. The figure below shows spherical and cylindrical capacitors. In both cases, the radius of the inner conductor is a, and the radius of the outer conductor is b. The region of interest is $a < r < b$, where r is a variable. In this case, the inner conductor has charge $+Q$, and the outer has charge $-Q$. Gaussian surfaces are shown as well. The calculus used to solve this problem is rather straightforward, and you should find the capacitance of the spherical shells to be

$$C = 4\pi\varepsilon_0 \frac{ab}{b-a}$$

and that of the coaxial cylinders to be

$$C = \frac{2\pi\varepsilon_0 l}{\ln\left(\dfrac{b}{a}\right)}$$

Notice that both expressions have the same units as the parallel plate. You should be able to show how to arrive at these answers. If you have problems, look at the free-response question for the C exam at the end of this chapter.

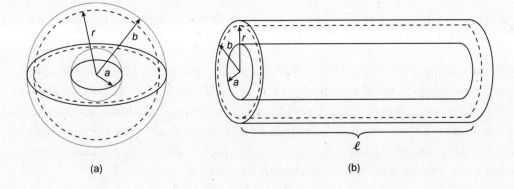

(a)   (b)

The above figure shows (a) a spherical capacitor and (b) a cylindrical capacitor. The dotted curve in each represents the Gaussian surface used to determine the field between the plates. Once you have an expression for this field, you need to integrate from the inner surface to the outer surface so that you can find the potential difference.

DIELECTRICS (C EXAM)

Dielectrics are insulating materials that react to an applied electric field. The reaction is called "polarization" because the charges separate slightly, creating dipoles all throughout the material. This polarization decreases the net electric field in the dielectric. The figure below should help you visualize what is happening here. Remember, in an insulator, charges are not free to move about, but they may be locally displaced by a small amount. The applied, or external, field causes the negative charges to shift one way and the positives to shift the other way. The dipole field thus established opposes the external field.

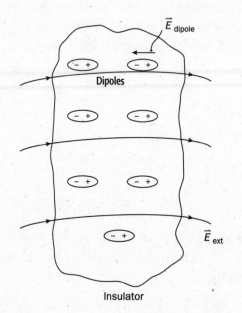

The figure above shows dielectric material in an external field. The dipoles formed inside the material decrease the field inside.

If a dielectric is placed between the plates of a capacitor, the capacitance will change. If we keep the geometry and the charge of the capacitor fixed, we can determine the effect of the dielectric. The field between the plates will decrease, which will cause the potential difference between the plates to decrease. This, in turn, leads to an increase in the capacitance. Thus, the capacitance of a capacitor with a dielectric is larger than an identical capacitor without the dielectric. The ratio between these two capacitors is called the dielectric constant, κ.

$$\kappa = \frac{C_{\text{dielectric}}}{C_{\text{vacuum}}}$$

The effect of dielectrics is explored more fully in the free-response question at the end of the chapter.

ELECTRIC CURRENTS

Electric currents are the main application of electricity in engineering and are responsible for generating *all* magnetic fields (see Chapter 9). The basic concepts of current, resistance, and power will be discussed in some detail in this section. Then we will develop the basics of electric circuits with resistors and capacitors in steady-state DC situations. We will close this section with a discussion of the transient phenomena associated with resistor/capacitor (RC) circuits for students taking the C exam.

CURRENT, RESISTANCE, AND POWER

Electric current is defined as the rate of flow of positive charge across a surface per unit time. Expressed as a formula on the B exam, this is

$$I_{\text{avg}} = \frac{\Delta Q}{\Delta t}$$

and on the C exam

$$I = \frac{dQ}{dt}$$

The two definitions are identical because I is a constant for steady currents. The units for current are coulombs per second, which is the **ampere** (A). We usually think of charge as being a fundamental unit, but in the SI system it is not. The ampere is the fundamental unit of electricity, and the coulomb is derived from the ampere. In Chapter 10, we'll see that the ampere's definition is based on the force between two parallel wires carrying electric currents.

Any flow of charge can be considered a current. A proton beam in a particle accelerator is one current. The current is defined by keeping track of the charges that cross the plane defined by line P every second (see figure below).

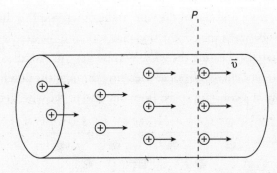

If there are 10^6 protons in every centimeter of length along the beam and each proton is moving at 10^4 m/s, we can calculate the current this beam represents. From the definition of current given above, we can see that we need to know how much charge crosses a plane every second. Each proton contributes 1.6×10^{-19} coulombs of charge. Slicing the beam with a plane perpendicular to its direction of motion, we can count the number of protons that pass this plane each second. Every centimeter of beam length corresponds to 10^6 protons; every meter therefore has 10^8 protons. In one second, 10^4 meters of beam pass our plane. We finally arrive at 10^{12} protons crossing our plane every second. Combining this information with the charge of each proton, we get a current of 1.6×10^{-7} A. Thus for a beam of charges, a tremendous number of individual charges is required for any appreciable current. When we deal with charges confined to a wire (as we do in circuits), we will find that we have plenty of charges to create these currents. Electron flow in wires is responsible for the current in electrical circuits; however, the "conventional" direction of current that we use assumes positive charges are flowing in a direction opposite to the actual electron flow.

To make charges move, we need an electric field. Rather than focusing on this field, we are typically more concerned with the **potential difference** that exists where the field is non-zero. This potential difference, V, is responsible for the motion of charges. When working with electric currents and circuits, it is common to refer to the potential difference as the "applied" or "impressed" voltage. So if we want an electric current, we need two things: free charges and a potential difference. Common sources of potential differences are batteries and power supplies. If there is nothing to oppose the motion of the charges, they all accelerate according to Newton's second law, where the magnitude of the force is given by Eq. In free space, this is the case. But in wires, the story is not yet complete.

When charges fly through a vacuum, there is little with which they can interact. Most importantly, there are few, if any, obstacles opposing the current established. In a physical conductor, this is not the case. As the charges ("free" or "conduction" electrons in conductors) move down a wire, they interact with the nuclei that make up the material in the wire. In the language of the physicist, the electrons interact with the "crystal lattice" of the conductor. Under normal circumstances, the electrons are confined to the wire, so they *must* stay fairly close to its atoms. When an electron collides with an atom, it usually bounces off in a direction that is different from its initial direction of travel. In the short term, then, no electron manages to go any great distance. However, there is a general drift of electrons from one end of the wire to the other, and this drift is what gives rise to the current. The collisions between the conduction electrons and the atoms result in a dissipation of energy, which is manifested as heating the wire. We also get what is called resistance. The net effect of resistance is to prevent charges from accelerating through the potential difference. Instead, the charges appear to move at a constant speed from one end of the wire to the other.

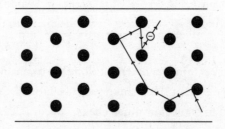

The previous figure is a simple picture of a conduction electron moving through a metal. The solid dots represent the atoms that make up the metal. The path of the electron is indicated.

Resistance is defined as the ratio of applied voltage to current produced:

$$R = \frac{V}{I}$$

The unit of electrical resistance is the **ohm** (Ω). One ohm is the resistance created when a 1-volt potential difference is applied to a sample and a 1-amp current is produced. This definition of resistance is not just **Ohm's law** in a slightly disguised form. It should be noted that while this definition is true for any voltage, current, and material, it does *not* necessarily mean that voltage and current are linearly related to one another. That is to say, resistance is not necessarily a constant for a given object. If the resistance is constant, then we say that it is an **Ohmic resistor**, and then Ohm's law does indeed apply. However, resistance often varies with applied voltage for materials. But to a good approximation, and for all problems on the AP Physics exams, it is acceptable to use Ohm's law as the relationship between voltage and current.

Another quantity associated with electrical resistance is resistivity (ρ). **Resistivity** is an intrinsic property of all materials. It is measured in ohm-meters ($\Omega \cdot m$). If you have a nicely shaped sample of material (e.g., a wire of uniform diameter), you can determine the resistance if you know the resistivity and geometry. The relationship between resistance and resistivity in such a case is given by

$$R = \frac{\rho l}{A}$$

where l is the length of the sample and A is the cross-sectional area (see the figure below, where resistivity ρ leads to a resistance R).

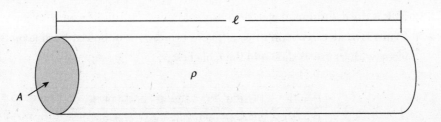

The relationship between resistance and resistivity makes sense if you think of electric current in a wire as analogous to fluid moving through a pipe. In a longer wire, there are more atoms along the path of the current flow with which to collide. The number of atoms increases as the area increases, but there are also more possible paths for an electron to cross a given plane. (This last claim is not strictly accurate because the conduction electrons move along the surface of the conductor, but we will ignore this complication here.) The resistivity of the current, then, would correspond to the viscosity of the fluid.

AP EXPERT TIP

Resistance also goes up as the temperature of the wire increases. So think of a low-resistance wire as being like a **penguin**—short, fat, and cold.

Because the current represents charges moving through a potential difference, the charges are losing potential energy. However, in a wire they do not accelerate, so they do not gain any kinetic energy. For energy conservation to hold, there must be some other place for the energy to go. Generally, it goes into heating the element carrying the current. In a sense, the resistance supplies a "force" to oppose the force that results from the potential difference. The work done by the resistive force and the potential difference must be equal (and opposite). How can we calculate the work done by the resistance if it is not strictly a force? Considering Ohm's law, we have the following:

$$V = IR$$

Bear in mind that V represents a potential difference. Thus, if we multiply V by the magnitude of a charge in the current, we will get the potential energy lost by that charge while moving across the potential:

$$Vq = IRq$$

If we divide both sides by the time required for the charge to complete its motion, we have the average power:

$$\frac{qV}{t} = IR\frac{q}{t}$$
$$P = IRI$$

This last step is possible because we recognize q/t as electric current. The quantity RI is nothing more than the potential V, so we have

$$P = IV$$

This is the definition of electric power. Note that by suitable substitution using Ohm's law, we can obtain three different expressions of power. Each expression, along with its use, is shown in the table below.

$P = IV$	Useful for batteries and solved circuits
$P = I^2R$	Useful with unknown voltages
$P = \dfrac{V^2}{R}$	Useful with unknown currents

The last equation in the table explains why electricity is transmitted from the power plant to distant locations at very high voltage. The wires carrying the current have extremely large resistances (due to length), decreasing the power lost in transmission.

STEADY-STATE DIRECT CURRENTS WITH BATTERIES AND RESISTORS

If you combine a potential source such as a battery, some conductive wires, and various electrical components (called "elements"), you can build a circuit. An **electric circuit** is a system through which electric current flows, causing changes in pieces of the system and quite often in its surroundings. As its name implies, a circuit is a closed path. Beginning at any point on the circuit, one can trace a route in one direction through the elements and return to the starting point. It is not necessary that a single path travel through all elements for a collection to qualify as a circuit; you just need to be able to return to your starting point without backtracking.

The simplest electric circuit consists of a single potential source, a single resistive element (called a "**resistor**"), and connecting wires. When making calculations regarding this and other circuits, we assume that the connecting wires are perfect conductors (i.e., they have zero electrical resistance). By "steady-state," we mean there is no change over time. Neither the electric current at any point nor the voltage between any two points in the circuit is changing with time. We diagram these so-called simple circuits using a straightforward scheme. Each type of element (resistor, capacitor, inductor, microphone, loudspeaker, light bulb, potential sources, etc.) has a plain line drawing associated with it. The pictures of the elements are connected with straight lines that represent the wires. Because the wires are assumed to be perfect conductors, they can be any length, as long as all elements are connected in the same way.

The figure below shows the most basic elements. You should know the meaning of these pictures by sight before test day.

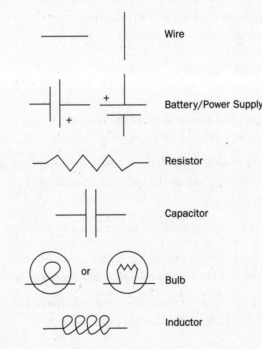

A circuit diagram consists of these symbols connected with straight line wires. The figure below shows three simple circuits. Notice that sometimes the battery is shown with the positive and negative terminals indicated and sometimes the locations are assumed based on the symbol. Just remember that the shorter line is the negative, or low, terminal.

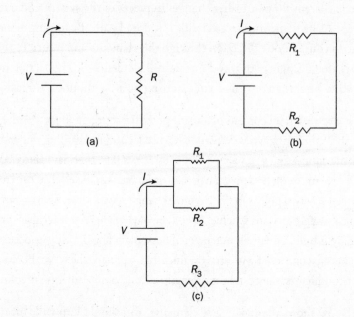

The above figures show a single resistor connected to a battery (a), two resistors connected in series with a single battery (b), and three resistors connected in a series/parallel combination (c). Despite the increasing complexity depicted here, all of these circuits are considered to be simple circuits.

The two important rules governing circuits, known as Kirchhoff's rules, are nothing more than conservation laws in disguise. Kirchhoff's rules state the following:

- The sum of all potential gains and losses around any closed loop in a circuit is always zero (Kirchhoff's loop rule, the loop rule, or KLR).

- The sum of all currents flowing into a junction is equal to the sum of all currents flowing out of that junction (Kirchhoff's junction rule, the junction rule, or KJR).

When using the loop rule, recall the directional conventions:

- If you cross a battery from *low to high* (short line to long line) you increase potential, but crossing from *high to low* decreases potential.

- If you cross a resistor *with* the current you decrease potential, while going *against* the current increases the potential.

KLR is nothing more than conservation of energy, while KJR is conservation of charge. We can use these two rules to develop the sum rules for resistors in series and parallel (and any other elements for which we know how the potential varies across the ends). Series resistors just add resistances. The formula is

$$R_S = \sum_i R_i$$

Parallel resistors add by reciprocals and the sum is then inverted. The formula is

$$\frac{1}{R_P} = \sum_i \frac{1}{R_i}$$

The result of combining resistors is often called the **effective resistance** or the **reduced circuit resistance**. Two resistors in series always have a *greater* effective resistance than either individually, while two resistors in parallel always have a *lesser* effective resistance than either individually. This fact allows a nice, quick check when doing calculations.

When you are given a circuit, the goal is to determine all currents, voltages, and resistances present. To interpret this statement, you need to know what the phrase "currents, voltages, and resistances" means. *Resistances* means the resistance of every element in the circuit, including possible effective resistances. We are dealing with simple circuits, so there are no capacitors or inductors present. *Voltages* means the electric potential difference between (or "across," as it is sometimes put) any two points in the circuit. *Current* means the electric current flowing through any point in the circuit. Finding what you need to know about a circuit requires using the loop rule, the junction rule, and Ohm's law to write a system of independent equations. Then you must solve the system for all of the unknown quantities. Generally, you will have more equations than variables, and some of the equations will not be independent. When this is the case, eliminate the superfluous equations.

To illustrate this approach, let's solve a circuit with seven resistors and two batteries. The circuit is shown in the figure below.

AP EXPERT TIP

- Resistors are only in series if passing through one resistor requires passing through the other, and vice versa. Series resistors form a chain.

- Resistors are only in parallel if they have the same endpoints as each other.

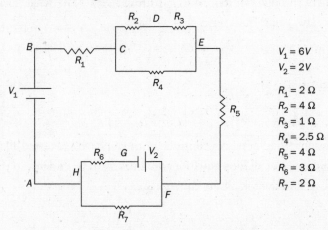

$V_1 = 6V$
$V_2 = 2V$

$R_1 = 2\,\Omega$
$R_2 = 4\,\Omega$
$R_3 = 1\,\Omega$
$R_4 = 2.5\,\Omega$
$R_5 = 4\,\Omega$
$R_6 = 3\,\Omega$
$R_7 = 2\,\Omega$

To keep track of everything we know as well as the things we need to find out, it's good to make a table like the one below.

Element	V (V)	I (A)	R (Ω)	P (W)
V_1	6		–	
V_2	–2		–	
R_1			2	
R_2			2	
R_3			3	
R_4			2.5	
R_5			4	
R_6			3	
R_7			2	

Notice how each element is given a name and how the known values are written in the table. This helps us identify the quantities we still need to know. Clearly, we need to know current and power for each element. Further, we must know the voltage across each resistor. In the absence of other information, we should assume that the batteries are ideal and have no internal resistance. Many conventions exist for naming the different currents in a parallel circuit. Here, we will call the current through the largest battery I and add Arabic numeral subscripts, starting with 1, to currents as we need them.

Now we must start tracing our paths using the loop and junction rules. The first path we can take is ABCDEFGHA. Through this closed path, the sum of all potential gains and losses must be zero (KLR). We also note that at point C, the current splits into I_1 (through R_2 and R_3, the current through our loop) and I_2 (through R_4). Then at point E, the current becomes I again. At F, it splits into I_3 (through R_6, the current through our loop) and I_4 (through R_7). Now, what does KLR say about our loop? From Ohm's law, you should verify that we have

$$4 - 6I - 5I_1 - 3I_3 = 0$$

A second loop ABCEFHA gives

$$6 - 6I - 2.5I_2 - 2I_4 = 0$$

This process of finding equations from loops can continue until you have exhausted all possible loops. Generally, you will find that not all loop equations are linearly independent, so you will have to resort to the junction rule to find the remaining equations.

In our case, KJR tells us:

$$I = I_1 + I_2$$
$$I = I_3 + I_4$$

This gives us plenty of information to solve our problems. Once we know all of the currents, we can use Ohm's law and the definition of power to solve the system completely. After some tedious algebra, we arrive at this table.

Element	V (V)	I (A)	R (Ω)	P (W)
V_1	6	0.6101	–	3.661
V_2	–2	–0.2255	–	0.451
R_1	1.220	0.6101	2	0.744
R_2	0.4066	0.2033	2	0.083
R_3	0.6099	0.2033	3	0.124
R_4	1.017	0.4067	2.5	0.414
R_5	2.440	0.6101	4	1.489
R_6	–0.6765	–0.2255	3	0.153
R_7	1.671	0.8355	2	1.396

All this calculating is well and good, but we need to compare these predictions with actual measurements. The measurements of voltage and current are performed with voltmeters and ammeters, respectively. **Voltmeters** are connected across the ends of an element, and **ammeters** are connected in series with an element. This helps cement the idea that the voltage is a difference in potential on either end of a circuit element and current is the flow of charge through a circuit element. The ideal voltmeter has an infinite resistance (because it is connected in parallel with your element), while the ideal ammeter has zero resistance (because it is connected in series).

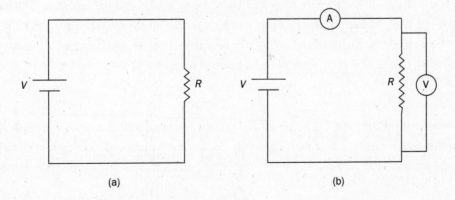

(a)   (b)

The previous figure (a) shows a circuit without any meters connected. Figure (b) shows the same circuit, but with a voltmeter and an ammeter connected to measure voltage and current. Notice that the voltmeter is in parallel with the resistor and the ammeter is in series. The symbol for each meter is a circle with a letter inside ("V" for voltmeter and "A" for ammeter). If you want to measure resistance, use an ohmmeter; you can construct an ohmmeter from a battery, a voltmeter, and an ammeter. You would then use Ohm's law to find the resistance of the element.

CAPACITORS IN CIRCUITS: STEADY STATE

In direct current circuits, capacitors act like breaks. After a sufficient amount of time, a capacitor prevents current from flowing across it. The charge that builds up on a capacitor always opposes the addition of more charge if the current does not change direction, as shown in the figure below. As charge accumulates on the capacitor plates, the current decreases until it stops completely. The amount of charge deposited on the plates depends on the capacitance and the voltage.

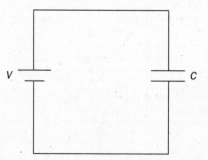

The figure above shows a basic capacitor-battery circuit. After a long time, the plates on the capacitor have sufficient charge to prevent the battery from depositing any more.

Recall that the definition of capacitance is

$$C = \frac{Q}{V}$$

where Q is the charge on the positive plate and V is the potential across the capacitor. Unless this potential is less than the battery potential, no current can flow in the circuit, thanks to Kirchhoff's loop rule. In a circuit with a battery of voltage V and a capacitor of capacitance C, a steady state will exist with a charge Q on the plates of the capacitor, where

$$Q = CV$$

Like resistors, multiple capacitors can be connected to a single battery in series, in parallel, or in a series/parallel combination. These multicapacitor circuits can be reduced to circuits with a single effective capacitance. For capacitors in series, the charge on each capacitor must be the same (think of Kirchhoff's junction rule). If you have a set of capacitors C_1, C_2, C_3..., all connected in series to a battery of voltage V, then you know that the charge on each capacitor must be the same; that is,

$$Q_1 = Q_2 = Q_3 = ... = Q$$

Q is also the charge on the effective capacitor that replaces the series group. The figure below shows a circuit with three capacitors in series and a circuit with the effective capacitor.

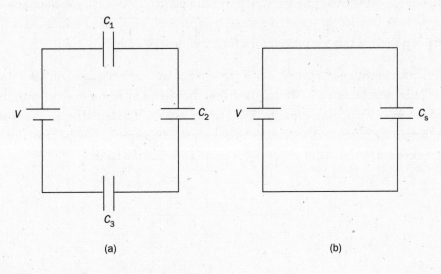

(a)   (b)

By applying KLR to both circuits, you should see

$$\frac{Q}{C_s} = \frac{Q}{C_1} + \frac{Q}{C_2} + \frac{Q}{C_3}$$

Canceling the common factor of Q, we arrive at the sum rule for capacitors in series:

$$\frac{1}{C_S} = \sum_i \frac{1}{C_i}$$

If we connect the same capacitors in parallel and take advantage of KLR with respect to parallel branches, we arrive at the sum rule for capacitors in parallel:

$$C_P = \sum_i C_i$$

You should note that the rules for combining capacitors are different from the rules for combining resistors; they are reversed. This is because capacitance acts like $\frac{1}{V}$, while resistance acts like V. It is important that you do not confuse these rules.

CAPACITORS IN CIRCUITS: TRANSIENTS IN RC CIRCUITS (C EXAM)

The previous section was concerned with DC circuits containing capacitors in the long term. After sufficient time has passed, the situation settles down to a steady state with no current flowing through the circuit. However, current does flow in the short term. For the C exam, you are expected to understand what happens in this short-term period. To do this, you must know the solution to the basic differential equation governing capacitors. These short-term behaviors are called **transients** because they are present for a time and then go away.

In an ideal circuit consisting of a battery and a capacitor, there is no resistance. Any real circuit, however, will have some resistance. We can model this by placing a resistor in series with the capacitor and treating the resistor-capacitor (RC) circuit as ideal. The figure below shows an ideal RC circuit. Before the switch is closed, there is no potential difference across the capacitor. After the switch is closed, a potential difference builds up and opposes the battery.

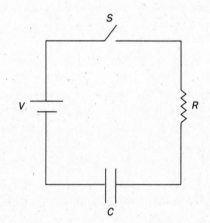

Let's remind ourselves of the qualitative aspects of this problem. When the switch is first closed, current will begin to flow through the resistor, and charge will accumulate on the capacitor. As charge accumulates, the current will decrease in magnitude because the capacitor will have a voltage difference built up to oppose the battery. Eventually, enough charge accumulates on the capacitor that the potential difference across it equals the potential supplied by the battery. At this point, no more current flows in the circuit.

Quantitatively, we can write an equation for this circuit that obeys Kirchhoff's loop rule. That equation (referring to the above figure) is

$$V - IR - \frac{Q}{C} = 0$$

If we recall that current is just the rate of flow of charge, we have

$$V - \frac{dQ}{dt}R - \frac{Q}{C} = 0$$

This is a differential equation for $Q(t)$. The solution for Q as a function of time is

$$Q = CV\left(1 - e^{-\frac{t}{RC}}\right)$$

This is sometimes called the charging equation; it shows charge building up on a capacitor as a function of time. The quantity RC is called the capacitive time constant. It is sometimes written as $RC = \tau_C$.

Differentiating the charging equation once with respect to time will give you the current. If you divide both sides of the charging equation by C, you obtain an expression for the potential across the capacitor. The figure below shows rough plots of (a) current and (b) potential across the capacitor during charging. Notice that as potential increases, current decreases.

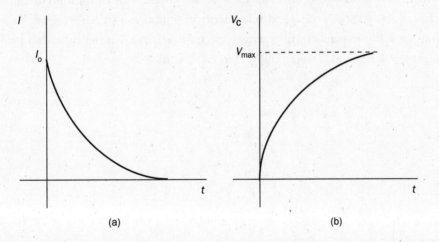

(a)   (b)

If you connect a charged capacitor to a resistor, you will get a current. In this case, the differential equation will be

$$\frac{dQ}{dt}R + \frac{Q}{C} = 0$$

This equation is much simpler to solve, and the solution for $Q(t)$ is given by

$$Q = Q_0 e^{-\frac{t}{RC}}$$

where Q_0 is the initial charge on the capacitor. This equation governs the discharging of a capacitor. Again, differentiation once by time will give you the current, and dividing each side by C will yield the potential across the capacitor.

Unlike the case of charging, current and potential follow each other in discharging, as shown in the figure below, where (a) shows rough plots of current and (b) shows rough plots of voltage.

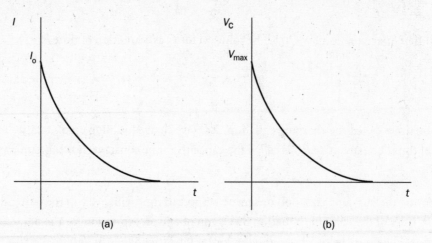

(a)   (b)

The transients are useful in setting up a timing mechanism. In some applications, it is desirable to have something happen at regular intervals; using an RC circuit with an appropriate time constant allows you to achieve a certain voltage with regularity in time after a switch is closed. You can then connect your capacitor in parallel with a sensing circuit that triggers at some threshold voltage. It is not the most elegant method of doing things, but it can work.

REVIEW QUESTIONS

B Exam

1. A cylindrical wire of radius 0.2 mm and length 5 m has a resistance of 0.5 Ω. Which of the following is the best estimate of the resistivity?

    (A) $3 \times 10^{-15}\, \Omega \cdot m$
    (B) $1.2 \times 10^{-15}\, \Omega \cdot m$
    (C) $3 \times 10^{-10}\, \Omega \cdot m$
    (D) $1.2 \times 10^{-8}\, \Omega \cdot m$
    (E) $3 \times 10^{-8}\, \Omega \cdot m$

2. Two identical insulating spheres each carry charge $+q$ and are separated by a distance r. If the charge on each sphere is halved and the distance doubled, what is the ratio of the magnitude of the new force to that of the original force?

    (A) 16
    (B) 4
    (C) $\dfrac{1}{2}$
    (D) $\dfrac{1}{4}$
    (E) $\dfrac{1}{16}$

3. A parallel plate capacitor is connected to a battery of potential V. The distance between the plates is doubled without disconnecting the battery. Which of the following is true?

    (A) Both the capacitance and the charge on the plates are decreased by one-half.
    (B) The capacitance is decreased by one-half, but the charge on the plates is unchanged.
    (C) The capacitance is unchanged, but the charge on the plates is decreased by one-half.
    (D) The capacitance is doubled, and the charge on the plates is unchanged.
    (E) Both the capacitance and the charge on the plates are doubled.

4. Which of the following is closest to the electric force between the proton and electron in a hydrogen atom?

    (A) $8 \times 10^{-15}\, N$
    (B) $8 \times 10^{-12}\, N$
    (C) $8 \times 10^{-10}\, N$
    (D) $8 \times 10^{-8}\, N$
    (E) $8 \times 10^{-6}\, N$

Use the following figure to answer question 5.

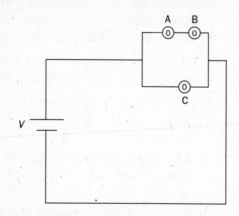

5. The circuit shown above consists of three light bulbs connected in the given series/parallel combination. If bulb A is disconnected, what can be said about the change in brightness of bulbs B and C?

(A) Both bulbs will become brighter.

(B) Bulb B will become brighter and bulb C will not be changed.

(C) Bulb B will not be changed and bulb C will become brighter.

(D) Bulb B will go out and bulb C will not be changed.

(E) Both bulbs will go out.

C EXAM

6. Three equal charges of magnitude q are placed at the corners of an isosceles triangle of side length a and base b. At what point inside the triangle might the net electric field be zero?

(A) At the intersection of the perpendicular bisectors of the sides of the triangle

(B) At the midpoints of the sides of the triangle

(C) At the reflection of one charge about the opposite side of the triangle

(D) Near one of the charges

(E) None of the above

7. A parallel plate capacitor of area A and plate separation d is made with two different dielectric materials in the gap. On the left half of the gap is material 1 with dielectric constant κ_1, and on the right half is material 2 with dielectric constant κ_2. The dielectrics sit side-by-side in the gap, and each dielectric occupies half the volume between the plates. An expression for the effective capacitance of this setup is

(A) $(\kappa_1 - \kappa_2)\dfrac{\varepsilon_0 A}{2d}$.

(B) $(\kappa_1 + \kappa_2)\dfrac{\varepsilon_0 A}{2d}$.

(C) $(\kappa_1 - \kappa_2)\dfrac{\varepsilon_0 A}{d}$.

(D) $(\kappa_1 + \kappa_2)\dfrac{\varepsilon_0 A}{d}$.

(E) $(\kappa_1 \kappa_2)\dfrac{\varepsilon_0 A}{2d}$.

Use the following figure to answer question 8.

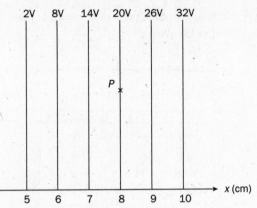

8. The above figure shows a region of equipotential curves. What can be said about the electric field at the point P?

(A) The field points to the right and has an approximate strength of 12 V/cm.

(B) The field points to the left and has an approximate strength of 12 V/cm.

(C) The field points to the right and has an approximate strength of 6 V/cm.

(D) The field points to the left and has an approximate strength of 6 V/cm.

(E) Nothing can be said because not enough information is given.

9. A capacitor of capacitance C stores a charge Q. If this capacitor is connected to a resistor R, what is the maximum current in the circuit created, and what is the maximum amount of energy converted to heat in the resistor?

(A) $I_{max} = \dfrac{QC}{R}$, $E_{max} = \dfrac{Q^2}{2C}$

(B) $I_{max} = \dfrac{Q}{CR}$, $E_{max} = \dfrac{Q^2}{2C}$

(C) $I_{max} = \dfrac{Q}{CR}$, $E_{max} = \dfrac{Q^2}{C}$

(D) $I_{max} = \dfrac{QC}{R}$, $E_{max} = \dfrac{2Q^2}{C}$

(E) $I_{max} = \dfrac{Q}{CR}$, $E_{max} = \dfrac{1}{2}Q^2C$

Use the following figure to answer question 10.

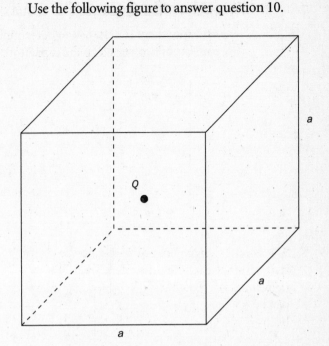

10. As shown in the above figure, a point charge Q is placed at the center of a cube of side length a. What is the electric flux through the top surface of the cube?

(A) $\dfrac{Q}{\varepsilon_0}$

(B) $\dfrac{Q}{3\varepsilon_0}$

(C) $\dfrac{Q}{6\varepsilon_0}$

(D) $\dfrac{Q}{9\varepsilon_0}$

(E) $\dfrac{Q}{18\varepsilon_0}$

FREE-RESPONSE QUESTION: B EXAM

An experiment is designed to determine the charge-to-mass ratio of subatomic particles. It consists of two parallel plate capacitors oriented as shown.

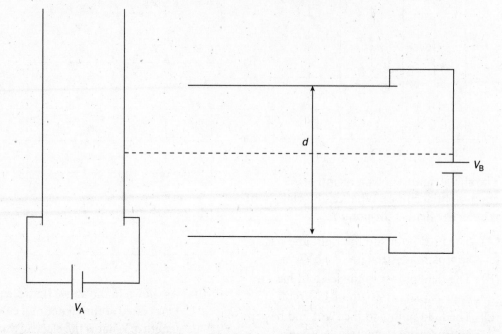

The particles start at the left plate of capacitor A and accelerate toward the right plate. The accelerated particles are able to pass through a small hole in the center of the right plate of capacitor A and fly into the space between the plates of capacitor B. The whole system is arranged so that if the particles travel undeviated through capacitor B, they are always equidistant from each plate.

(a) If the particles have charge q and mass m and capacitor A has a potential difference V_A (as shown), at what speed will the charges exit capacitor A?

(b) Capacitor B has a potential difference V_B across its plates, as shown in the figure. What is the force (magnitude and direction) on the particles while they are in capacitor B?

(c) How far from the left side of capacitor B will the particles strike a plate?

(d) Will this setup allow one to determine q/m for the particles?

FREE-RESPONSE QUESTION: C EXAM

Two concentric cylinders carry uniform linear charge densities λ_1 and λ_2, respectively. Cylinder 1 has radius a, and cylinder 2 has radius b ($a < b$). Each cylinder is a conductor with length l.

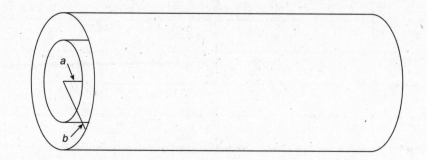

(a) Find the electric field for all values of r measured from the axis of the two cylinders.

(b) If $\lambda_1 = -\lambda_2$, find the capacitance of the cylinders. What is the potential difference between the cylinders?

(c) Suppose the space between the cylinders was filled with a dielectric material with dielectric constant $\kappa > 1$. What would the new potential difference be?

(d) Finally, imagine that the dielectric material was slowly pulled out from between the cylinders. Describe the source of any resistive forces to removing the dielectric.

ANSWERS AND EXPLANATIONS

B EXAM

1. D

Resistivity can be found by $\frac{RA}{\ell}$. To determine the area, you can approximate π to be 3. This introduces an error of less than 5%, which is accurate enough for estimations. Also, don't forget to convert all units to meters before you calculate the area.

2. E

Because the changes in the configuration are such that the force will definitely decrease, you can eliminate choices (A) and (B). The initial force is proportional to $\left(\frac{q}{r}\right)^2$. The charge decreases to half its initial value, and the radius increases to twice its initial value. Thus, the new force is proportional to $\left(\frac{q}{4r}\right)^2$. This gives the new force as $\frac{1}{16}$ the strength of the old force. It is much easier to solve comparison problems like this one if you realize that you can neglect the constants.

3. A

Here you must recall that the formula for the capacitance of a parallel plate capacitor acts like $\frac{1}{d}$. When d gets larger, C gets smaller by the same factor. Thus, you know right away that the capacitance goes to $\frac{1}{2}$ its initial value. The potential across the plates remains constant because the capacitor is connected to a battery. This means that if the capacitance changes, the charge on the plates must change as well. From $Q = CV$, you can see that charge acts like capacitance. This question is designed to test your ability to apply two different definitions of the same concept, as well as whether you understand what is happening when one quantity is held constant.

4. D

This question requires you to know several important physical constants. The first is the charge on the electron, 1.6×10^{-19} C. The second is the size of the hydrogen atom, 1 angstrom or 10^{-10} m. You also must have an idea of the value of $\frac{1}{4\pi\varepsilon_0}$. For most estimating purposes, this combined constant is about 9×10^9 Nm^2/C^2. Combine all of this with Coulomb's law and remember to use half the size of the hydrogen atom for the distance between the electron and proton. This yields 8×10^{-8} N as a result.

5. D

When bulb A is disconnected, the current through bulb B is cut off. Therefore, bulb B is no longer lit, and it goes out. To determine what happens to bulb C, you must consider its loop with the battery. Bulb C is all by itself with the battery, so what happens with A and B does not affect C, and bulb C therefore remains unchanged.

C EXAM

6. E

If the answer to any question is "None of the above," you must be *completely* sure of the answer before you select it. Choice (B) is easy to eliminate, since two of the forces would completely cancel out, leaving the third force to dominate the net force. Choice (C) is likewise easy to rule out, as the three force vectors would all be pointing outside the triangle and would reinforce each other. At a point near one of the charges, choice (D), the force from the closer charge would dominate and could not be canceled out by the others. Choice (A) looks like it could be the right answer, but after some analysis, you can see the point created by the intersection of the three perpendicular bisectors is the center of the *circumscribed* circle and would not result in a net force of zero (draw out a 45-45-90 right triangle to confirm this for yourself). With all the answers ruled out, we are left with choice (E). A little further analysis shows that the actual point of zero net force is the intersection of the 3 *medians* of the triangle, which forms the center of the *inscribed* circle. A great way to confirm any of these conclusions is to draw out diagrams with force vectors. You can use simple examples (equilateral triangles and 45-45-90 right triangles are both isosceles), but make sure you test each case on more than one type. On an equilateral triangle, the point described in choice (B) is the center of both the inscribed and circumscribed circles, but that is the only case where it is correct.

7. B

Since the two dielectrics fill equal volumes in the capacitor, you may treat the dielectric constant as an average of both: $\kappa_{avg} = \dfrac{\kappa_1 + \kappa_2}{2}$. Therefore,

$$C = (\kappa_1 + \kappa_2)\frac{\varepsilon_0 A}{2d}$$

8. D

Based on the definition of E as $-\dfrac{dV}{dx}$, you can determine the strength and direction of the field. All you have to do is remember that the derivative is nothing more than the slope. The spacing of the equipotentials is nice and even, and they step up by 6 V each line. This makes a graph of V versus x a straight line whose slope is 6 V/cm. Because the high potential is on the right and low is on the left, the field must point to the left. Students often forget that the derivative of a linear function is just the slope, but there is no reason to get tripped up by a simple problem like this one.

9. B

Determining the current is fairly easy if you recall that current is charge per time. So you know right away that Q must be in the numerator and t must be in the denominator. If you recall that the capacitive time constant is RC, you know that this quantity has units of time and therefore $I = \dfrac{Q}{CR}$. The maximum amount of heat produced in the resistor is given by the total energy stored in the electric field of the capacitor. This is just $\dfrac{Q^2}{2C}$.

10. C

This type of question is a classic because it tests whether you really understand the concept of flux. If you take a point charge and calculate the electric flux through a sphere containing the point charge, you will find that it is $\dfrac{Q}{\varepsilon_0}$ regardless of the size of the sphere. Indeed, one may take this as the definition of Gauss's law. The subtle issue here is that this is the flux through *any* closed surface containing charge Q. Therefore, the cube has a total flux of $\dfrac{Q}{\varepsilon_0}$. This flux is split evenly among all 6 faces of the cube. So the total flux through any one face is just $\dfrac{Q}{6\varepsilon_0}$. A popular variant of this problem places the charge at one corner of the cube and asks you to determine the flux through a face that does not share that corner. All you have to do is construct cubes so that the charge is at the center. You should then be able to see that the flux in this case is $\dfrac{Q}{24\varepsilon_0}$.

FREE-RESPONSE QUESTION: B EXAM

(a) Based on the definition of electric potential we have $W = qV_A$, where W is the work done on the charge. From the work-kinetic energy theorem, $W = \Delta K$. So we are led to

$$\Delta K = qV_A$$

$$\frac{1}{2}mv^2 = qV_A$$

$$v = \sqrt{\frac{2qV_A}{m}}$$

(b) Once the particle enters capacitor B, it is in another electric field, this time perpendicular to its initial motion. The force on the charge is given by the definition of the electric field, $F = qE$.

To determine the field, we need to recall the basic idea behind the parallel plate capacitor, namely that there is a uniform electric field between the plates if there is a potential difference across those plates. From the relationship between field and potential, we have $E = -\dfrac{\Delta V}{\Delta x} = -\dfrac{V_B}{d}$. The minus sign reminds us that the field points in the direction of decreasing potential. The force is then given by $F = -q\dfrac{V_B}{d}$, and it points down in the picture.

(c) This is a two-dimensional kinematics problem. We know the following quantities:

$$v_{x0} = \sqrt{\frac{2qV_A}{m}} \qquad v_{y0} = 0$$

$$x_0 = 0 \qquad y_0 = \frac{d}{2}$$

$$x_f = ? \qquad y_f = 0$$

$$a_x = 0 \qquad a_y = \frac{qV_B}{md} \text{ down}$$

$$t = ?$$

Once we have this information, we can approach it just like a projectile motion problem. First, find the time of flight:

$$\Delta y = v_{y0}t + \frac{1}{2}a_y t^2$$

$$t = \sqrt{\frac{2\Delta y}{a_y}} = \sqrt{\frac{d}{\left(\dfrac{qV_B}{md}\right)}}$$

$$t = d\sqrt{\frac{m}{qV_B}}$$

Now use the time of flight to determine the change in x:

$$\Delta x = v_{x0}t + \frac{1}{2}a_x t^2$$

$$\Delta x = d\sqrt{\frac{2qV_A}{m}}\sqrt{\frac{m}{qV_B}}$$

$$\Delta x = d\sqrt{\frac{2V_A}{V_B}}$$

(d) This setup is not a useful way to determine q/m because the total distance traveled through capacitor B is independent of the desired quantity. Any experiment designed to measure a quantity must be sensitive to changes in the quantity.

FREE-RESPONSE QUESTION: C EXAM

(a) The charge distributions suggest using cylinders as Gaussian surfaces. If the Gaussian surfaces are concentric with the charge distributions, then there are three regions of space identified by $r < a$, $a < r < b$, and $b < r$, where r is the radius of the Gaussian cylinder. If we label these regions I, II, and III, respectively, and use Gauss's law in each region, we will find the following:

REGION I:

$$\oint_c E\,dA = \frac{Q}{\varepsilon_0}$$

$$\oint_c E\,dA = 0 \Rightarrow E = 0$$

REGION II:

$$\oint_c E\,dA = \frac{Q}{\varepsilon_0}$$

$$\oint_c E\,dA = \frac{\lambda_1 l}{\varepsilon_0}$$

$$E(2\pi rl) = \frac{\lambda_1 l}{\varepsilon_0} \Rightarrow E = \frac{\lambda_1}{2\pi\varepsilon_0 r}$$

REGION III:

$$\oint_c E\,dA = \frac{Q}{\varepsilon_0}$$

$$\oint_c E\,dA = (\lambda_1 + \lambda_2)\frac{l}{\varepsilon_0}$$

$$E2\pi rl = \frac{(\lambda_1 + \lambda_2)l}{\varepsilon_0} \Rightarrow E = \frac{\lambda_1 + \lambda_2}{2\pi\varepsilon_0 r}$$

(b) Now $\lambda_1 = -\lambda_2 = \lambda$. In region II the field is given by $E = \dfrac{\lambda}{2\pi\varepsilon_0 r}$. The definition of

capacitance is $C = \dfrac{Q}{V}$, so we need the potential difference. From $E = -\dfrac{dV}{dx}$, we can find V by

integrating both sides with respect to a path. If we choose our path to be radial from $r = a$ to $r = b$, we have this:

$$V = \int_a^b E\,dr = \int_a^b \frac{\lambda}{2\pi\varepsilon_0 r}\,dr$$

$$= \frac{\lambda}{2\pi\varepsilon_0}\int_a^b \frac{1}{r}\,dr = \frac{\lambda}{2\pi\varepsilon_0}(\ln b - \ln a)$$

$$= \frac{\lambda}{2\pi\varepsilon_0}\ln\left(\frac{b}{a}\right)$$

Using the definition of capacitance and the fact that the charge on the inner cylinder is λl, we find

$$C = \frac{\lambda l}{\left(\dfrac{\lambda}{2\pi\varepsilon_0}\right)\ln\left(\dfrac{b}{a}\right)}$$

$$= \frac{2\pi\varepsilon_0 l}{\ln\left(\dfrac{b}{a}\right)}$$

(c) When the dielectric is inserted, the capacitance increases by a factor of κ because $C_{\text{dielectric}} = \kappa C_{\text{vacuum}}$. Since the charge does not change, the potential must decrease by a factor of κ because

$V = \dfrac{Q}{C}$. This means the new potential is

$$V = \frac{\lambda\ln\left(\dfrac{b}{a}\right)}{2\pi\kappa\varepsilon_0}$$

(d) The potential energy is lower with the dielectric material in place because the capacitance is higher. Thus, removing the material increases the potential energy. Therefore, we should expect there to be a force resisting the removal of the dielectric because work would be required to increase the potential energy of the system.

CHAPTER 9:   MAGNETISM

IF YOU MEMORIZE ONLY 15 EQUATIONS IN THIS CHAPTER . . .

$$F_M = qvB\sin\theta \qquad \Phi_M = BA\cos\theta \qquad \varepsilon = B\ell v$$

$$F = I\ell B\sin\theta \qquad B = \frac{\mu_0\,I}{2\pi\,r} \qquad \varepsilon_{avg} = -\frac{\Delta\Phi_M}{\Delta t}$$

C Exam Equations:

$$\Phi_M = \int_S \vec{B}\cdot d\vec{A} \qquad \varepsilon = -\frac{d\Phi_M}{dt} \qquad \varepsilon = -L\frac{dI}{dt}$$

$$\vec{F} = q\vec{v}\times\vec{B} \qquad d\vec{B} = \frac{\mu_0}{4\pi}\frac{I\,d\vec{\ell}\times\vec{r}}{r^3} \qquad U_L = \frac{1}{2}LI^2$$

$$\oint \vec{B}\cdot d\vec{l} = \mu_0 I \qquad F = \int I\,d\vec{\ell}\times\vec{B} \qquad B_s = \mu_0 nI$$

INTRODUCTION

The study of magnetism is intimately related to electric currents, a relationship that we will explore in detail in this chapter. The connection between electricity and magnetism went unnoticed for a long time, but once that connection was made, it allowed great strides in the field of magnetics. The fundamental definition of a **magnetic field** is that it produces a force on a moving charge at right angles to the motion of the charge. This definition will be refined as we explore the topics in this chapter. We will see how magnetic fields affect single moving charges and electric currents in a wire.

We will also explore how magnetic fields are produced by electric currents. Once this theory is established, we will discuss the production of electric currents with time-varying magnetic fields (which play a role in generating more than 90 percent of our electric power). We will also cover the concept of inductance and the use of inductors in simple circuits, a discussion that will reveal the similarity between an inductor-resistor (LR) circuit and a resistor-capacitor (RC) circuit. A closer look at the inductor-capacitor (LC) circuit will show how it mimics a simple harmonic oscillator. The discussion part of the chapter will end with Maxwell's equations, named for James Clerk Maxwell, a 19th-century Scottish physicist who brought together a set of equations that allowed him to show that electromagnetic waves exist and travel at the speed of light.

The chapter will end with a set of practice exam questions and their answers and explanations. Also, experimental techniques and classic experiments will be discussed whenever possible to help you develop a sense of what sorts of things can be done with the (nearly) complete electromagnetic theory.

FORCES ON MOVING CHARGES IN MAGNETIC FIELDS

We've already mentioned that magnetic phenomena are closely tied to moving electric charges. Before we go into the details of this connection, it is helpful to refresh our knowledge of basic magnetic properties.

A **magnet** is an object that has a magnetic field (**B**) around it. This is a circular definition—one needs to know what a magnet is to understand the idea of a magnetic field—but we will break it apart in a moment. The magnetic object is characterized by having two ends, or **poles**. One pole is called **north**, and the opposite pole is called **south**. The magnetic field points from north to south. The figure below shows the magnetic field around (a) a bar magnet and (b) a horseshoe magnet. Just like electric charges, opposite magnetic poles attract, and like magnetic poles repel.

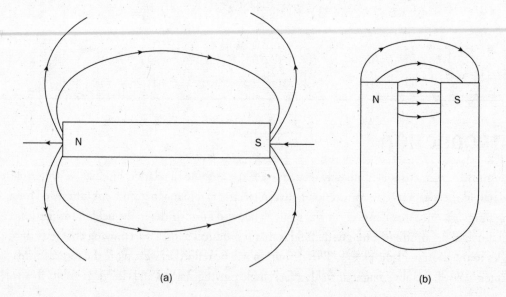

(a)   (b)

Unlike with electric charges, however, one cannot isolate a north pole from a south pole (at least, not according to current scientific knowledge). If you take an electric dipole, it is possible to separate the constituent charges so that you have an isolated negative charge and an isolated positive charge. If, however, you try to separate the north pole of a magnet from its south pole (e.g., by breaking a simple bar magnet), you will end up with two whole magnets. Each is complete, with its own north-south pair of poles (see following figure). We will see many similarities between electric and magnetic phenomena, but this fundamental difference persists.

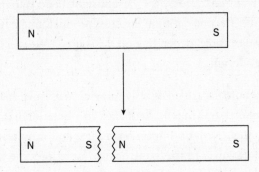

Experimentally, we never see a north pole without a south pole, and vice versa. Stated another way, there are no **magnetic monopoles**. As far as scientists can tell, this is a fact dictated by nature. There are searches under way to detect a magnetic monopole, which emphasize putting lower limits on the mass of the monopole and upper limits on the magnetic charge. To date, no magnetic monopoles have been found. The absence of magnetic monopoles can help you to determine whether a particular field can be a magnetic field. The figure below shows three fields. The first two are possible magnetic fields, but the third is not. To understand why, compare the fields with those of the electric dipole, parallel plate capacitor, and isolated electric charge.

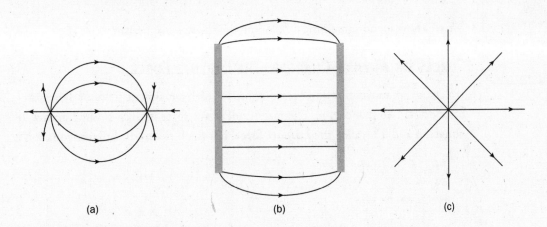

(a)   (b)   (c)

Just as electrical phenomena have a constant associated with them, ε_0, magnetic phenomena have a constant as well. **The magnetic constant** is known as the permeability of free space, and its symbol is μ_0 (pronounced "mu naught" or "mu zero"). **The permeability of free space** is exactly equal to $4\pi \times 10^{-7}$ T · m/A (tesla-meter per amp). One interesting thing to notice here is that if you calculate the quantity $\dfrac{1}{\sqrt{\varepsilon_0 \mu_0}}$, you will get the speed of light. James Clerk Maxwell showed why this is when he collected the known equations governing electricity and magnetism.

The nature of the magnetic field is such that it is created by moving electric charges and it exerts a force on a moving charge. The spatial relationship among field, force, and motion is summarized by the various right-hand rules. (There is really only one right-hand rule, the one defining the cross product, but it is common practice to talk about two or three, depending on the source.) We will identify these rules as they become relevant.

Magnetic phenomena are necessarily three-dimensional. It is difficult to render three-dimensional drawings on flat paper, so a notation has been developed to assist us. There are slight variations from text to text and from teacher to teacher, so we will set our notation as in the figure below. Sometimes the vector for "into page" is shown with a circle around it (like the "out of page" vector), but we will not do that here.

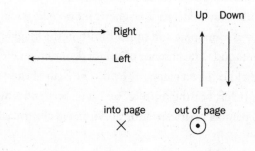

We will now spend some time developing the rules of magnetism.

FORCES ON MOVING CHARGES: THE LORENTZ FORCE

If a charge, q, moves through a region of space with a magnetic field, B, there is a force exerted on that charge. The magnitude of the force is given by the **Lorentz equation** and it is called the **Lorentz force**. For the B exam, the Lorentz equation is

$$F_M = qvB \sin \theta$$

For the C exam, it is

$$\vec{F} = q\vec{v} \times \vec{B}$$

q is the charge including the sign, v is the speed (or velocity in the case of the C exam), B is the magnitude of the magnetic field, and θ is the angle between the velocity and the field. To determine the direction of the force, we need the first right-hand rule. Take the index finger of your right hand and point it in the direction of the velocity of the charge. Then orient your hand so that your remaining three fingers naturally close toward the direction of the magnetic field. Your thumb points in the direction of the force on a positive charge; if the charge is negative, your thumb points opposite the direction of the force on the charge. The following figure should help you visualize this idea.

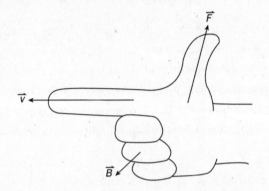

Another way to think of this rule is to point all of the fingers on your right hand in the direction of the velocity. Then sweep the fingers into the magnetic field through the smaller angle. Your thumb points in the direction of the force on a positive charge. Again, if the charge is negative, your thumb points opposite the direction of the force on the charge. The next figure shows the first right-hand rule in action. A proton is moving to the right through a magnetic field that points into the page. By the right-hand rule, the force on the proton is pointing up the page.

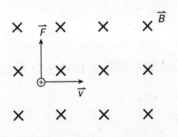

The Lorentz force is used as the actual definition of the magnetic field. It is the field that exerts a force always perpendicular to the velocity of an electric charge.

$$B = \frac{F}{qv \sin\theta}$$

If F is in newtons, q is in coulombs, and v is in meters per second, then B should have teslas (T) as its units. One tesla is a very large field. A more common unit for magnetic fields is the gauss (G). One tesla is equal to 10,000 gauss. The Earth's

AP EXPERT TIP

Note that this rule assumes the flow of a positive charge. If a question asks about a moving electron, just use your left hand to get the answer.

magnetic field is on the order of 50 milligauss, and a common refrigerator magnet has a field of a few gauss near one of its poles. Even though the tesla is a very large unit, it is still common on the AP Physics exams because it is the accepted SI unit and derives from the basic units of force: charge and velocity. However, you should still beware of answers that appear to be a large number of teslas. If, for example, you have to find a magnetic field and your answer comes out to be 1,000 T, you should probably check your work, because this is about 10 times larger than the largest fields produced in research laboratories. Most research is done at the 5 to 15 T level.

Because the magnetic force is always perpendicular to the motion of the charge, this force can do no work on the charge. That means the kinetic energy of a moving charge does not change due to the magnetic field. This claim is not strictly true because accelerating charges emit radiation and therefore lose energy. However, this happens even if the acceleration is colinear with the velocity; it is not due to magnetic fields alone. At this level in the subject, we can happily ignore this complication.

Another aspect of this force is that it results in centripetal acceleration. If a charged particle (charge q, mass m) is injected into a uniform magnetic field of strength B with velocity v, there will be a Lorentz force on the charge. Since the force is perpendicular to the velocity and the speed does not change, the charge moves in uniform circular motion (see following figure). It is relatively straightforward to show that the radius of the motion is given by

$$r = \frac{mv}{qB}$$

and the frequency of the orbit is

$$f = \frac{qB}{2\pi m}$$

This is done by setting the Lorentz force equal to the centripetal force and using the fact that—in circular motion—the time required to complete an orbit (the orbit period) is

$$T = \frac{2\pi r}{v}$$

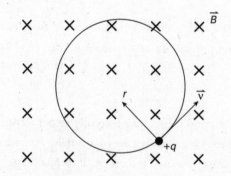

AP EXPERT TIP

Notice that since the force is perpendicular to the direction of motion, no work is done by the magnetic field.

This figure shows a charged particle moving with speed v in a uniform magnetic field B. The particle's path is a circle.

Given the equations on the previous page, knowing the speed of the charged particle allows us to determine the charge-to-mass ratio. What follows is a brief description of the $\frac{e}{m}$ for the electron experiment in a typical college undergraduate lab. Electrons are boiled off a hot filament and subjected to a known potential difference. This gives them a known kinetic energy, and thus speed, as a function of charge and mass. It is fairly easy to show that this speed is

$$v = \sqrt{\frac{2eV}{m}}$$

where v is the speed, V is the potential difference, e is the magnitude of the charge on the electron, and m is the mass of the electron. (See Chapter 8, Free-Response Question: B exam, part (a).) Once you have this speed, the electron is allowed to travel through a region of known uniform magnetic field. You can then measure the radius of the electron's path and use that measurement to determine $\frac{e}{m}$. Typically, a single potential is selected, and the magnetic field strength is changed from one trial to the next. Then a plot of r^2 versus $\frac{1}{B^2}$ is made. From the slope of the straight line generated, you can extract the value of $\frac{e}{m}$.

The Lorentz force is exploited in many devices. Any cathode ray tube (CRT) screen, such as an older model television or computer monitor, steers the electrons by using magnetic fields. A microwave oven uses a magnetic field to accelerate electrons, thus producing the radiation used for cooking. In particle accelerators, magnetic fields direct atomic and subatomic particles down tubes and toward targets.

A particularly novel use of the Lorentz force is the homopolar generator. In this device, a thick conducting disk is rapidly rotated like a high-speed record or a CD. Electrical contacts are placed at the center of the disk and at one point along the edge. These contacts are connected to a circuit. If a magnetic field is introduced perpendicular to the disk along the line between the two contacts, an extremely large current can be generated. The faster the disk rotates, the higher the current. The trade-off is that this is a rather low-voltage device.

The Lorentz force is directly linked to many other magnetic phenomena that will be discussed in this chapter. You should try to spot this force at work in as many different topics as possible.

FORCES ON CURRENT-CARRYING WIRES IN MAGNETIC FIELDS

Because a current is nothing more than the flow of electric charges, a current-carrying wire in a magnetic field experiences a force. You can imagine this force to be the result of the sum of the Lorentz force on each individual charge that makes up the current. Strictly speaking, this is not quite accurate because the motion of the charges through a wire is not as uniform as we might imagine. However, the net effect is the same.

Imagine that there is a long wire carrying a current I to the right. The wire is in a region of space with a uniform magnetic field B pointing into the page. Each charge in the current (remember that current is the flow of positive charge) feels a force directed up the page, according to the first right-hand rule. Because the charges are confined to the wire, the wire itself feels the force. The following figure shows a current-carrying wire in a magnetic field. The direction of the force is given by the first right-hand rule.

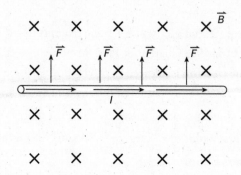

To determine the magnitude of the force, we need to think about the Lorentz force equation and the relationship between current and charge. The Lorentz equation $F = qvB\sin\theta$ can be rewritten as follows:

$$F = q\frac{\Delta L}{\Delta t}B\sin\theta, \text{ where } L \text{ is the length of the wire segment}$$

If we think of the current in the wire as a uniform line of moving charge, we can move the change in t over to q, resulting in

$$F = \frac{\Delta q}{\Delta t}LB\sin\theta$$

We recognize the quotient on the right-hand side of this last equation as the current I. Therefore, we claim that the force on a current-carrying wire in a magnetic field is

$$F = ILB\sin\theta$$

where θ is the angle between the direction of the current and the magnetic field. In the last figure, this angle was $\frac{\pi}{2}$ radians. In general, though, the angle can have any value between 0 and π. Often, we are interested in the force per length rather than the total force.

You can see an example of this by setting up a system like that shown in the following figure. It can be constructed from stiff wire, a 6 V lantern battery, and some strong magnets. When a current flows through the wire in one direction, the wire swings out from its support. If you reverse the direction of the current, the wire swings the other way. You can also reverse the direction of swing by flipping over the magnets.

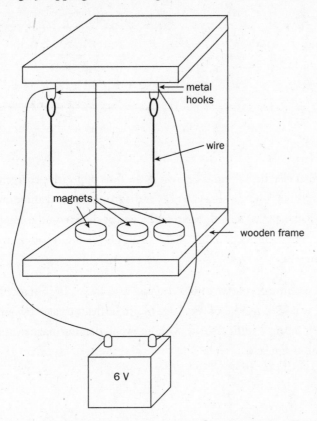

FIELDS OF LONG CURRENT-CARRYING WIRES

So far, we have talked about how moving charges are affected by magnetic fields, but we have yet to describe how the magnetic fields are created. *All* magnetic fields are produced by moving charges, from the magnetic field associated with the electron's magnetic moment to the permanent magnets on a refrigerator to the magnet used to lift scrap iron at a junkyard. Furthermore, every moving charge produces a magnetic field around it.

Because a current is nothing more than moving charges, a wire carrying a current, I, will have a magnetic field around it. The field lines are composed of circles all centered on the wire. The direction of the field can be determined by a second right-hand rule: if you grasp a wire carrying a current with your right hand such that your thumb points in the direction of the current, then your fingers naturally curl in the direction of the magnetic field loops.

The following figure shows (a) a three-dimensional perspective view of the magnetic field around a current and (b) an end view of (a).

AP EXPERT TIP

Sometimes it helps to hold your pencil in your right fist to execute this rule. Direct your thumb toward the pencil point with the current, and your fingers will circle just like the magnetic field lines.

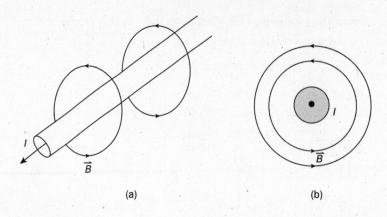

(a)   (b)

It is important to remember that electric current is defined as the flow of positive charge when applying the second right-hand rule. You can see evidence for the magnetic field around a current-carrying wire by bringing a small compass near the current. North, as shown on the compass, will revolve around the wire according to the second right-hand rule. A 6 V lantern battery with some bell wire works well for this experiment.

The magnetic field around a current-carrying wire must decrease with increasing distance from the wire. If this were not the case, there would be regions of very large magnetic fields created by our electrical power grid. At the level of the B exam, it suffices to say, without proof, that the magnetic field around a current-carrying wire obeys this formula:

$$B = \frac{\mu_0}{2\pi} \frac{I}{r}$$

where r is the perpendicular distance from the wire. In the next section, we will derive this result from Ampere's law for the benefit of those taking the C exam.

We will now address an important application of the last two sections. If two parallel wires are both carrying currents, each of them exerts a force on the other. To see why this is so, look at the following figure.

You can see that wire 1 carries current I_1, so it has a magnetic field given by

$$B_1 = \frac{\mu_0}{2\pi} \frac{I_1}{r}$$

Wire 2, with current I_2, is a distance d away from wire 1. At wire 2, the magnetic field produced by 1 is

$$B_{21} = \frac{\mu_0}{2\pi} \frac{I_1}{d}$$

A current in a magnetic field feels a force, as described in the previous section. We can now write the equation for the force per length on wire 2 as follows:

$$\frac{F}{l} = I_2 B_{21} = \frac{\mu_0}{2\pi} \frac{I_1 I_2}{d}$$

The direction of the force is determined by using the first and second right-hand rules as appropriate.

By Newton's third law, we can see that wire 2 exerts an equal and opposite force per length on wire 1. If one or the other of the currents (but not both) is reversed, the force reverses direction as well. As a general rule, parallel currents attract one another and antiparallel currents repel.

The material presented up to this point has three very interesting applications. The first is the Hall effect, the second is the rail gun, and the third is the mass spectrometer. Rail gun-type problems and mass spectrometer-type problems are likely to appear on the AP Physics exam; the mass spectrometer is especially likely to show up because it involves physics concepts that cross the boundaries among mechanics and electricity and magnetism. The Hall effect is less likely to be on the free-response part of the exam, but if you understand the basic physics involved, you will be able to tackle more difficult problems as well as multiple-choice questions in which it appears.

The Hall effect is connected with a stationary conductor carrying a current in a magnetic field. The conductor itself does not react noticeably to the magnetic field, but the moving charges do. Imagine there is a flat square conductor, as shown in the next figure (sometimes a flat object like this one is called a "lamina"). The conductor carries a current, I, from left to right. This means the conduction electrons are moving from the right to the left. Imagine that there is a magnetic field directed perpendicularly through the conductor as well. Because the electrons are moving across the conductor, they experience a Lorentz force deflecting them to one side.

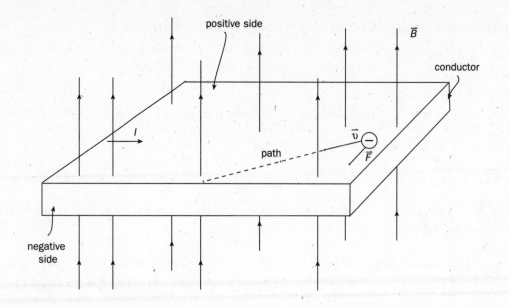

The deflection results in a charge density gradient. That is, there are more electrons on one side of the conductor than on the other. The electrons that pile up on that one side create an electric field that repels additional electrons. Provided there are enough electrons in the metal, which is nearly always the case, at some point there will be equilibrium between the Lorentz force from the magnetic field and the electrostatic force from the piled-up electrons. When this equilibrium is achieved, there will be a measurable potential difference perpendicular to the flow of the current. The potential difference will be directly proportional to the strength of the perpendicular magnetic field through the conductor. A device based on this principle is called a Hall probe and is a very sensitive magnetic field detector.

The rail gun is shown in the figure below. It consists of two parallel conducting rails with a potential difference between them. Oriented perpendicularly to the plane of the rails is a large magnetic field. A conducting rod joins the rails and is free to slide along them. If everything is set up properly, there will be a large force on the rod, causing it to accelerate down the rails until it flies off the end. The force on the conducting rod depends on its length, the magnetic field, and the current through the circuit.

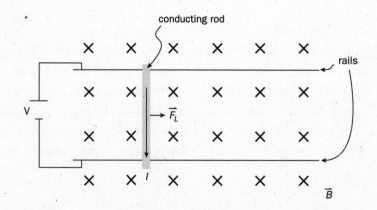

A rail is theoretically capable of accelerating a massive conductor to several times the speed of sound. However, there are problems with implementation. To generate the magnetic fields needed, you must have a very large electromagnet. Also, the current through the rod must be reasonably large and sustainable over the whole length of the rails. Most home-built attempts at a rail gun consist of massive banks of capacitors connected in parallel to provide the tremendous current needed to accelerate the rod. Tackling the problem of generating the magnetic field usually stops most builders.

Regardless of the engineering difficulties, you should be able to show that the initial force on the rod is given by

$$F_{\text{rod}} = ILB$$

In rail gun problems, students are commonly asked to determine the current via Ohm's law and then to determine the "muzzle velocity" of the rod as it leaves the rails. To do these calculations, you need to resort to basic dynamics.

The final major application of all the material we have reviewed thus far is the mass spectrometer. This is a classic problem because, like the rail gun, it crosses the boundaries among mechanics, electricity, and magnetism. You should be very comfortable with the mass spectrometer before you take the AP Physics exam.

A mass spectrometer has two regions. In the first region (sometimes referred to as the velocity selector), there are an electric field and a magnetic field at right angles to each other. Mutually perpendicular to both fields is the velocity vector of a beam of mixed particles each with the same charge, q. Each particle may have a different mass. The second region contains only a magnetic field parallel to the field in the first region. Usually it is just an extension of the magnetic field from the first region. The next figure lays out the entire mass spectrometer. Region I is the velocity selector, and region II is the actual mass spectrometer.

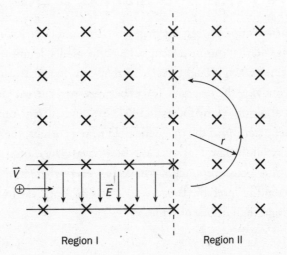

In the velocity selector region of the spectrometer, the electric and magnetic fields are situated such that they exert oppositely directed forces on moving charges. Thus if a positive (or a negative) charge is moving to the right in the previous figure, the electric field would be directed down, and the magnetic field would be directed into the page. Because the Lorentz force is velocity dependent, there is a particular velocity at which the electric and magnetic forces exactly cancel each other. It is trivial to show that this velocity is

$$v = \frac{E}{B}$$

If the beam of particles is aimed at a small hole between regions I and II, only those particles with this velocity will be able to pass through the hole. Thus, when particles enter region II, they have a known velocity. In region II there is only a magnetic field, so the particles travel in circular paths. Because each particle is traveling at the same speed through the whole path, each experiences the same force. Differences in mass, however, result in different accelerations, which lead to differences in the radii of the paths. More massive particles have larger radii for their paths than less massive particles. If the magnetic fields in the two regions are the same, you can see that the radius of the path for a particle with mass m is given by

$$r = \frac{mE}{qB^2}$$

The mass spectrometer is very useful in determining the isotopic concentrations of certain elements, which is necessary for processes such as carbon dating.

Your textbook should discuss these three applications in some detail, and you are encouraged to spend some time thinking about them.

THE BIOT-SAVART LAW AND AMPERE'S LAW (C EXAM)

The discovery of the relationship between the magnetic field and electric currents is generally credited to Oersted. It is said that during a lecture, he performed a demonstration showing how a magnetic needle is deflected when brought near a current-carrying wire. Ampere then performed a series of experiments studying the magnetic forces between two current-carrying wires of various shapes. Biot and Savart also studied the forces and fields around current-carrying wires. They developed a formula that determined the magnetic field near a current; the **Biot-Savart law** is the magnetic analogue of Coulomb's law. However, the Biot-Savart law is quite tedious to use because it requires you to determine a cross product of two vectors everywhere in space. While we can ignore this complication at this level, the Biot-Savart law is the only way to determine magnetic fields in situations lacking a high degree of symmetry.

Ampere's law—the magnetic analogue of Gauss's law—is directly derivable from the Biot-Savart law. Ampere's law deals with the contour integral of the magnetic field in the direction of the path of integration and relates this result to the enclosed current. Ampere's law is written as follows:

$$\int_C \vec{B} \cdot d\vec{l} = \mu_0 I$$

Notice that you must integrate around a closed path. This is similar to the requirement that you integrate over a closed surface with Gauss's law.

In general, the result of $\vec{B} \cdot d\vec{l}$ depends on where you are in the path. However, in certain highly symmetrical cases, this dot product is the same everywhere along a particular path. That means that $\vec{B}$ is independent of $d\vec{l}$ and may come out of the integral. If this is the case, the result of the left-hand side of Ampere's law is just the magnetic field times the perimeter of the closed path. The most basic use for Ampere's law is the long, straight wire carrying a current I.

The procedure for using Ampere's law is similar to that for using Gauss's law. The main difference is that Gauss's law requires you to integrate the flux through a closed (i.e., Gaussian) surface, whereas Ampere's law requires you to integrate the magnetic circulation around a closed path, called an **Amperian loop**. Consequently, you want to choose loops that make the integrand as simple as possible.

For a long, straight wire carrying a current, we know that the field is composed of concentric circles around the wire. We also know that the strength of the field must be the same at all points that are a given distance from the wire. These two facts suggest that we should use a circle for our Amperian loop. The following figure shows the loop for a current I in a long, straight wire.

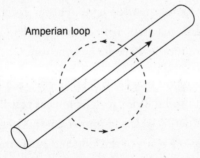

With this loop, $\vec{B}$ and $\vec{B} \cdot d\vec{l}$ are parallel everywhere, and Ampere's law becomes

$$B \oint_{circle} dl = \mu_0 I$$

The contour integral is just the circumference of a circle of radius r. We immediately arrive at

$$B = \frac{\mu_0 I}{2\pi r}$$

as was previously stated.

Ampere's law can also provide evidence for the fringing of magnetic field lines near the edges of magnets. The Amperian loop in the figure below contains no current, so $\oint_c \vec{B} \cdot d\vec{l} = 0$. For this to be true, the field on the left half of the loop cannot be zero as indicated. It must fringe out somewhat to cancel the effect of the right half of the loop.

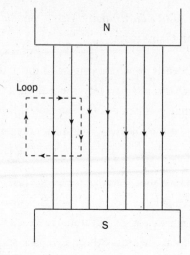

FARADAY'S LAW

We have seen how electric currents generate magnetic fields and how electric currents are influenced by magnetic fields. At this point, it is logical to ask whether magnetic fields can generate electric currents, and the answer to this question is yes. However, certain requirements must be met before this can happen. A static magnetic field cannot generate a current in a static conductor. Either the field or the conductor must be dynamic in some way.

If a conducting wire is pulled through a magnetic field, an effect similar to the Hall effect occurs; current flows for a brief moment until the electrostatic force is equal and opposite to the Lorentz force. If you make a loop of conducting wire and immerse it in a magnetic field, some of the magnetic field will probably cross the plane of the loop and create magnetic flux. If this magnetic flux changes with time, there will be a current induced in the loop. This induced current is much more useful because we can extract the current and use it to light bulbs and cook food. The two laws that govern this phenomenon are **Faraday's law** and **Lenz's law**.

Faraday's law relates the change in magnetic flux through a surface bounded by a path to an induced potential called the **electromotive force**, or emf, ε. This is an unfortunate name because it is not a force at all—it is a potential. However, history and inertia win out, and we are stuck with the emf, which takes volts as its units. Magnetic flux is completely analogous to electric flux. For the benefit of students taking the B exam, let's spend some time reviewing this concept.

Magnetic flux is defined as the dot product of the magnetic field and the area vector for a surface. Mathematically, it is written as follows:

$$\Phi_M = \vec{B} \cdot \vec{A} = BA \cos \theta$$

The following figure shows the geometric relationships among the field, area, and area vector. Notice that the area vector is perpendicular to the area it defines. The length of the vector represents the area. For the C exam, magnetic flux has a slightly different definition, which is consistent with the definition of electric flux:

$$\Phi_M = \int_S \vec{B} \cdot d\vec{A}$$

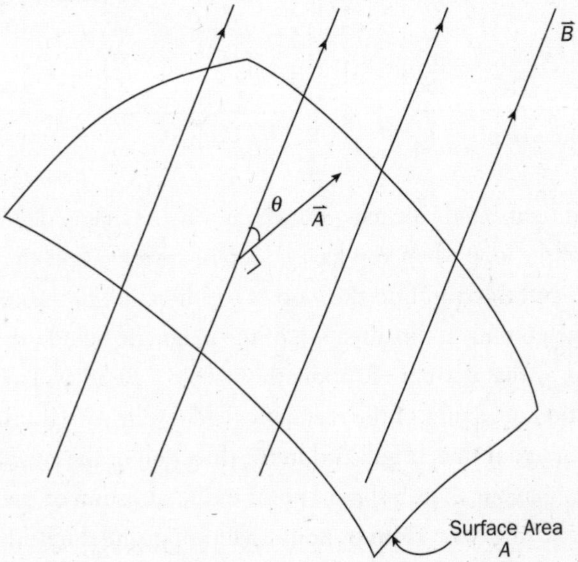

Surface Area
A

When the magnetic flux enclosed by a loop changes, the loop experiences an emf. This emf is such that if a current were produced, it would create a magnetic flux that would oppose the change in magnetic flux. The phrase "if a current were produced" is necessary because the emf is induced *regardless* of the presence or absence of a conducting path. However, if a conducting path *is* present, the emf produces an induced current in that path such that the magnetic field created by that induced current opposes the change in the enclosed magnetic flux. This is Faraday's law. As an equation it is written

$$\varepsilon_{avg} = -\frac{\Delta \Phi_M}{\Delta t}$$

on the B exam; on the C exam it is

$$\varepsilon = -\frac{d\Phi_M}{dt}$$

The negative sign indicates that the emf opposes the change in magnetic flux.

To understand how this works, refer to the next figure. A rectangular loop of wire encloses a uniform magnetic field. The field is perpendicular to the plane of the loop, so the magnetic flux is just BA. If the field strength is decreased over time, there will be an induced emf around the loop resulting in an induced current in the wire. This current flows such that it produces a magnetic field that tends to increase the magnetic flux enclosed. A third right-hand rule can be used here.

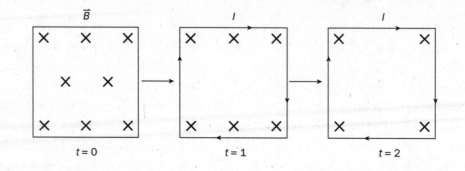

$t = 0$ $t = 1$ $t = 2$

AP EXPERT TIP

Imagine that your thumb can produce its own magnetic field lines and direct those new lines so that they fight against the change in flux. Then your curled fingers will indicate the current.

The third right-hand rule tells you which way the magnetic field points in the interior of the loop when you have a loop of current. Take the fingers of your right hand and curl them around the loop in the direction of the current. Your right thumb then points in the direction of the magnetic field in the interior of the loop generated by that current. To apply this rule to a Faraday's law problem, determine the direction of change of the magnetic field. Point your thumb opposite this change, and your fingers will curl in the direction of the emf. If the magnetic flux is decreasing, your thumb should point in the direction of the field. If the magnetic flux is increasing, your thumb should point opposite the field.

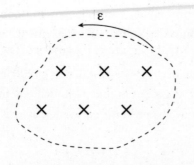

Increasing flux, thumb out (opposite field)

If there is a conducting loop around the region of changing magnetic flux, there will be a current in that loop. To determine the magnitude of the current, recall that emf is really an electric potential. The wire making up the loop has a certain resistance; the induced emf is what causes the current to flow. Rather conveniently, in cases like this the emf can be considered to be the voltage supplied by an appropriate battery. Therefore, if the resistance is R and the emf is ε, the induced current as given by Ohm's law is

$$I = \frac{\varepsilon}{R}$$

LENZ'S LAW

Lenz's law deals with the negative sign in Faraday's law when currents are produced. Ultimately, Lenz's law is nothing more than conservation of energy. Imagine a coil of wire, perhaps a spring. Place a long, thin bar magnet—one thin enough to fit inside the coil—near the coil. Then give the magnet a little push toward the coil. According to Faraday's law, there will be an induced emf in the first loop of the coil because as the magnet draws closer, the magnetic field strength (and thus the flux) increases. If Faraday's law did not have the minus sign, this emf would be such that the current in the loop would work *with* this increase in field. The result would be the magnet being drawn toward the coil, which would, in turn, increase the flux at a greater rate, thereby increasing the current. Indeed, this would be quite favorable because you would be able to extract a great deal more energy from this system than you put in, meaning that energy would not be conserved.

Lenz's law tells you that the induced current must *oppose* the change in flux. A striking example of Lenz's law is shown by suspending a conducting ring from a thin string. If a strong bar magnet is slowly inserted into the ring and then rapidly withdrawn, the ring swings toward the magnet. The induced current in the ring creates a magnetic field that is attracted to the bar magnet. If the slightest gap is left in the ring, though, this swinging does not happen. If there is no complete conducting path, there is no induced current and, thus, no induced magnetic field.

Typical problems involving Faraday's law and Lenz's law are concerned with loops being pulled out of or pushed into magnetic fields, rail gun-like loops that have the rod being pulled in one direction, and rotating loops in a static magnetic field. This last example is most likely C exam material, but the qualitative aspects are easily understood at the B exam level. Let's spend some time discussing each of these three problems and how to go about solving them.

First, consider the situation shown in the following figure. A square loop of wire with area A is in a region of constant magnetic field B. The wire has resistance R that is modeled by creating a loop from ideal wire and placing a resistor in the loop. Suppose the loop is pulled at a constant speed v out of the region containing the magnetic field. A typical exam question would be "What are the magnitude and direction of the current induced in the wire?"

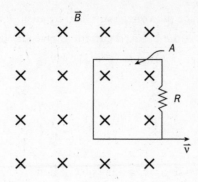

The first thing you need to do is determine the emf in the loop, which can be accomplished with Faraday's law. You can see that you are going to need the rate of change of magnetic flux over time. Before the leading edge of the loop reaches the boundary between field-filled space and field-free space, the magnetic flux is given by

$$\Phi_M = BA$$

As soon as the leading edge enters field-free space, the flux begins to decrease. Because the magnetic field does not change, the change in flux is related to the change in the area that the magnetic field pierces. This can be stated as

$$\frac{\Delta A}{\Delta t} = -v\sqrt{A}$$

Thus, the change in magnetic flux with respect to time is

$$\frac{\Delta \Phi_M}{\Delta t} = B\frac{\Delta A}{\Delta t} = -Bv\sqrt{A}$$

You can see that the emf will be equal to:

$$\varepsilon = -\frac{\Delta \Phi_M}{\Delta t} = Bv\sqrt{A}$$

The fact that ε is positive indicates that the induced current should create a magnetic field that will work with the existing field. In this case, the current will be clockwise. The magnitude of the current is given by Ohm's law, which leaves us with

$$I = \frac{Bv\sqrt{A}}{R}$$

In general, you should find currents from this type of problem to be magnetic field times speed times length (the square root of area is length) divided by resistance. If you can remember this relationship, you may be able to eliminate choices on the multiple-choice section of the exam.

Next, let's consider a rail gun-type problem in which the projectile rod is pushed by an external agent to create an electric current. The situation is shown in the following figure.

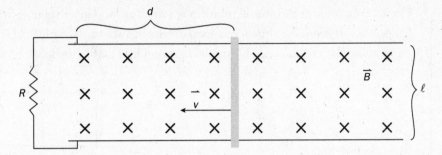

Two parallel rails separated by a distance ℓ lie in a region of uniform magnetic field B. The field is perpendicular to the plane containing the rails and points into the page. At one end, the rails are connected by a wire having resistance R. There is a rod at the other end that is pushed toward the resistor with constant velocity v. Our goal is to find the current generated through R.

As usual, we need to find the rate of change of flux in the loop bounded by the resistor, rails, and rod. Suppose the rod is a distance d from the resistor at some moment in time. The magnetic flux would then be

$$\Phi_M = B\ell d$$

After a short time, Δt, the rod moves a small distance equal to $v\Delta t$ so that the new flux is given by

$$\Phi'_M = B\ell(d - v\Delta t)$$

The change in flux is just the difference between these two expressions, or

$$\Delta\Phi_M = -B\ell v\Delta t$$

To find the induced emf, we must divide this by a small time interval and take the negative. It is fairly easy to see that the induced emf is

$$\varepsilon = B\ell v$$

Notice the similarity between this expression for ε and that found for the loop being removed from the magnetic field. The current is given by

$$I = \frac{\varepsilon}{R} = \frac{B\ell v}{R}$$

The direction of the current is determined by Lenz's law (or the third right-hand rule). In this problem, it is clockwise because the flux is into the page and decreasing. You should not be surprised if you find these two problems very similar. The physics behind them is identical.

Our final example of this sort of problem is a rotating loop in a magnetic field. The figure below lays out the situation: a loop of wire in a static magnetic field is made to rotate about the axis shown. This results in an induced emf in the loop.

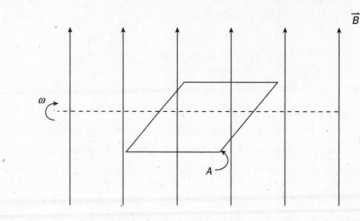

A loop of area A is in a region of constant, uniform magnetic field B. To make things more definite, we take B as pointing up the page. Imagine that the loop is made to rotate with constant angular speed ω around the axis shown in the figure. The magnetic flux will change as time passes because the angle between the field and the area vector will change. The goal is to determine *how* this flux changes.

Let's suppose at time $t = 0$, the angle between the area vector and the magnetic field is ϕ. Then at time $t = 0$, the magnetic flux through the loop is

$$\Phi_M = BA\cos\phi$$

At a later time, the loop has rotated through an angle given by $\omega\Delta t$, and the new flux is

$$\Phi'_M = BA\cos(\omega\Delta t + \phi)$$

The average time rate of change of flux is then

$$\frac{\Delta\Phi_M}{\Delta t} = BA\frac{\cos(\omega\Delta t + \phi) - \cos\phi}{\Delta t}$$

If we let Δt go to zero, we have the definition of the derivative of $\cos(\omega t + \phi)$ with respect to time. This means we get

$$\frac{d\Phi_M}{dt} = -\omega BA\sin(\omega t + \phi)$$

The negative of this is the induced emf:

$$\varepsilon = \omega BA\sin(\omega t + \phi)$$

Notice that in this case, the induced emf changes sign with periodicity $\frac{\pi}{\omega}$. This means that any induced current will first go one way around the loop, then switch and go the other way around the loop (i.e., it will alternate). Also notice that when the magnitude of the flux is at its minimum, the induced emf is at its maximum magnitude, and vice versa. This is a general property of this type of problem.

You can build some extremely useful devices by exploiting Faraday's law properly. For example, if you use a triple beam balance in your lab, it is likely that there is a magnetic damper on it. The end opposite the pan probably has an aluminum flag that sits between two closely spaced magnets. As the flag moves up and down through the field, tiny loops of current are induced. The electrical resistance of the aluminum causes heat to be generated when these currents flow. This heat bleeds off the energy of oscillation, allowing the balance to come to rest more quickly. The tiny loops of current are called **eddy currents**.

Most common demonstrations that show the interaction of electricity and magnetism exploit these eddy currents. Perhaps your teacher has performed a demonstration in which she holds a long, narrow copper tube vertically and drops a strong magnet down it. It takes a very long time for the magnet to reach the bottom because at each point along the way, currents are induced to oppose the change in magnetic flux. This means that regions below the falling magnet are pushing up on the magnet and regions above the falling magnet are pulling up on it. In very short order, the magnet is falling with constant speed. What is even more interesting is that you can measure the effect of the magnet on the tube by hanging the tube from a spring scale. While the magnet is falling, the scale will register both the tube and the magnet, thanks to Newton's third law.

It is unlikely that you have seen the following demonstration in your classroom, but it is extremely impressive and really must be seen to be believed. A large, thick copper slab is cooled to 77K by immersion in liquid nitrogen. Once it is cold, a strong magnet is dropped so that the same pole faces the slab the whole way down. The cold copper is such a good conductor that the magnet can bounce *without* striking the copper. This happens because the current induced by the falling magnet opposes the increase in magnetic flux through the copper. Thus, there is a magnetic force upward on the magnet as it falls.

INDUCTANCE: MAGNETIC FIELDS IN CIRCUITS (C EXAM)

The material in the remainder of the discussion portion of this chapter is solely for those taking the C exam. However, everyone with an interest in physics is encouraged to keep reading.

If you take a length of wire and wrap it into a coil, you build what is known as an **inductor**. An inductor acts much like a capacitor in that it stores energy in a field. A capacitor can be charged, disconnected from the charging circuit, and carried around with little loss of electric field; an inductor cannot do this. The reason is that an inductor stores energy in a magnetic field.

The magnetic field requires a current to exist, so once the inductor is removed from the circuit, its field disappears. Inductors have a property called inductance, abbreviated L in formulas. The unit of inductance is the henry, abbreviated as H. The following figure shows an inductor and the symbol used to represent it in a circuit diagram.

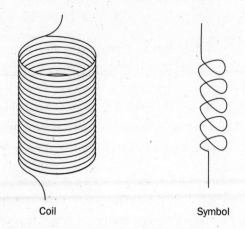

Coil   Symbol

If we think about what an inductor does in a circuit, we can develop a better understanding of what inductance actually is. The figure below shows a simple battery/switch/inductor circuit. Before the switch is closed, there is no current in the circuit, and there is no magnetic field inside the inductor. In the language of Faraday's law, the magnetic flux in the inductor is zero at this point. Immediately after the switch is closed, however, current begins to flow. This is when things get interesting.

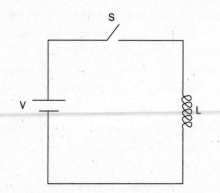

As the current starts to flow, a magnetic field is generated inside the loops of the inductor, causing an increase in the magnetic flux. By Faraday's law, an emf is induced in the loops to oppose this increase in magnetic flux. As long as the current is changing in time, the inductor is producing an emf to oppose this change. This is the defining characteristic of an inductor: it is a circuit element that opposes the changes in currents running through it, via Faraday's law. Here we assume inductance is a given quantity and just write the induced emf across the ends of an inductor as

$$\varepsilon = -L\frac{dI}{dt}$$

After a sufficiently long time, the current will reach a maximum value, and the inductor will cease to have an emf across its ends. As long as the current remains steady, the inductor will behave like a simple straight length of wire. This is the steady state for an inductor in a circuit.

There is a geometric expression for the inductance of an inductor. It is related to the total magnetic flux and the current. Specifically, the inductance is the integrated flux volume divided by the current. If a coil of N turns has a magnetic flux Φ_M when a current I passes through it, then the inductance of that coil is

$$L = N\frac{\Phi_M}{I}$$

In fact, this is the actual definition of **inductance**. However, we take inductance as a given quantity at this level, so don't spend too much time fretting over this equation.

When a current is flowing through an inductor, there is a magnetic field inside the inductor. If the current begins to decrease, the inductor has an induced emf across its ends to oppose the decrease in current. In essence, the inductor is using its stored energy to maintain the current. The magnitude of that stored energy is given by

$$U_L = \frac{1}{2}LI^2$$

LR Circuits

We can build a circuit with inductors and resistors. If an inductor is in a circuit with a resistor, we call that circuit an inductor-resistor circuit, or an LR circuit. LR circuits are very similar to the RC circuits discussed in Chapter 8.

Let's consider the circuit shown in the following figure. Here we have a battery connected to a resistor, R_1, and a switch. Across the switch and battery are an inductor, L, and another resistor, R, connected in parallel. If we close the switch and wait a while, the inductor will act like a short circuit and have a current flowing through it. After this time, no current will flow through R and the total current can be found from Ohm's law to be $I = \dfrac{V}{R_1}$. What happens when we open the switch?

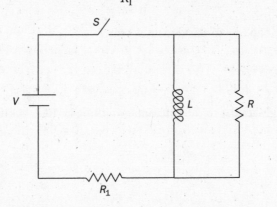

When the switch is opened, the battery and R_1 are removed from the circuit, causing the current through the inductor to decrease. By Faraday's law, the inductor has an induced emf across it to oppose this decrease. Because the inductor is connected to the resistor R, the induced emf will result in a current through R. However, this current is short-lived because it is created by the decay of the magnetic flux in the inductor. Eventually this flux will reach zero, and it will not increase again. The magnitude of this decaying current in the LR circuit shown is given by

$$I = I_0 e^{\frac{-tR}{L}}$$

This equation governs the behavior of the decrease of current through an inductor in an LR circuit. The quantity $\frac{L}{R}$ is called the inductive time constant and has the symbol τ_L. In this example, I_0 is given by $\frac{V}{R_1}$, the current at the time the switch is opened.

If we remove the resistor R from the circuit in the above figure, we get the circuit in the following figure. Here we have the inductor, resistor, battery, and switch all in series. When the switch is closed, the battery will try to push current around the circuit. The inductor will resist this current because prior to closing the switch, there is no magnetic flux in the inductor. As the current begins to flow, magnetic flux is generated in the inductor. By Faraday's law (yet again), the inductor has an induced emf across its ends.

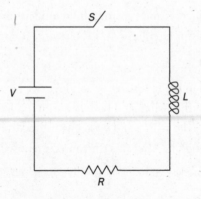

If we apply Kirchhoff's loop rule to this circuit, we see

$$V - L\frac{dI}{dt} - RI = 0$$

This equation is strikingly similar to the equation governing an RC circuit. As a reminder, that equation is

$$V - R\frac{dQ}{dt} - \frac{Q}{C} = 0$$

All that is different here is that Q is replaced by I, R becomes L, and $\dfrac{1}{C}$ becomes R. This means the solution to the RC equation is mathematically identical to the solution to the LR equation. Specifically, we have

$$I = I_{max}\left(1 - e^{-\frac{t}{\tau_L}}\right)$$

This equation governs the behavior of the increasing current through an inductor in an LR circuit. Again, noting that after a sufficiently long time inductors behave like short circuits, we can use Ohm's law to obtain $I_{max} = \dfrac{V}{R}$.

If we need to find the emf across the inductor, then in situations of both increasing and decreasing current, all we need to do is take the first time derivative of the current and multiply by the negative of the inductance. For the decreasing current case, we get

$$\varepsilon = \frac{LI_0}{\tau_L}e^{\frac{-t}{\tau_L}}$$

For the increasing current case, we get

$$\varepsilon = -\frac{LI_{max}}{\tau_L}e^{\frac{-t}{\tau_L}}$$

Notice that in the decreasing current situation, the current is large when the emf is large, and the current is small when the emf is small. For the increasing current situation, the emf and current are opposite: when the current is large the emf is small, and vice versa.

We have now seen how the LR circuit closely mirrors the behavior of the RC circuit. For an RC circuit, the capacitor starts out like a short circuit and evolves into a break in the circuit. For an LR circuit, the inductor starts out like a break in the circuit and evolves into a short circuit. The next section explores the LC circuit's behavior and compares it to a spring-mass mechanical system.

LC CIRCUITS

The following figure shows a simple LC circuit. The capacitor is initially charged to a potential difference of V volts. When the switch is closed, the capacitor will discharge through the inductor, but the inductor will resist this discharging. Eventually, the capacitor will discharge enough so that the current drops off. When this happens, the inductor will oppose the decrease in current. This will result in a recharging of the capacitor with the potential reversed. The process then repeats itself, with the current flowing in the other direction. In the absence of resistive heating, nothing will stop the oscillation of charge back and forth on the capacitor.

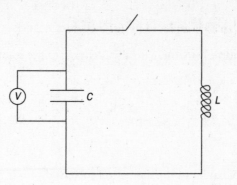

If we pick some arbitrary time and analyze the circuit using Kirchhoff's loop rule, we arrive at the following equation:

$$\frac{Q}{C} + L\frac{dI}{dt} = 0$$

From the definition of current, we can rewrite this equation as follows:

$$\frac{1}{C}Q + L\frac{d^2Q}{dt^2} = 0$$

If we compare this equation with that of a mass on a spring, we gain tremendous insight. As a reminder, the equation of motion for a mass on a spring is

$$kx + m\frac{d^2x}{dt^2} = 0$$

These two equations are mathematically identical with the following substitutions:

$$x \rightarrow Q$$
$$m \rightarrow L$$
$$k \rightarrow \frac{1}{C}$$

This means the solution to the spring-mass oscillator is the solution to the LC oscillator as well; namely,

$$Q = Q_0 \cos\left(\frac{1}{\sqrt{LC}}t\right)$$

This equation shows how the charge on the plates of the capacitor varies with time. The quantity $\frac{1}{\sqrt{LC}}$ is the natural frequency of the oscillator.

This frequency is fairly sensitive to changes in either the inductance or the capacitance. If the inductance increases somehow, the natural frequency of oscillation will decrease. You can increase the inductance by inserting a magnetic needle into the inductor. This fact is exploited to measure how deeply magnetic fields penetrate different materials, with a resolution of approximately 1 angstrom.

MAXWELL'S EQUATIONS (C EXAM)

The so-called Maxwell's equations are really just the culmination of all the scientific insights into electricity and magnetism. Maxwell himself did only two things. First, he collected all of the equations together. Second, and most importantly, he made them all internally consistent. As you probably know from your physics classes, Maxwell's equations are the following:

$$1. \oint_S \mathbf{E} \cdot d\mathbf{A} = \frac{Q}{\varepsilon_o}$$

$$2. \oint_S \mathbf{B} \cdot d\mathbf{A} = 0$$

$$3. \oint_C \mathbf{E} \cdot d\mathbf{l} = -\frac{d\Phi_M}{dt}$$

$$4. \oint_C \mathbf{B} \cdot d\mathbf{l} = \mu_0 \varepsilon_0 \frac{d\Phi_E}{dt} + \mu_0 I$$

Equation 1 is Gauss's law for electric charges. Equation 2 is Gauss's law for magnetic charges. It states the experimental fact that no isolated magnetic charges are known to exist. Equation 3 is Faraday's law. If you think about the left side of this equation, what you have is an electric potential. The right side is the negative rate of change of magnetic flux. Equation 4 deserves a closer look.

Equation 4 is essentially Ampere's law, but Maxwell added the term $\mu_0 \varepsilon_0 \frac{d\Phi_E}{dt}$ to make Ampere's law entirely consistent with Gauss's law for magnetic charges. Maxwell did not stop here; he went on to show that these equations, taken together, predict the existence of traveling electromagnetic waves. The speed of the waves is given by $\frac{1}{\sqrt{\mu_0 \varepsilon_0}}$, which—when evaluated in SI units—has a value of nearly 3×10^8 m/s. Maxwell recognized this as the speed of light in a vacuum. He thus succeeded in proving that light is nothing more than an electromagnetic wave.

As soon as scientists started thinking about light as a wave, they wanted to know how the wave was propagated—what did it move through? Physicists initially postulated the existence of a substance called "luminiferous aether." This was thought to be a very tenuous material that permeated all space. If this "aether" existed, scientists argued that we should be able to measure our speed relative to it. The first major experiment designed to measure this speed was the Michelson-Morely experiment. It found that the speed of the Earth through the aether was practically zero. All other experiments designed to measure this speed found it to be zero as well. That meant either the Earth drags the aether with it or there is no aether at all. Various ad hoc explanations were created to account for this famous null result, but each made predictions that were at odds with what is seen in nature. Einstein put forth the final explanation in his 1905 paper "On the Electrodynamics of Moving Bodies." This was the birth of the special theory of relativity.

REVIEW QUESTIONS

B EXAM

1. Of the fields shown below, which is not a possible magnetic field?

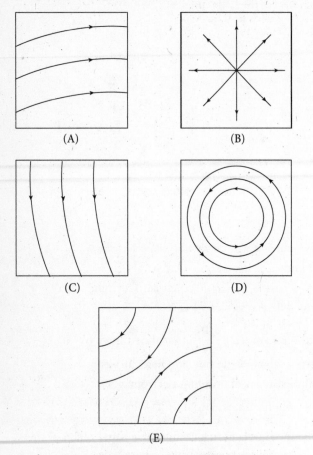

(A)

(B)

(C)

(D)

(E)

2. An electron is traveling in the plane of the paper from left to right. Which direction must a magnetic field point so that the electron is pushed out of the plane of the paper by a Lorentz force?

   (A) Toward the top of the paper in the plane of the paper

   (B) Toward the bottom of the paper in the plane of the paper

   (C) Into the paper

   (D) Out of the paper

   (E) From the right to the left in the plane of the paper

3. A rectangular (non-square) loop of conducting wire lies in the plane of the paper. A uniform magnetic field is directed into the paper. If the loop is made into a square, what direction does the induced current flow in the loop, and why?

   (A) Clockwise, because the magnetic flux is increasing

   (B) Clockwise, because the magnetic flux is decreasing

   (C) Counterclockwise, because the magnetic flux is increasing

   (D) Counterclockwise, because the magnetic flux is decreasing

   (E) There is no current because the magnetic flux is constant.

4. Two long, parallel conducting wires, each carrying a current I, are separated by a distance d. If d and I are both simultaneously made twice as large, what happens to the magnitude of the force per length between the two wires?

   (A) The force per length increases by a factor of 4.

   (B) The force per length increases by a factor of 2.

   (C) The force per length remains unchanged.

   (D) The force per length decreases by a factor of 2.

   (E) The force per length decreases by a factor of 4.

5. Two identical pendulums are made of large, flat copper sheets attached to very long, thin rods. Each rod is free to swing in a plane about the end away from the copper sheet. Both pendulums are drawn aside by the same angle and released. Pendulum A passes through a region of uniform magnetic field such that the field is perpendicular to the plane of the copper sheet. Pendulum B passes through field-free space at all times. Which of the following is true?

(A) Pendulum A will swing for a longer time and remain cooler than pendulum B.

(B) Pendulum A will swing for a longer time and get hotter than pendulum B.

(C) Pendulum A will swing for a shorter time and remain cooler than pendulum B.

(D) Pendulum A will swing for a shorter time and get hotter than pendulum B.

(E) Pendulum A will swing for the same amount of time as pendulum B and will not get hotter or cooler than pendulum B.

C Exam

6. A conducting rod of length l and width w passes through a region of uniform magnetic field (magnitude B) with speed v as shown below. Which of the following gives the best estimate for the potential difference across the ends of the rod?

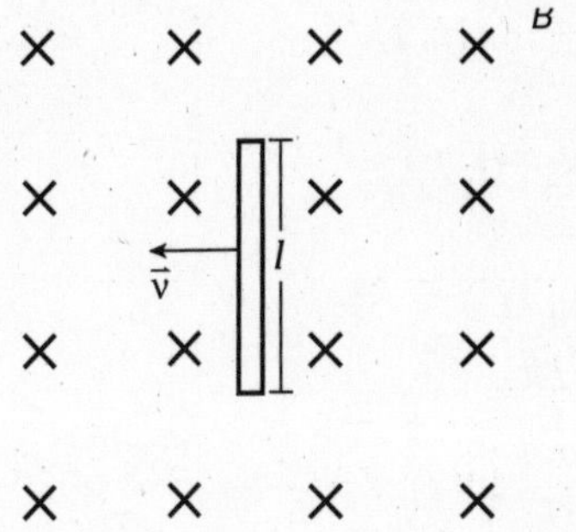

(A) Bvw

(B) Bvl

(C) $\dfrac{Bl^2v}{w}$

(D) $\dfrac{Bw^2v}{l}$

(E) Bv^2

7. A proton and an electron each travel with velocity v to the right. If a uniform magnetic field is directed out of the page, which of the following statements is/are true?

I. The proton is deflected up the page.

II. The electron is deflected up the page.

III. The radius of the electron's path is greater than the radius of the proton's path.

(A) I only

(B) II only

(C) I and II only

(D) II and III only

(E) None of the statements is true.

8. A square loop of side length a is near a long wire carrying current I. Two sides of the loop are parallel to the wire. If the side of the loop closest to the wire is a distance a away, what is the magnetic flux through the loop?

   (A) $\dfrac{\mu_0 I a^2}{2\pi}$

   (B) $\dfrac{\mu_0 I a^2}{2\pi}\ln(2)$

   (C) $\dfrac{\mu_0 I a}{2\pi}\ln(2)$

   (D) $\dfrac{\mu_0 I a}{2\pi}$

   (E) $\dfrac{\mu_0 I}{2\pi a}\ln(2)$

9. A conducting rod of length a and mass m is free to slide down rails with constant speed v. The rails are inclined at an angle θ with respect to the horizontal, and they are connected to one another by a wire at their high ends. If there is a uniform magnetic field of strength B directed vertically downward, what is the magnitude of the induced emf in the rod?

   (A) $Blv\cos\theta$

   (B) $Blv\sin\theta$

   (C) Blv

   (D) $Blv\tan\theta$

   (E) $Blv\sec\theta$

10. Which of the following graphs best represents the current through an inductor in an LC circuit if at time $t = 0$ the capacitor is charged?

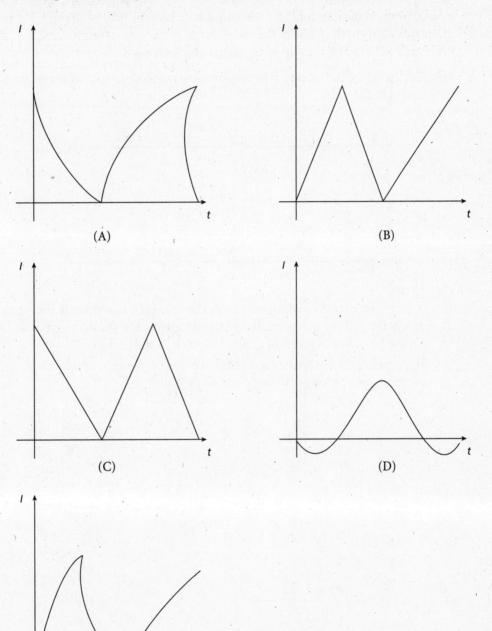

(A)

(B)

(C)

(D)

(E)

FREE-RESPONSE QUESTION: B EXAM

A rail gun is constructed of two parallel rails of length l separated by a distance d. The region between the rails is filled with a uniform magnetic field of strength B directed perpendicularly to the plane of the rails. The projectile rod is length d, mass m, and resistance R. Consider the rails to be perfect conductors.

(a) On the figure below, sketch in the directions of B and I so that the projectile is accelerated to the right.

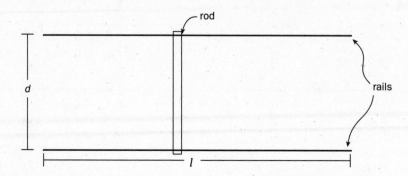

(b) A potential source is connected across the rails such that there is always a constant V across the ends of the projectile. Derive an expression for the magnitude of the force on the projectile in terms of quantities given in the problem.

(c) If the projectile starts from rest at the left end of the rails, what will its "muzzle velocity" be as it leaves the right end of the rails?

FREE-RESPONSE QUESTION: C EXAM

A circular loop of wire with a resistance of 200 Ω is in a region of uniform magnetic field as shown below.

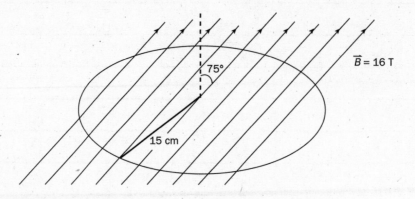

(a) Calculate the magnetic flux through the loop in T · m².

(b) If the magnetic field (in units of tesla) obeys the equation $B(t) = 16\cos(t)$, calculate the induced current in the loop as a function of time. (Note: The direction of $\vec{B}$ is always parallel to the same line. Also, we have assumed a frequency of $\omega = 1$ rad/s for B.)

(c) How much energy is dissipated by the loop from the time $t = 1$ second to the time $t = 1.5$ seconds?

(d) When is the induced current at a maximum in the loop for the first time after $t = 0$?

ANSWERS AND EXPLANATIONS

B Exam

1. B

The only nonmagnetic field possible is the magnetic monopole. You should recognize at once that choice (B) represents a monopole field. Choices (A) and (C) could both be the fringing fields of a dipole. Choice (D) is the magnetic field around an electric current and choice (E) could be the fields of two interacting magnets. This leaves choice (B) as the correct answer.

2. B

Unthinking application of the first right-hand rule would indicate the answer should be choice (A). However, you must remember that the charge on the electron is negative. Therefore, any result you get with the right-hand rule applied to an electron must be reversed. Choices (C) and (D) can be eliminated because both of these choices set the magnetic field parallel or antiparallel to the force. Choice (E) can be eliminated because it sets the magnetic field antiparallel to the velocity. Recall that for the Lorentz force to work, velocity, magnetic field, and force must all have mutually perpendicular components.

3. C

When dealing with rectangular shapes of constant perimeter, it is crucial to know that the square has the largest area. Therefore, the area bounded by the wire increases when the rectangle becomes a square, which results in an increase in flux. By Faraday's law and Lenz's law, the induced current opposes the increase in flux.

4. B

You should know that the force per length on parallel wires is $\frac{F}{l} = \frac{\mu_0}{2\pi} \frac{I^2}{d}$ if both wires carry the same current. When d doubles, the force is decreased by a factor of 2. When I doubles, the force is increased by a factor of 4. The net effect is an increase by a factor of 2.

5. D

As pendulum A swings through the magnetic field, eddy currents are established in the copper. These currents flow through a material with some finite resistance. Therefore, some energy is used to heat the copper. The source of this energy is the kinetic energy of the pendulum as it moves. Thus, there is a reduction in the energy of pendulum A that is not present in pendulum B. This means pendulum A will stop before pendulum B. This question requires you to apply concepts from several branches of physics. It tests how well you "think like a physicist."

C Exam

6. B

As the bar moves through the field, the charge carriers in the bar experience a Lorentz force that tends to deflect them to one end. This results in an electric field being established in the bar. When the electric repulsion from the charge separation has a magnitude equal to the Lorentz force, no more charge carriers will flow. This condition will give an estimate of the potential difference between the ends. Mathematically, your reasoning should be as follows:

$$F_E = F_M$$
$$Eq = qvB$$
$$E = vB$$

But since we know that $E \approx \dfrac{V}{l}$, we arrive at choice (B) after substituting.

7. B

By the first right-hand rule, the force on a positive charge is directed down the page. This allows us to eliminate any answers with statement I. Because the electron is negative, though, it feels a force directed up the page. Our answer must therefore include statement II. The path radius for a charged particle in a magnetic field is $\frac{mv}{qB}$. Both particles have the same magnitude charge and the same speed. The mass of the proton, however, is about 1,800 times the mass of the electron, and its path radius will be an equal factor larger.

8. C

We must solve the integral $\int_S \vec{B} \cdot d\vec{A}$. The following figure depicts the situation.

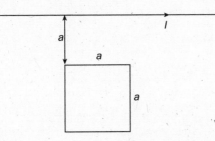

Everywhere in the loop, $\vec{B} \| d\vec{A}$. This simplifies our integral a bit to $\int_S B dA$. In a direction parallel to the wire, the magnetic field strength is constant. This lets us evaluate that portion of the surface integral immediately. Further, we can insert the formula for the magnetic field strength near a long, conducting wire. We get

$$\int_a^{2a} \frac{\mu_0 I}{2\pi} \frac{a}{r} dr$$

Moving all of the constants out, this becomes a standard integral of $\frac{1}{r}$. The answer is

$$\frac{\mu_0 I a}{2\pi}(\ln(2a) - \ln(a))$$

which, after using the subtraction/quotient rule for logarithms, gives us choice (C).

9. A

This problem is largely a geometry problem. You need to realize that the angle of incline is also the angle between the vertical and the line normal to the incline. This type of situation should be familiar from the numerous inclined plane problems in mechanics. Solving the problem then becomes a matter of recognizing the rail gun-type nature of this Faraday's law problem. It is a good idea to remember that Blv has the unit of volt and it is almost always going to be an induced emf.

10. D

In an LC circuit, the current is zero when the maximum charge q is on the capacitor (such as at $t = 0$ in this problem). As the charge decreases, the current $I = \frac{dq}{dt}$ decreases from zero. When the charge on the capacitor is zero, the current is at its minimum. This current causes a buildup of charge on the capacitor, eventually leading to the point where the capacitor is fully charged but in the opposite direction. At that point in time, the capacitor discharges again, and current now flows in the opposite direction. Therefore, current oscillates sinusoidally. Nature does not usually produce straight-line growth and decay graphs, so they should always be initially suspect.

FREE-RESPONSE QUESTION: B EXAM

(a)

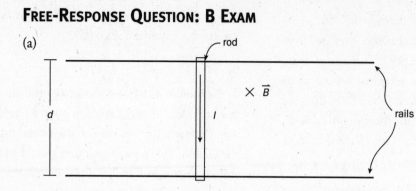

We have a choice in doing this. If both the current and the magnetic field were reversed, we would get the same force, so this is not a unique answer.

(b) The magnetic force on a current-carrying wire (or conducting rod) of length d is

$$F = IdB \sin \theta = IdB \sin 90° = IdB$$

To find the current in the rod, we use Ohm's law:

$$V = IR$$

$$I = \frac{V}{R}$$

Substituting this in the force equation gives

$$F = \frac{VdB}{R}$$

which is in the form requested.

(c) This is a one-dimensional dynamics problem starting from rest. From the work-kinetic energy theorem, we have

$$\Delta K = W$$

$$\frac{1}{2} m \left(v_f^2 - v_i^2 \right) = F\Delta x \cos \theta$$

The force is parallel to the displacement, so the cosine term is 1. Since the initial velocity is zero, we can write

$$v_f^2 = \frac{2F\Delta x}{m}$$

$$v_f = \sqrt{\frac{2F\Delta x}{m}}$$

Substituting our expression for F from part (b) and the value of $\Delta x = l$, we get

$$v_f = \sqrt{\frac{2VdlB}{mR}}$$

FREE-RESPONSE QUESTION: C EXAM

(a) $\Phi_M = \displaystyle\int_S \vec{B} \cdot d\vec{A} = \int_S B dA \cos\theta$

Here, $\theta = 75°$ because it must be measured from the normal to the surface. Because B is constant over the whole circle, it can come out of the integral, leaving

$$B \cos\theta \int_S dA = \pi r^2 B \cos\theta$$

Substituting $B = 16$ T, $r = 0.15$ m, and $\theta = 75°$, we find

$$\Phi_M = 0.293 \quad \text{T} \cdot \text{m}^2$$

(b) The constant B is replaced by a time varying B. This changes the magnetic flux to

$$\Phi_M = \pi r^2 B \cos t \cos\theta$$

The induced emf is just the negative of the time derivative here. The only term that varies with time is the cosine term. The induced emf is then

$$\varepsilon = -\frac{d\Phi_M}{dt}$$

$$= -0.293 \frac{d}{dt} \cos t \quad \longleftarrow \text{radian}$$

$$= 0.293 \sin t$$

in volts. To find the current, we just use Ohm's law

$$\varepsilon = IR$$

$$I = \frac{\varepsilon}{R} = \frac{0.293}{200} \sin t$$

$$= 1.47 \times 10^{-3} \sin t \, \frac{\text{T} \cdot \text{m}^2}{\Omega \cdot \text{s}} = 1.47 \times 10^{-3} \sin t$$

(c) From the two definitions of power, we have

$$\frac{dW}{dt} = P$$

$$= IV = I\varepsilon = (0.29)(1.47 \times 10^{-3}) \sin^2 t$$

$$= 4.26 \times 10^{-4} \sin^2 t$$

Because work is the change in energy, or the energy dissipated, we have

$$\Delta E = \int_{1.0}^{1.5} 4.26 \times 10^{-4} \sin^2(t)\, dt = 4.26 \times 10^{-4} \int_{1.0}^{1.5} \sin^2(t)\, dt$$

This integral is actually outside the scope necessary for the exam; however, it can be found in tables to be of this form:

$$\int \sin^2(ax)\, dx = \frac{x}{2} - \frac{\sin(2ax)}{4a}$$

If we remember to keep our calculators in radian mode for this calculation, it is a straightforward process to find that the energy dissipated is about 1.8×10^{-4} J.

(d) I is first a maximum when $\sin t$ is first a maximum. This is when $t = \dfrac{\pi}{2}$ s. The current is first at maximum at 1.57 seconds.

CHAPTER 10: WAVE MOTION

> ## IF YOU MEMORIZE ONLY THREE EQUATIONS IN THIS CHAPTER . . .
>
> $$v = \sqrt{\frac{T}{\mu}} = \sqrt{\frac{lT}{m}} \qquad f' = f\frac{v}{v + v_s} \qquad v = f\lambda$$

NOTE: This topic appears *only* on the B exam.

INTRODUCTION

Any student of physics must understand waves and wave phenomena. Waves help describe many natural phenomena, from the pure note produced by a world-class whistler to tall grass undulating in the wind. Even more striking, perhaps, is the fundamental role waves play in processes occurring at the atomic and subatomic levels.

GENERAL PROPERTIES OF WAVES

We all have a qualitative understanding of what a wave is: the thing that the surface of water does when a rock is dropped into it. The peaks (**crests**) and dips (**troughs**) that spread out from the rock are waves. Less clear, however, is the question of what is actually moving in a wave. Does the water move along with the wave, or does it just go up and down as the wave passes? Provided the water waves are not "rolling" (this condition is met on the AP Physics exam), the water just moves up and down as the wave passes. So the wave moves through the water but does not move the water itself.

The question of what is actually spreading out from the point where the rock hit the water is yet to be answered. The short answer to this question is that nothing is moving. That is, no object with mass is conveyed from one point to another by a wave. The only thing that happens when a wave moves is that the wave carries energy and momentum through the water. As a wave in water passes some point, it causes the water to rise (increasing gravitational potential energy). The wave also causes the momentum of the water to change as it passes. The energy and momentum come from whatever is driving the wave.

A **wave** is defined as a disturbance that moves through a mass, or field, transferring energy and momentum but not mass. A "medium" is just something that can transmit a wave. Note that we are neglecting the effects of the special theory of relativity (which relates energy and mass) in our discussion. As long as we keep our energies small, this is not a problem. Any object you see can be a medium for a wave; we will have more to say about wave media shortly.

If two people each hold one (opposing) end of a long, weak spring and move a good distance apart, a wave can be demonstrated quite easily. It is very likely you have seen this demonstration done in class at some point. This type of wave (a wave on a spring) is a physical wave. Other types of wave are optical, or light, waves and the probability waves or wave-packets associated with quantum mechanics. As you see here, we can classify waves by their medium.

If the medium is massive (i.e., made of something with mass), then the wave is mechanical. **Mechanical waves** include sound waves, water waves, and waves on a string. If the medium is not massive, then the wave is not physical. Waves that are not mechanical include light (which travels in an electromagnetic field) and the probability wave-packet mentioned earlier. No one is quite sure what the medium for a probability wave-packet is.

Beyond classification by medium, we can further distinguish waves by the direction of the disturbance they create relative to the direction in which the wave is moving. This is sometimes called the wave's "mode of oscillation." If the disturbance is perpendicular to the direction that the wave moves, the wave is said to be **transverse**. Waves on a string are generally transverse waves, as are light waves. If the disturbance is parallel to the direction that the wave moves, the wave is said to be **longitudinal**. Sound waves in air are typical of longitudinal waves. Interestingly, the mode of oscillation does not matter when it comes to the mathematics used to analyze wave motion. Even though there is a fundamental difference between sound waves moving through air and light waves moving through air, they both behave in precisely the same way when they encounter a narrow opening, once you account for the length-scale difference. In fact, all of physical optics (covered in Chapter 11) could be studied using sound.

What are the mathematics of wave motion, then? We tend to represent a wave by a sine or cosine curve (sometimes called a "sinusoidal curve"). All waves have five properties that can be measured. These measurable quantities are wavelength, frequency, amplitude, speed, and phase. The first four of these quantities are relevant to all waves. The fifth (phase) is relevant only when there are two or more waves of approximately the same wavelength in the same medium at the same location at the same time. Phase has a lot of restrictions before it comes into play. The following figure shows a typical wave.

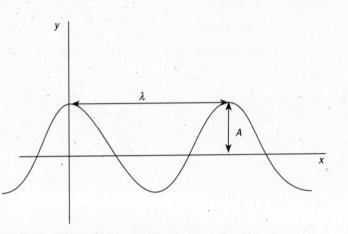

We usually represent a single wave with the following equation:

$$y(x,t) = A\cos\left[2\pi\left(\frac{x}{\lambda} - ft\right) - \phi\right]$$

This equation describes a wave moving with a velocity in the positive x direction. Notice that the expression for y depends on the x position and time. This indicates that there may be different y displacements at a single location, depending on when you look at it.

In the equation above, A represents the **amplitude** of the wave. This is the maximum value that the medium is disturbed from equilibrium. The units on A depend on the type of wave. For waves on a string, the amplitude is measured in meters (or length in general). For sound waves, the sound level, which is related to the amplitude, is measured in decibels; decibels are related to pressure variations over time, which suggests that sound is a pressure wave. For the purposes of a wave, **equilibrium** is defined as the condition of the medium when there is no wave present. For a wave on a string (e.g., on a guitar), equilibrium occurs when the string has not been plucked. You can disturb the equilibrium by pulling the string to one side. When the string is released, a wave (really two waves traveling in opposite directions) travels up and down the string. The further the string is displaced initially, the greater the amplitude of the wave. The change in amplitude can be detected in this case by the change in the volume coming from the guitar.

The parameter λ is the **wavelength**, measured in meters. In general, this is the length between two points on the wave that have the same y-value and are on the same slope as each other at the same time. Usually it is easiest to talk about the wavelength as the distance between two successive maxima (crests) or two successive minima (troughs). This does not need to be the case, though. The following figure should help you visualize wavelengths. (Line segment L does not represent a wavelength, even though both end points have the same y value at the same time. Line segments M and N do represent wavelengths.)

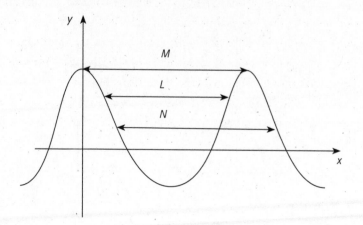

The **frequency** of the wave is denoted by f. In its simplest form, frequency represents the number of wave crests that pass a point in 1 second (i.e., it is the number of wave cycles occurring per unit of time). It is the inverse of period. The unit of frequency is the hertz (Hz); 1 hertz is equal to 1 s^{-1}, or one count per second. To determine the frequency of a wave, all one has to do is count the number of waves that pass by each second. This may prove a more difficult task than it seems, however. It is generally easier to measure the wavelength and then determine frequency via the wave speed.

Speed does not show up explicitly in the expression for a wave, but it is there nonetheless. From observing the rock dropped into the still pond, we know waves move with some speed. Without the aid of calculus, we must take on faith the statement that a wave's speed is given by $v = f\lambda$.

We can see that this equation is correct dimensionally, at least. This is the true definition of the speed of a wave in any medium. The speed depends on the medium, and wavelength then depends on v and frequency. There are other expressions for the speed of a wave, but all of them are subject to the $v = f\lambda$ equation. It is a classic experiment to measure the wavelength of a wave on a string under tension. Then you use a formula relating wave speed to tension and the string properties. From this, you can determine the frequency of oscillation of the wave.

For the most part, we will be concerned with waves in one dimension. That is, the wave travels only along the x-axis. Certain phenomena require the consideration of two-dimensional waves, particularly diffraction. When we come to that topic, we will introduce the concepts of plane and circular waves.

TRAVELING WAVES

Traveling waves are just what they seem to be: waves that travel or move through a medium. All waves are actually traveling waves, but we make a distinction between waves that can be seen to move and waves that appear to be stationary (which are called **standing waves**). When we deal with traveling waves, we are generally concerned only with a wave pulse and not with an entire wave. However, you should not automatically assume that this is the case; consider the exam question carefully to determine whether you are dealing with a wave pulse or an entire wave. The following figure shows the difference between a wave pulse (a) and a wave (b). A wave pulse is usually only half of a wave.

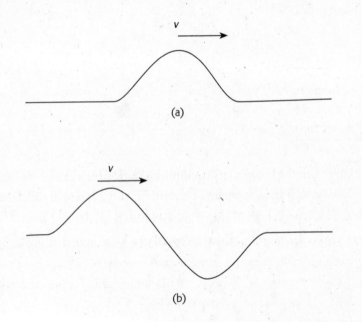

The way that a wave travels through a medium depends on both the wave and the medium. We will discuss three different media here: a string under tension, a stretched spring, and air. After that we will discuss what happens to a traveling wave when it encounters a boundary or a change in the characteristics of its medium.

TRAVELING WAVES IN VARIOUS MEDIA

To understand waves in a string under tension, there are several basic concepts we must grasp. First, the wave is a transverse displacement wave. That is, the dimension of the wave equation is length. So we measure $y(x, t)$ in meters. Recall that

$$y(x,t) = A\cos\left[2\pi\left(\frac{x}{\lambda} - ft \right) \right]$$

It is not a good idea to assume that the string is massless. If the string *were* massless, the wave would have an infinite speed. Mass restrains the speed of the wave because the mass of each section resists the acceleration imparted by the wave as the wave propagates down the string. This resistance forces the wave to move at a finite speed. So the medium's mass is one important factor in determining how fast the wave moves in this case. The other significant factor here is the tension in the string. If the tension is high, there is a large restoring force on a displaced portion of the string; in this situation, wave speed should increase as tension increases. The following figure shows a string with a wave pulse. The shaded region has a small mass Δm. The tension in the wave tends to lift this mass element up. If this mass is large, the acceleration is small, and hence the wave travels slowly. If the tension is high, however, the acceleration of the mass element can be large, leading to faster wave speed.

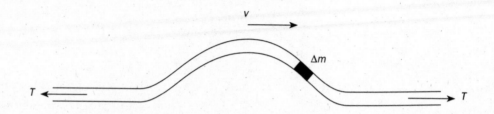

The tension tends to return the string to its equilibrium state, which causes the wave to move along the string. Because each small segment of the string is accelerated at different times, it is best to talk about the mass per length of the string, or linear mass density, $\mu = \dfrac{m}{L}$. If a string has a tension T in it and a mass per length μ, then the speed of a wave in that string is

$$v = \sqrt{\frac{T}{\mu}} = \sqrt{\frac{LT}{m}}$$

Here L is the total length of the string, and m is the total mass of the string. You should check the units to verify that this is, in fact, a speed.

Waves on a stretched spring are also displacement waves. However, these waves may be transverse or longitudinal, and in general, the transverse wave and the longitudinal wave have different speeds because of the way a spring reacts to the different displacements. There is no general formula that you are expected to know for the speed of a wave on a spring.

REFLECTION AND DIFFRACTION OF A WAVE

Imagine a string tied to a hook in a wall:

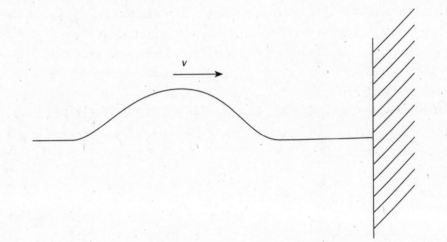

If you send a wave pulse down this string toward the wall, it should reflect back at you. But what will the reflected pulse be like? If, as the above figure suggests, the end of the string is securely tied to the wall, then this point on the string cannot move at all. It is fixed in space, and any attempt the wave makes to lift the point will therefore fail. Recall Newton's third law, though: if the wave exerts an upward force on the hook, the hook must exert a downward force on the wave. The magnitudes of the forces are the same, but the accelerations of the masses are different. The hook essentially has an infinite mass and no acceleration. If we assume that the mass is very large (but not infinite), then we can say the wave on the string experiences a very large (but not infinite) acceleration downward. This follows from Newton's third law.

$$\frac{a_{string}}{a_{hook}} = -\frac{m_{hook}}{m_{string}}$$

Technically, we should talk not about the acceleration of the wave but rather about the acceleration of the string elements, hence, the subscripts above.

This amounts to claiming that the section of string adjacent to the hook has such a large downward acceleration that it flips the wave over. The reflected wave pulse travels in the opposite direction on the opposite side of the string. In the language of waves, we say the reflected wave's phase has been shifted by 180° or π radians. The following figure depicts a wave pulse movement. Going down the page, the wave pulse is first traveling toward the fixed point and is then reflected and traveling away from the fixed point. The reflected wave is on the opposite side of the string.

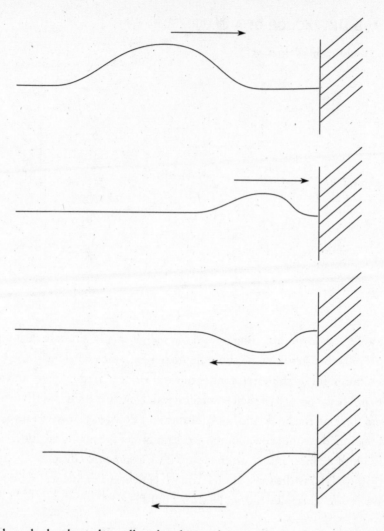

Suppose we replace the hook on the wall with a device that simulates a free end: a nearly massless ring that is free to slide along a frictionless rod. In this case, the ring will rise with the pulse when the wave pulse arrives at the end point. The restoring tension will cause the ring and the string to return to equilibrium, and the reflected wave pulse will be on the same side of the string as the incident wave. As a result, there is no phase shift in this case. The following figure shows how the wave pulse reacts to a free end. As the wave pulse is reflected, it experiences no phase shift and is on the same side of the string as the incident wave pulse.

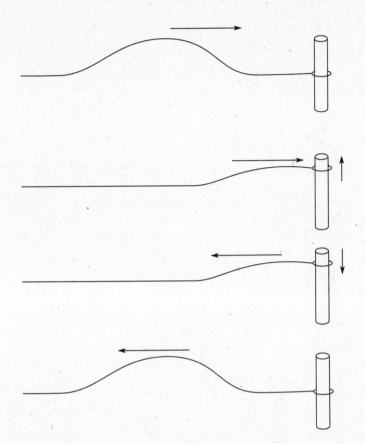

In this discussion of different ways the end of the string can respond, we have described the behavior of fixed and free boundaries, the two most common boundary conditions. The reflected wave is either phase shifted by π radians in the case of a fixed boundary or not phase shifted at all in the case of a free boundary.

You may occasionally encounter a third boundary condition known as a knot. In this case, two strings of different linear mass densities are tied together. A wave pulse travels toward the knot, and you need to determine what will happen. Part of the wave pulse is reflected, as before, and part of the pulse is transmitted through the knot. As you would expect, the transmitted part continues on in the same direction as the incident pulse with no phase shift. The behavior of the reflected wave depends on which string received the incident pulse. If the pulse is traveling along the lighter string to the heavier string, the reflected wave acts is if the knot is fixed. That is, the reflected pulse is phase shifted. If the pulse is traveling on the heavier string toward the lighter string, however, the reflected pulse acts as if the knot is free. That is, the reflected pulse is *not* phase shifted. You can generalize the knot problem to sound waves by considering the wave speed relations.

In the following figure, a wave pulse incident from the lighter string results in a phase-shifted reflected pulse (a). If the pulse is incident from the heavier string, the reflected pulse has no phase shift (b).

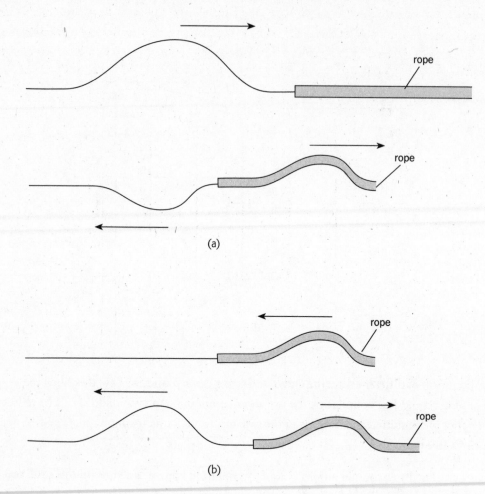

(a)

(b)

Both strings must have the same tension due to Newton's third law; the wave speed is less on the heavier string than it is on the lighter string. So, without any proof, we generalize to the following two statements:

1. If a wave is incident on a boundary from medium A to medium B, then the wave that is reflected back into medium A is phase shifted by π radians if the wave speed in B is less than the wave speed in A.

2. If the wave speed in B is greater than the wave speed in A, the reflected wave is not phase shifted.

These statements apply to sound waves moving from one medium to another (e.g., from air to water) and to light waves reflected off a mirror. These facts are particularly useful for determining how laser cavities work.

Diffraction of waves occurs only when the waves are at least two-dimensional. We are largely concerned here with two types of two-dimensional waves: plane waves and circular waves. Plane waves are represented by a set of parallel lines, called **wave fronts**, with a vector perpendicular to these lines to indicate the velocity of the wave; the parallel lines represent the wave crests. Circular waves are represented by a set of concentric circles; the waves are considered to be traveling out from the center of the circle. Circular waves are familiar—they are the waves you see when you toss a rock into a pond. The following figure shows plane (a) and circular (b) waves.

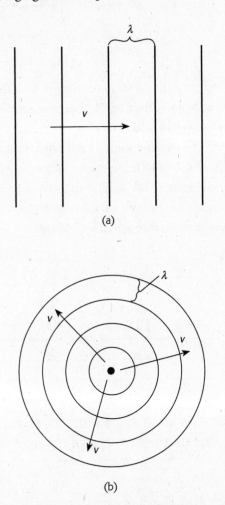

(a)

(b)

An interesting thing happens when a wave encounters a sharp corner: the wave bends, or diffracts, around the corner. This is why you can hear the sounds made by someone who is standing around the corner of a building. The corner acts like the center of a circular wave that propagates around the edge. If you extend a line from the corner parallel to the motion of the wave, the plane wave fronts are not affected. The following figure shows this effect.

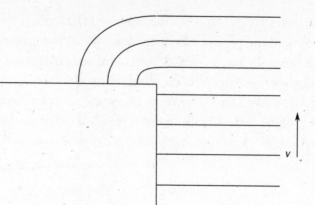

If the plane waves encounter a small opening in an impenetrable barrier, then there is an even more striking diffraction; the opening acts like the center for circular waves. This can be seen in a wave tank or a large bathtub. Fill the tank or tub with about two inches of water and shine a bright light straight down on the water. If you dip the edge of a long board in the water, you can produce plane waves. By placing two metal blocks in the water, you can make a small opening for the waves to diffract through. The shadows are the wave crests. The following figure shows the diffraction of plane waves through a small opening.

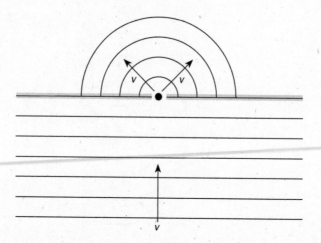

AP EXPERT TIP

The greater the wavelength, the more a wave diffracts. So for a given opening, you will notice diffraction much more easily for sound than for light.

What counts as a sharp corner or a small opening depends on the size of the wavelength. If the radius of the corner is much less than the wavelength, then the corner is sharp. If the opening is about the same size as the wavelength, then the opening may be considered small.

THE DOPPLER EFFECT

When an ambulance drives past you with its siren on, you hear the sound's pitch change from high to low. This change in pitch is due to the Doppler effect or Doppler shift. In the case of sound waves, the nature of the Doppler shift depends on whether the source is moving or the observer is moving. In the case of light waves, the Doppler effect has a different cause and is independent of whether the source or observer is moving. In this chapter, our discussion of the Doppler effect will only deal with sound. The following figure shows a source of sound moving toward an observer with speed v_s for speed of the source.

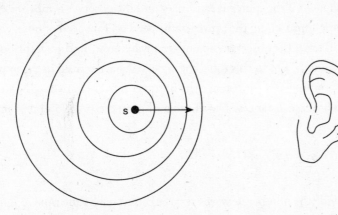

We will take the speed of sound to be v for this discussion. At one instant, the source emits a wave crest; before the next wave crest is emitted, the source moves a distance x_1 given by the source speed divided by the emitted frequency, f. Recall from our discussion of oscillation that frequency is the inverse of period. The first wave crest has moved a distance x_2 given by the sound speed divided by the emitted frequency. This results in a decreased distance between wave crests as the observer hears them, creating an increase in the frequency of the observed sound when compared with the emitted sound. If we call the observed wavelength λ', then the observed frequency is

$$f' = \frac{v}{\lambda'}$$

The observed wavelength is the reduced distance between successive wave crests. This is given by

$$\lambda' = \frac{v}{f} - \frac{v_s}{f}$$

Making this substitution and performing the algebra, we arrive at

$$f' = f\frac{v}{v - v_s}$$

where f' is the observed frequency and f is the emitted frequency. This is for the source moving toward the observer. If the source is moving away from the observer, the wavelength is increased, and the frequency is decreased. So for a source moving away from a stationary observer,

$$f' = f \frac{v}{v + v_s}$$

If you get confused on the exam, remember this point: if the source is approaching the observer, the observed frequency must increase.

What happens if the source is stationary and the observer is moving? Suppose the observer is approaching the source with speed v_o, for speed of observer. At some instant in time, the observer encounters a wave crest. In time t, the observer will move a distance $v_o t$. In this time, the wave will have moved in the opposite direction by an amount vt. So the distance the observer moves relative to the wave fronts is the sum of these distances. The number of wave crests the observer encounters is given by

$$N = \frac{vt + v_o t}{\lambda}$$

If we divide the number of wave crests encountered by the time required to encounter them, we have the **observed frequency** of the wave. In this case,

$$f' = \frac{N}{t} = \frac{v + v_o}{\lambda}$$

If we use the fact that $v = f\lambda$, then we can show

$$f' = f \frac{v + v_o}{v}$$

If the observer is moving away from the source, then the observed frequency is

$$f' = f \frac{v - v_o}{v}$$

Here, like above, f' represents the observed frequency, and f is the emitted frequency.

If the source and observer are both moving, the observed frequency is given by

$$f' = f \frac{v \pm v_o}{v \mp v_s}$$

The upper signs are used when the objects are approaching, and the lower signs are used when objects are moving away.

SUPERPOSITION

When two physical waves in the same medium are at the same place at the same time, they will **interfere**. The net effect of the two waves is just the sum of their displacements; the nature of the interference may be constructive or destructive. **Constructive interference** occurs when the crest of one wave passes through the crest of another wave (or the trough of one wave passes through the trough of another wave), resulting in a wave that is larger than either individual wave. **Destructive interference** occurs when the crest of one wave passes through the trough of another wave, resulting in a wave that is smaller than one of the other waves. The following figure shows constructive (a) and destructive (b) interference.

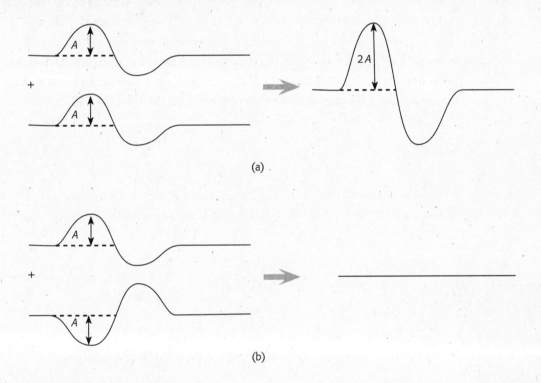

(a)

(b)

STANDING WAVES

A standing wave can occur when two waves of the same frequency move in opposite directions through the same bounded medium; the fact that the medium is bounded means that one of the waves is a reflected wave. Normally, we picture standing waves as if they are on a string. However, it is possible to set up standing waves using sound, water, or light—you can even see standing waves on a soap film held in a rigid frame.

As noted earlier, one of the two waves at work here is a reflected wave. As you will recall, reflections occur only when the medium changes characteristics (i.e., at a boundary). We will ignore knots as boundaries in this discussion and confine ourselves to fixed and free boundary conditions.

The most commonly diagrammed standing wave is the fixed-fixed wave. This is a standing wave that has its medium fixed at both ends. If we imagine an incoming wave train, rather than a single pulse, we can determine what will happen at the fixed endpoints of the medium. The first full wave to be reflected from the fixed point will have a phase shift of π radians. When the reflected wave interacts with the next incoming wave, there will be interference. Under certain conditions, the interference will be totally constructive. These are the conditions that create standing waves. The following figure shows how we represent the first three standing waves on a string with both ends fixed.

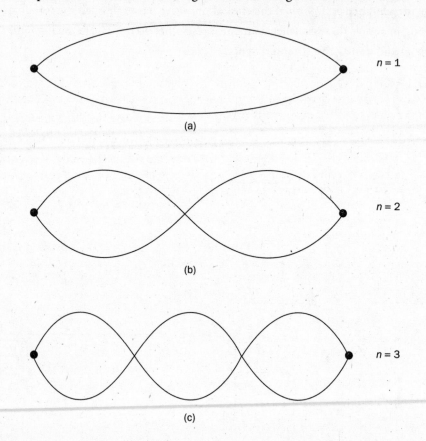

$n = 1$

(a)

$n = 2$

(b)

$n = 3$

(c)

Each standing wave is numbered, starting with 1 and increasing from there. The first standing wave in a standing wave pattern is called the fundamental wave, or the first harmonic. The fundamental wave is the standing wave with the longest wavelength. The other standing waves are ordinal harmonics (second, third, fourth, etc.) based on decreasing wavelength.

When you look at a standing wave on a string, you will see something like the previous figure, which is why we draw standing waves in this manner. Notice that certain points along the wave do not deviate from equilibrium at all, while other points experience maximum disruption. Those points that remain in equilibrium are called **nodes**. In the fixed-fixed case, both ends of the medium are nodes. Those points that have maximum disruption or displacement are called **antinodes**. There is always exactly one antinode halfway between two consecutive nodes; similarly, there is always exactly one node halfway between two antinodes. These two facts allow us to determine which waves are going to produce standing waves in a given configuration. Suppose that in the last figure, the distance between the endpoints is L. Because both ends are fixed, we know that they must be nodes (i.e., points along the medium that do not deviate from equilibrium). The rule that tells us there is an antinode halfway between any two consecutive nodes allows us to construct the first standing wave. What is the wavelength of this wave? Moving from left to right along the top path (a), we traverse one half of a sine curve. Thus, L is one half of a wavelength for the fundamental standing wave, or

$$\lambda_1 = 2L$$

If we want to discover the second harmonic, all we have to do is add one node between the end points such that the distance between any two consecutive nodes is the same. For the second harmonic, the added node is in the middle; we can now draw the wave, recalling the rule regarding the placement of antinodes. We thus arrive at (b). Following the sine curve along the top path from left to right, we move up to the first antinode and then down to the second node. We continue down to the second antinode and then up to the third node. The path is shown in the figure below. This path is an entire sine curve, so now the distance between the endpoints L is a full wavelength, or

$$\lambda_2 = L$$

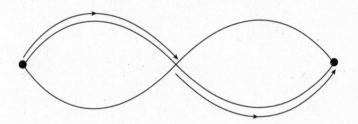

It follows that the third harmonic has a wavelength given by

$$\lambda_3 = \frac{2}{3}L$$

Notice that the wavelength of harmonic number n appears to be given by

$$\lambda_n = \frac{2}{n}L$$

This is the case for a fixed-fixed situation such as the strings on a guitar, which are all fixed-fixed strings of the same length.

What happens when we have a fixed-free case, as shown in the following figure? One end is a node, but what about the other end? From the rules about wave reflection, we know that the free end is an antinode. It is important to realize that a free end is *always* an antinode and a fixed end is *always* a node. This will allow you to determine which wavelengths are going to create standing waves. Remember that wavelengths are the determining factor here.

The figure below shows the fixed-free condition modeled on a string. The right side of the string is tied to a ring that slides without friction along a rod.

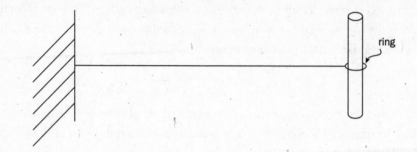

The following figure shows the first three harmonics for the fixed-free condition.

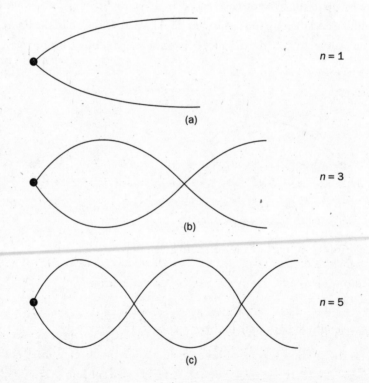

Notice here how the fundamental goes through only one-quarter of a wavelength between the endpoints (the string is one-quarter of a wavelength long). This should be clear because the distance between two nodes is half a wavelength, and an antinode is halfway between two consecutive nodes. We can write the following equation for the wavelength of the first harmonic in the fixed-free condition:

$$\lambda_1 = 4L$$

If we examine the next harmonic we can trace the following path. Go up from the left node through the first antinode. Then go down through the second node and finish on the bottom side of the second antinode. This path is three-quarters of a wavelength, so

$$\lambda_3 = \frac{4}{3}L$$

For the third standing wave, we see that

$$\lambda_5 = \frac{4}{5}L$$

In the fixed-free case, it appears the formula for the n-th harmonic wavelength is

$$\lambda_n = \frac{4L}{n}, \text{ where } n \text{ is odd integers only.}$$

To determine the variable n from the harmonic number, use the following equation:

$$n = 2n_h - 1, \text{ where } n_h \text{ is the harmonic number.}$$

For example, n for the fifth harmonic ($n_h = 5$) is 9.

A fixed-free situation is most easily modeled by using a long, narrow tube that is closed at one end. Blowing across the open end in a certain way results in an audible tone because a standing sound wave has been established in the tube. Exactly which end is fixed and which is free is open to some debate, as we will discuss later.

The final situation to be examined here is the free-free case. Mathematically, the results for the fixed-fixed and free-free cases are identical because there is exactly one node between two consecutive antinodes. The fundamental wavelength for the free-free condition is therefore twice the distance between the ends, just as for the fixed-fixed condition. One might think that we should abandon the whole "fixed-fixed/fixed-free/free-free" labeling system in favor of a "same end/different end" system, but we stick with the free/fixed distinction because the reference to specific types of end conditions helps us to recall the rules of wave reflection.

Once we know the wavelengths that will generate standing waves, we usually want to find their associated frequencies. Resonant frequencies are generally of more interest than the standing wavelengths. We need a method for determining these frequencies, but fortunately there is an easy relationship between wavelength and frequency. For a given medium, their product is a constant known as the wave speed. If we have some independent method for determining the wave speed, we can find the frequencies associated with the standing wavelengths. There are two media commonly used for standing wave experiments: the string under tension and air (for sound waves). Light waves in a medium are employed less often. These common techniques are broadly applicable.

Recall that for a wave on a string, the speed of the wave is given by

$$v = \sqrt{\frac{LT}{m}} = \sqrt{\frac{T}{\mu}}$$

Once we have the wave speed, we can determine the resonant frequencies. Suppose we have a fixed-fixed string with a distance L between the endpoints. We know the first standing wavelength will be $2L$. We also know that

$$v = f\lambda$$

and

$$\lambda_1 = 2L$$

Therefore, we can write

$$f_1 = \frac{v}{2L}$$

But based on the formula for the speed of a wave on a string, we have

$$f_1 = \frac{1}{2L}\sqrt{\frac{T}{\mu}}$$

This is the first harmonic frequency for a fixed-fixed (or free-free) standing wave on a string. You should feel confident in writing the following equation for the n-th harmonic frequency for the same string.

$$f_n = \frac{n}{2L}\sqrt{\frac{T}{\mu}}$$

Notice that as wavelength decreases, frequency increases. For a high harmonic, the wavelength is quite small, but the frequency is large. Without further discussion, we can write the formula for the nth harmonic for a fixed-free standing wave on a string as follows:

$$f_n = \frac{n}{4L}\sqrt{\frac{T}{\mu}}$$

For sound waves in air, the speed depends on many factors, including the air density, pressure, and temperature. The most straightforward speed variation, however, is dependent on temperature. The speed of sound in air at sea level is approximated by

$$v_{\text{sound}} = 331 + 0.6T_{\text{C}}$$

T_C is the temperature in degrees Celsius; the units for this speed are meters per second. Now we come to the tricky part of a standing sound wave. As stated earlier, sound is a pressure wave; therefore, we should think of sound in terms of pressure variation and not in terms of the displacement of air molecules or the local density of the air.

A narrow tube with one open end, as shown in the following figure, has a node at the open end (pressure is fixed at the atmospheric pressure) and an antinode at the closed end. Some textbooks adopt an alternative convention, placing an antinode at the open end and a node at the closed end, but this alternative convention treats sound as a displacement wave rather than a pressure wave. However, one's choice of convention does not affect the harmonics of the situation—it affects only the locations of the nodes and antinodes.

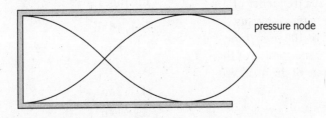

pressure node

pressure antinode

Notice that in the above figure, the displacement nodes and antinodes are reversed. Note also that the equations given above are approximate for sound waves due to the diameter of the pipe.

When there is a pressure node in a standing sound wave, there is a displacement antinode at the same location, and vice versa. In any case, the harmonic wavelengths for the fixed-free standing sound waves are given by $\lambda_n = \dfrac{4L}{n}$, where n is an odd integer and L is the length of the tube. To determine the harmonic frequencies, we exploit the relationships among speed, wavelength, and frequency to get $f_n = \dfrac{n}{4L}(331 + 0.6T_C)$.

These arguments can be extended to waves of any type, provided you have the proper wavelength and wave speed formulas. Once you have that information, all you need to do is solve for the frequencies using

$$f = \frac{v}{\lambda}$$

If a given wavelength does not satisfy the boundary conditions, then no standing wave will be established. Take care to note which boundary conditions are relevant to the problem you are working so that you can employ the wavelength formulas accurately.

REVIEW QUESTIONS

1. A wave of length 2 m is found to have a frequency of 140 Hz. What is the speed of this wave?

   (A) 0.142 m/s

   (B) 70 m/s

   (C) 138 m/s

   (D) 142 m/s

   (E) 280 m/s

2. A siren is emitting sound at a frequency of 2,000 Hz, and an observer is moving away from the source at a speed of 100 m/s. If the speed of sound in air is 340 m/s, which of the following is the best estimate of the observed frequency of the siren?

   (A) 1,410 Hz

   (B) 1,550 Hz

   (C) 2,000 Hz

   (D) 2,590 Hz

   (E) 2,830 Hz

3. A sound wave in air moves at a speed of 340 m/s. Which is closest to the wavelength of middle C ($f = 256$ Hz)?

   (A) 87,000 m

   (B) 596 m

   (C) 84 m

   (D) 1.33 m

   (E) 0.753 m

4. A string is tied between two walls 8 m apart. Which of the following is *not* an allowed standing wavelength for this string?

   (A) 32 m

   (B) 16 m

   (C) 8 m

   (D) 2 m

   (E) 0.5 m

Use the following figure to answer questions 5 and 6.

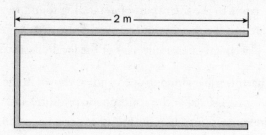

5. The tube above is closed at one end and open at the other. If the tube is 2 m long, what is the fundamental harmonic wavelength for the tube?

   (A) 16 m

   (B) 8 m

   (C) 4 m

   (D) 2 m

   (E) 1 m

6. If the closed end of the tube is opened, and all other things remain the same, what is the ratio $\dfrac{f'}{f}$ between the new fundamental frequency (both ends open) f' and the old fundamental frequency (one end open and one end closed) f?

   (A) 4

   (B) 2

   (C) 1

   (D) 0.5

   (E) 0.25

Use the following figure to answer question 7.

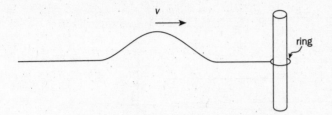

7. One end of a horizontal string is tied to a ring that is free to slide along a frictionless rod. A transverse wave pulse is generated at the other end, moves toward the ring as shown, and is reflected at the ring and rod. Which of the following are properties of the reflected wave?

  I. Its speed is greater than that of the incident pulse.

  II. Its amplitude is smaller than that of the incident pulse.

  III. It is on the same side of the string as the incident pulse.

  (A) I only
  (B) III only
  (C) I and II only
  (D) II and III only
  (E) I, II, and III

8. A sound wave traveling through air encounters the surface of a pond. How are the speed and phase of the transmitted wave related to the speed and phase of the reflected wave?

  (A) The transmitted wave moves more quickly and is in phase with the reflected wave.

  (B) The transmitted wave moves more slowly and is in phase with the reflected wave.

  (C) The transmitted wave moves more quickly and is out of phase with the reflected wave.

  (D) The transmitted wave moves more slowly and is out of phase with the reflected wave.

  (E) The transmitted wave moves at the same speed and is out of phase with the reflected wave.

9. A sound source is moving away from an observer at the speed of sound. What is the ratio $\dfrac{f'}{f}$ of the observed frequency f' to the emitted frequency f?

  (A) 4
  (B) 2
  (C) 1
  (D) 0.5
  (E) 0.25

10. Two strings, each fixed at both ends, have different linear mass densities. String 1 has a linear mass density μ_1, string 2 has a linear mass density μ_2, and $\mu_1 > \mu_2$. If both strings have the same length and are under the same tension, which of the following statements describes the relationships between the harmonics of the strings? Subscript 1 refers to string 1 and subscript 2 refers to string 2.

  I. $f_1 > f_2$
  II. $\lambda_1 > \lambda_2$
  III. $v_1 < v_2$

  (A) I only
  (B) II only
  (C) III only
  (D) I and II only
  (E) I, II, and III

FREE-RESPONSE QUESTION

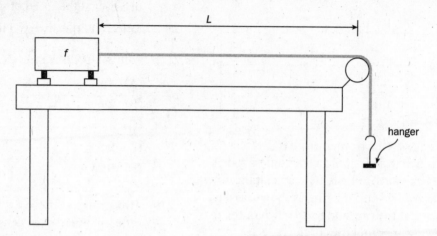

The figure shows a system designed to study the properties of standing waves. A string is attached to a vibrator of unknown fixed frequency f. At a distance L away from the vibrator is a pulley, over which the string runs. A mass hanger is hanging from the end of the string, on which various masses may be hung.

(a) Treating the place at which the string attaches to the vibrator and the pulley as fixed points, write an expression for the harmonic wavelengths for this string. Between the sets of points, sketch the first three harmonics for the system. Identify nodes and antinodes.

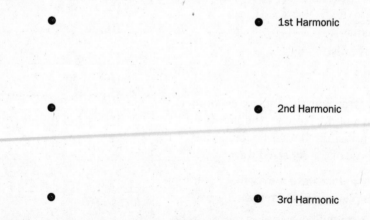

(b) Assume a standing wave corresponding to the third harmonic exists on the string for some initial value of the tension. If the tension in the string is increased sufficiently, which harmonic wavelength next appears on the string for the constant, but unknown, vibrator frequency? Explain your answer.

(c) If the string has a total mass of m and a total length of l, derive an expression for harmonic wavelength as a function of tension.

(d) If you plot λ_n^2 (the square of the successive harmonic wavelengths) versus T_n (the tension necessary to produce the nth harmonic wavelength for the unknown but constant vibrator frequency) (y versus x), show how you would find the frequency of the vibrator.

ANSWERS AND EXPLANATIONS

1. E

Because the speed of sound is $f\lambda$, all you have to do here is multiply 2 m by 140 Hz. Notice that the other choices express incorrect ways that one might combine the given numbers (e.g., 138 is 140 minus 2).

2. A

Because the observer is moving away from the source, the observed frequency should be *less* than the emitted frequency. Knowing this allows you to eliminate the last three choices immediately. Recalling the formula for the Doppler shift for the moving observer as

$$f' = f\frac{v - v_0}{v}$$

you can make a quick estimate. Substituting the given values, you can see that the fraction $\dfrac{v - v_0}{v}$ is about $\dfrac{12}{17} = \dfrac{3}{4.25}$. This a little less than three-quarters, so the answer must be less than three-quarters of 2,000, or less than 1,500 Hz.

3. D

Here you must recall the relation among speed, frequency, and sound. Because $\lambda = \dfrac{v}{f}$, you know the numerical value of the wavelength must be less than 340. Choices (C), (D), and (E) are the only choices that satisfy this requirement. You can also estimate the fraction $\dfrac{v}{f}$ as $\dfrac{34}{25}$, which is about $\dfrac{7}{5}$. This is much smaller than 84 but greater than 1. Through the process of elimination, only choice (D) survives.

4. A

This is a standing wave with fixed ends. The harmonic wavelengths for the fixed-fixed condition are given by $\lambda_n = \dfrac{2}{n}L$. Because n is an integer, λ_n must be a fraction of 16 (2×8). Nonallowed wavelengths must therefore be greater than 16, leaving choice (A) as the only viable choice. You can also eliminate the other answers by showing that $16 = \dfrac{16}{1}$, $8 = \dfrac{16}{2}$, $2 = \dfrac{16}{8}$, and $0.5 = \dfrac{16}{32}$.

5. B

A quarter of a sound wave exists in the tube during the vibration (node at closed end and antinode at open end). Therefore, $\lambda = 4L = 4$ (2 m) = 8 m.

6. B

When the closed end of the tube is open, the situation becomes fixed-fixed (if you think of it in terms of pressure waves) or free-free (if you think of it in terms of displacement waves). Regardless of your choice, the new harmonics are given $\lambda_n = \dfrac{2}{n}L$, which means that the fundamental wavelength is cut in half. Because the speed of the wave is unaffected, you can equate the product of the old wavelength and frequency with the new wavelength and frequency:

$$f'\lambda' = f\lambda$$
$$\frac{f'}{f} = \frac{\lambda}{\lambda'}$$

where the primes indicate new quantities. Because $2\lambda' = \lambda$, we can see that $\dfrac{f'}{f}$ must be 2.

7. **B**

The speed of the wave will not change because the linear mass density and tension do not change—any choice that includes statement I should be eliminated. The amplitude of the pulse will not change because the system is frictionless—any choice that includes statement II should be eliminated. The only choice left is (B), statement III only.

8. **A**

As you probably know, sound travels faster through solids and liquids than it does through gases. Because this is the case, the reflected wave is not phase shifted relative to the incident wave. Further, the transmitted wave is always in phase with the incident wave, which means that the transmitted and reflected waves must be in phase with each other.

9. **D**

The Doppler formula for a sound source moving away predicts a decrease in the observed frequency when compared with the emitted frequency. This allows you to eliminate the first three choices. If you recall, the formula for the shifted frequency in this case is $f' = f \dfrac{v}{v + v_s}$. Since the source is moving at the speed of sound, the fraction is equal to $\dfrac{1}{2}$, or 0.5.

10. **C**

Because the two strings are the same length between two fixed points, we know immediately that the harmonic wavelengths will be the same. This is because the wavelengths depend only on the distance between the two points. Any choice that includes statement II is thereby eliminated. The strings are under the same tension, so the wave speeds must depend on linear mass density. As mass density goes up, speed goes down. Since $\mu_1 > \mu_2$, we know $v_1 < v_2$. Thus, any choice that does not include statement III must be eliminated, which leaves only choice (C).

FREE-RESPONSE QUESTION

(a) This is a fixed-fixed condition, so the wavelengths are given by

$$\lambda_n = \frac{2}{n} L$$

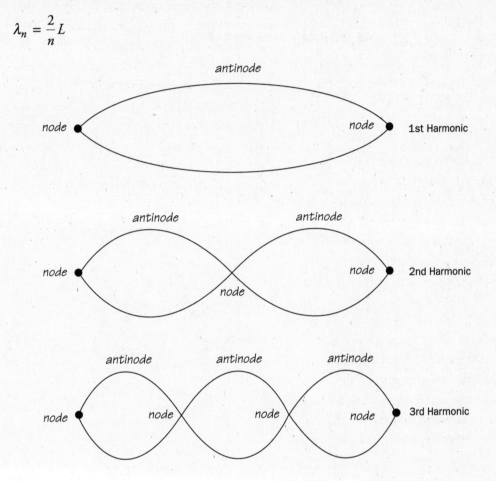

(b) As the tension is increased, the wave speed increases because $v = \sqrt{\dfrac{TL}{m}}$. Because the frequency is fixed, the next harmonic wavelength that should appear would be longer because $v = f\lambda$. As speed increases, wavelength increases as well. This longer wavelength would then correspond to the next *lower* harmonic, or the second harmonic.

(c) From the answer to (b), the speeds are going to be given by

$$f\lambda = \sqrt{\frac{TL}{m}}$$

so

$$\lambda_n = \frac{1}{f}\sqrt{\frac{T_n L}{m}}$$

(d) From the answer to (c), λ_n^2 is given by

$$\lambda_n^2 = \frac{1}{f^2}\frac{T_n L}{m}$$

A plot of λ_n^2 versus T_n (y versus x) should be a straight line with slope $\dfrac{L}{f^2 m}$ and intercept zero. If the slope is s, then

$$s = \frac{L}{f^2 m}$$

$$f^2 = \frac{L}{sm}$$

$$f = \sqrt{\frac{L}{sm}}$$

CHAPTER 11: OPTICS

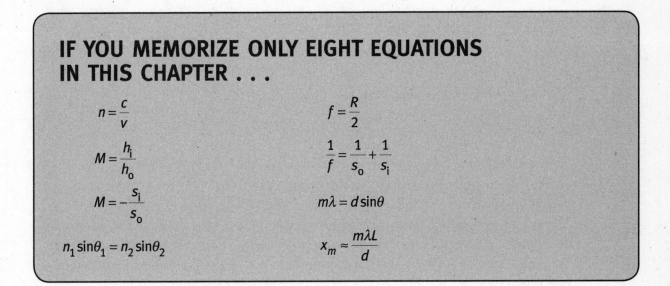

IF YOU MEMORIZE ONLY EIGHT EQUATIONS IN THIS CHAPTER . . .

$$n = \frac{c}{v}$$

$$M = \frac{h_i}{h_o}$$

$$M = -\frac{s_i}{s_o}$$

$$n_1 \sin\theta_1 = n_2 \sin\theta_2$$

$$f = \frac{R}{2}$$

$$\frac{1}{f} = \frac{1}{s_o} + \frac{1}{s_i}$$

$$m\lambda = d\sin\theta$$

$$x_m \approx \frac{m\lambda L}{d}$$

NOTE: This topic appears *only* on the B exam.

INTRODUCTION

Optics is the study of how light travels through various media. Light can be studied as a stream of particles or as a wave. Newton promoted the idea of treating light as a stream of particles, or corpuscles; if light were composed of particles, then Newton's laws of mechanics should have governed the way light moves. For many phenomena it was, and is, convenient to treat light as a stream of particles or a ray; geometric optics is the branch of optics that takes this approach. However, Huygens promoted the idea of light as a wave. Huygens's principle showed that considering light as a wave allowed all of geometric optics to be reproduced and other phenomena, like the diffraction of light around a boundary, to be explained. For more than 200 years, light was treated only as a wave. In the late 1800s and early 1900s, this view was called into question. Einstein showed that light acted as a particle with wavelike properties when he published his Nobel prize-winning paper on the photoelectric effect.

In one sense, light is neither a wave nor a particle, and in another sense, it is both. Physicists tend to believe that if you are not surprised by this, you don't really understand it. Even if you are surprised by it, you still may not understand it, but that's okay. It is reasonably safe to say that no one really understands this fact in the sense that we understand Newtonian mechanics. So we will take it on faith that light has some unusual properties and satisfy ourselves with learning the rules that it follows.

The first rule light follows is that in a vacuum, it travels with a constant speed of approximately 3×10^8 m/s. If light is traveling through anything other than vacuum, it slows down. In air under normal atmospheric conditions, it slows only very slightly (less than 0.1 percent). In other materials, such as water or glass, it slows significantly. For a given material, we define the ratio of the speed of light in vacuum to the speed of light in the material as the index of refraction, n. This is a dimensionless number. If we let c equal the speed of light in vacuum and v equal the speed of light in a material, then n is given by

$$n = \frac{c}{v}$$

Note that n will always be greater than or equal to 1. Sometimes the phrase "optically dense" is used to describe media. An optically dense medium is one in which light travels slower than c. Clearly, this describes everything except a vacuum. However, when we deal with multiple media in refraction, we may find it convenient to discuss *more* optically dense mediums and *less* optically dense mediums. More optically dense mediums have a higher index of refraction. We will also be concerned only with positive n values. Recent developments have shown that some exotic materials have negative n values, but this is far beyond the scope of the AP Physics exam.

We can further classify objects by how they transmit or fail to transmit light. Transparent objects are made of media that let light pass through without causing a narrow beam to spread. If you place a transparent object flat on a table and shine a laser parallel to the table through the object, you will not be able to see the beam from above. This is because none of the laser light enters your eye. More than likely, though, you will see a very faint beam in any "transparent" object subjected to this experiment. This happens because most "transparent" objects are actually translucent, meaning that they allow light to pass through them but tend to scatter some part of the light in all directions. The final type of object, optically speaking, is an opaque object. Opaque objects do not transmit any light through their bulk. A brick is an opaque object.

The rest of the rules we develop will fall under two broad headings. The first heading, geometric optics, is the branch of optics that treats light as a ray. The ray travels in a straight line through any given medium; it may bend at the boundary between two media. Geometric optics is largely concerned with image formation using lenses and mirrors. The second heading, physical optics, is the branch of optics that treats light as a wave. Physical optics allows us to do things like measure the spacing between adjacent tracks on a CD or DVD with a meter stick and a laser pointer. Physical optics also helps us understand why light spreads out, or disperses, when it passes though a prism.

GEOMETRIC OPTICS

We will begin our discussion with the technique known as ray tracing. Ray tracing allows us to predict image formation in lenses and mirrors; it also helps explain how objects like prisms work on monochromatic light (i.e., light of one color, like a laser). We then will discuss reflection and refraction in general terms as well as the specifics of flat-surface refraction. We will look at the commonly assigned problem of apparent depth, which students often find difficult.

We will then shift our attention to plane and spherical mirrors and the way that ray tracing is used to determine image placement and magnification with these optical elements. We'll touch briefly on the spherical aberration that can occur with mirrors and end this section with a discussion of thin lenses.

RAY-TRACING DIAGRAMS

When we treat light as a ray, we are assuming that light travels in a straight line through a continuous, constant medium. Actually, this assumption is a by-product of the way that light really moves. In going from one place to another, light always takes the least time-consuming path available, even if that path is crooked and long. Thanks to Euclidian geometry, we generally see this path as a straight line. Therefore, when we draw a picture of light being transmitted from one object to another, we draw a straight line with an arrow pointing from the source of light to the lit object. The following figure details this.

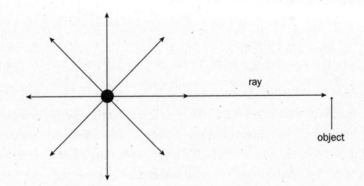

When creating ray-tracing diagrams, we generally want to know what effect a particular optical element (usually a mirror or lens) or set of elements will have on the image of an object. An image is nothing more than the convergence or apparent convergence of light. Images have three basic characteristics: an image may be real or virtual, it may be upright or inverted, and it may be enlarged or reduced. Under certain circumstances, an object may be virtual. However, problems involving virtual objects generally take more than 15 minutes to solve. Therefore, we will confine ourselves to real objects, which are always understood to be upright. This last bit is redundant because, as we will see shortly, the judgment that something is upright or inverted is based on the orientation of the object.

The figure below shows the basic setup for a ray-tracing diagram designed to determine the image formed by an object placed near a lens or mirror.

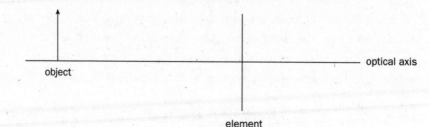

No rays are shown, but this is typically how you want to begin ray tracing for a lens or mirror. For a prism or slab, the diagrams are much easier; we'll discuss them in detail when we talk about refraction. In the figure above, an optical axis passes through the center of the optical element and is perpendicular to the element's face or faces. By convention, the object is placed to the left of the optical element pointing straight up. The bottom of the object is placed on the optical axis. When doing this sort of tracing, we assume that ambient light is shining on the object from all directions, which means that the very top of the object has reflected light coming off of it in all directions. We are thus free to pick any rays that might be convenient. We use an arrow for the object because we need some asymmetry to determine the orientation of the image.

It is good practice to draw this kind of bare-bones sketch first when you are told to make a ray-tracing diagram. It is also good practice to use a straightedge for these diagrams. If you forget to bring a ruler to the AP Physics exam, a spare pencil or a piece of folded paper could both work quite well as straightedges.

To determine the image's properties, you must first successfully complete the ray tracing. If the light rays that form the image actually cross, then the image is real, and it can be projected onto a screen. If the light rays only appear to cross after they are traced backward, the image is virtual and cannot be projected onto a screen. If the image points the same direction as the object, then the image is upright. If the image points in a direction opposite that of the object, then the image is inverted. You can also distinguish upright from inverted by examining the image's relationship to the optical axis: If the image is on the same side of the optical axis as the object, the image is upright. If not, it is inverted. A magnified image is larger than the object, and a reduced image is smaller. Regardless of whether the image is magnified or reduced, the ratio of the image height to the object height is called the magnification, M.

$$M = \frac{h_i}{h_o}$$

It can be shown that this equation is equivalent to

$$M = -\frac{s_i}{s_o}$$

On the free-response section of the exam, if you are dealing with optics, h refers to height, s refers to distance along the optical axis from the element, i as a subscript refers to the image, and o as a subscript refers to the object. Thus, s_o is the object's distance from the optical element.

Another important convention that applies here regards the mathematical signs on distances. Real objects always have positive distances. If an image is real, it has a positive distance. If an image is virtual, it has a negative distance. You can avoid a great deal of confusion by committing this convention to memory.

REFLECTION AND REFRACTION

When a ray of light encounters the boundary between two different materials, it may be reflected or refracted. In reality, part of the ray is always reflected and part is always refracted, but we are only concerned with what happens to the brighter portion. Reflection can be thought of as light bouncing off a surface, while refraction bends the ray as it enters the new material. There is one law of reflection and one law of refraction. The law of reflection states that the angle of incidence equals the angle of reflection. The figure below shows light incident on a reflecting surface. The incident angle, θ_i, is equal to the reflected angle, θ_r. By convention, the important angle here is the angle between the ray of light and the line perpendicular (normal) to the surface. This convention is consistent with the law of refraction.

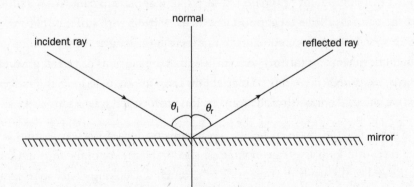

AP EXPERT TIP

It's always a good idea to draw a normal line to the surface first, if one has not already been provided.

Generally, we classify reflection as specular or diffuse. Under specular reflection, parallel incident rays result in parallel reflected rays. Specular reflection is what happens with typical bathroom mirrors. Under diffuse reflection, parallel incident rays do not result in parallel reflected rays. Diffuse reflection is what happens when light shines on any surface that is not flat and shiny. We depict a flat surface when we draw specular reflection, while a surface that creates diffuse reflection is drawn rough. The following figure shows the two reflections: specular reflection (a) and diffuse reflection (b).

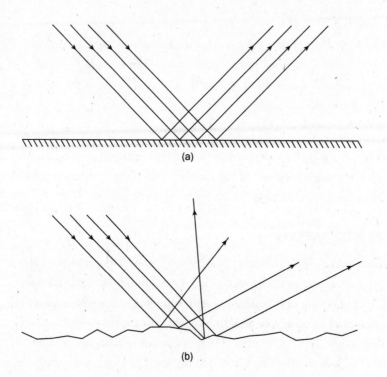

(a)

(b)

MIRRORS

A **mirror** is an object designed to reflect light in a particular way. **Plane mirrors** are intended to cause specular reflection over their entire surfaces. **Curved mirrors** are designed so that incoming parallel light is reflected to a single point or is reflected such that it appears to come from a single point. In either case, the point is called a **focal point** or a focus. Mirrors with surfaces that are portions of a sphere, or spherical mirrors, approximate this focusing as long as the ray is not too large in diameter. Parabolic mirrors (or mirrors with surfaces that are portions of a parabola rotated about its symmetry axis) can *truly* achieve this no matter how large the ray is, as long as it strikes the surface. For now, we will deal only with plane mirrors, leaving focusing mirrors for later.

It is usually simple to determine the image created by a plane mirror. The easiest way to do this is to imagine viewing the world so that you are looking parallel to the surface of the mirror. One side of the mirror is the real world, while the other side is "through the looking glass," so to speak.

Suppose we have an arrow like the one shown in the following figure. What will its reflection look like in the mirror shown? The easiest way to find out is to fold the paper along the mirror and trace the object.

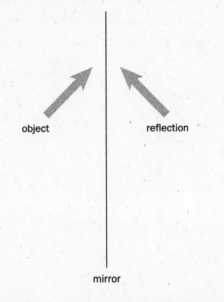

For a plane mirror, the image distance is always equal to the object distance. Further, the image produced by a plane mirror is always virtual and upright. Plane mirrors are the easiest optical elements to deal with, but they can present some difficult problems when it comes to certain shapes. The way around these problems is to use the fact that $s_i = s_o$ and to locate the images of key points in the object. The figure below gives an example of this.

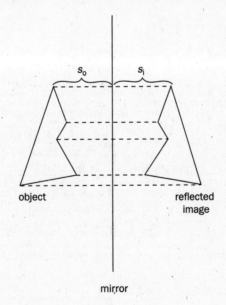

The **law of refraction**, often called Snell's law, is slightly more complicated than the law of reflection. Recall that light travels at different speeds in different materials. The figure below shows light traveling in air, incident on a glass surface.

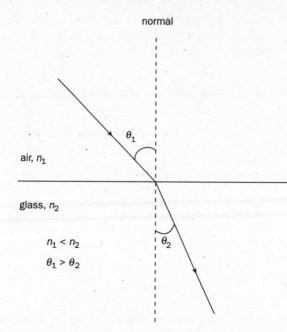

When the light enters the glass, it bends toward a line drawn normal to the interface. The exact relationship between the angle of incidence and the angle of refraction is Snell's law, $n_1 \sin \theta_1 = n_2 \sin \theta_2$. The subscript 1 refers to the first medium (air in this case), and the subscript 2 refers to the second medium (glass in this case). Remember, the angles are measured from the normal to the ray.

From Snell's law, you can see that if $n_1 < n_2$, then $\theta_1 > \theta_2$, and vice versa. This means that if light goes from a less optically dense medium to a more optically dense medium, it bends *toward* the normal. If the light moves from a more optically dense medium to a less optically dense medium, it bends *away* from the normal, which leads to an interesting phenomenon. Suppose light is shining on an air-glass interface from the glass side. Now imagine the angle of incidence is increased. As θ_1 gets larger, θ_2 gets larger at a faster rate. For some θ_1, θ_2 will equal 90 degrees. This means the light is refracted parallel to the interface. If the incident angle increases by a small amount beyond this critical angle, then the law of reflection takes over. This is called **total internal reflection** and is the basis for all of fiberoptic communication. The angle at which total internal reflection first occurs is called the **critical angle**. It is given by using Snell's law and setting θ_2 equal to 90 degrees. The result is

$$\theta_{\text{critcal}} = \sin^{-1} \frac{n_2}{n_1}$$

For this equation to be valid, $n_1 > n_2$ because the range of the inverse sine function is −1 to 1. Any incident angle greater than the critical angle results in total internal reflection.

With flat surfaces, refraction will not produce focusing. However, with curved surfaces, we can make parallel light rays focus to a point or spread out as if they are coming from a point, just as we can do with mirrors. Refraction has two more interesting applications as well: prisms and the apparent depth of objects in water.

Prisms are generally transparent triangular slabs. The angles of the triangle and the index of refraction, along with the incident direction of the ray, determine how a light ray will behave through the prism. Figure (a) below shows a typical prism problem setup. Figure (b) includes the angles you typically have to determine to solve for what is called the angle of deviation.

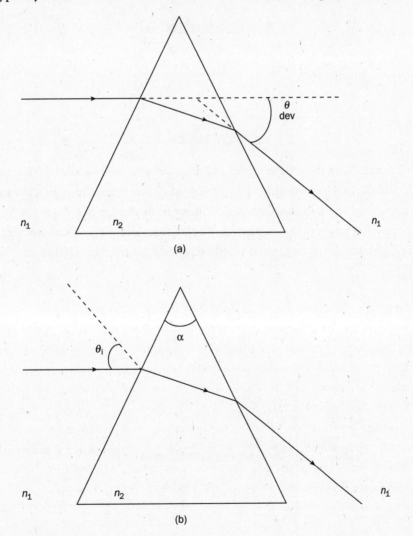

Prism problems require a good deal of geometry. Fortunately, it is comprised mostly of triangle and parallel line relationships, along with Snell's law.

The apparent depth problem takes on many guises. The basic idea is that some object, such as a quarter, is at the bottom of a pool of depth d. You are looking straight down on the quarter from a height h above the surface of the water. Solving this problem requires you to determine how deep in the pool the quarter appears to be. The figure below sets up the apparent depth problem.

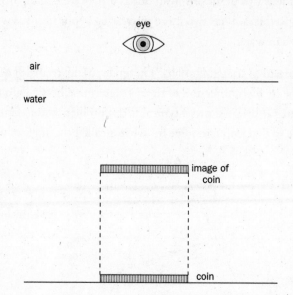

The following figure shows the various angles and distances you need to solve the apparent depth problem. The solution has two key steps. First, you need to determine the angle a refracted ray makes with the normal when the incident ray is coming from an edge of the quarter. Second, you need to realize that the image of the quarter must be the same size as the actual quarter. The first step is based on physics; the second step is based on how the human brain judges distance.

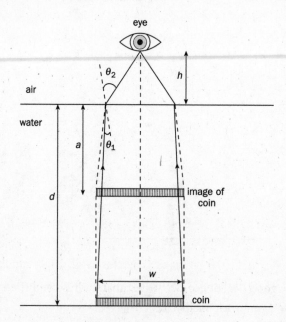

Reflection and refraction have broad applications with respect to flat surfaces, but optics really becomes interesting when we begin to talk about curved surfaces. We will confine our discussion here to simple spherical surfaces, which necessitates the use of the paraxial approximation. The paraxial approximation requires that all incoming rays be near the optical axis and also that they be nearly parallel to the axis. You should strive to remain aware of this approximation, even though we will exaggerate the physical situation so that this approximation is not apparent when we create ray-tracing diagrams. We will treat mirrors first and then move on to lenses. The discussion of mirrors will include a short section on spherical aberration, a defect commonly seen when looking at a shiny mirrored ball.

CONVERGING SPHERICAL MIRRORS

Converging spherical mirrors are also called concave mirrors; you can think of them as being part of the curved interior surface of a ball. Any spherical mirror, whether converging or diverging, has two important lengths: the radius of curvature and the focal length. The radius of curvature, R, is the radius of the sphere that the mirror is taken from. The focal length, f, is the distance from the reflecting surface to the point where parallel rays converge. There is a simple relationship between R and f under the paraxial approximation:

$$f = \frac{R}{2}$$

Because R and f are both lengths, they have units of length, commonly centimeters. The figure below shows the ray-tracing diagram for parallel rays incident on a converging mirror (notice that they cross at the focus). In the diagram, c represents the center of curvature of the mirror; the distance from c to the mirror is equal to R.

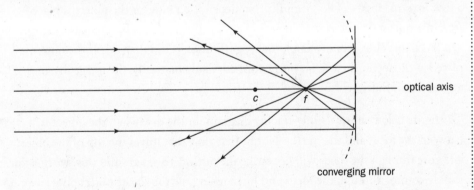

converging mirror

If you place a point source of light at the focus, the rays reflected off the mirror will be parallel to the optical axis. This is because the angles of incidence and reflection must be the same for any incident/reflected pair. Therefore, if the light suddenly reverses direction, it should just backtrack. Note that for converging mirrors, the focal length is a positive number.

To determine the location, orientation, and size of the image of an object placed near the mirror, you need to make a ray-tracing diagram. You may choose any rays you like to complete the diagram, but three are particularly useful. All three originate at the top of the object. The first goes into the mirror parallel to the optical axis, then out through the focus. The second goes into the mirror through the focus and comes out parallel. The third goes into the intersection of the mirror and the optical axis; it comes out at the same angle relative to the optical axis as the incident ray. The intersection of the three reflected rays marks the location of the top of the image of the object. The base of the object is on the optical axis, so you just draw an arrow perpendicular to the axis out to the intersection point. The figure below illustrates this technique. Some texts and teachers prefer to draw the mirror as a curved surface; others draw it as a straight line. Here we opt for the latter, because under the paraxial approximation the mirror is very nearly flat.

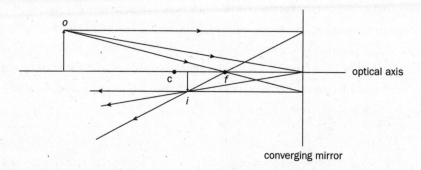

By making a rough ray diagram, you can extract qualitative information such as the relative location of the image compared to the object. In the above figure, the object distance is greater than the focal length. In this case, a converging mirror produces a real, inverted image. The magnification depends on how far the object is placed from the mirror. If s_o is greater than R, the magnification is between 0 and -1. If s_o is less than R but greater than f, the magnification is less than -1. If the object is placed at the focus, there is no image formed; mathematically, the image distance approaches infinity.

The figure below shows the ray-tracing diagram for the case when the object is between the focus and the mirror. Notice that a ray that leaves the top of the object and goes through the focus will not strike the mirror. To get around this, we imagine the ray coming through the focus and just missing the top of the object on its way to the mirror. Notice, too, that the reflected rays do not actually cross. The reflected rays must be traced backward from the mirror until we find the point where it appears that they have crossed; this is a virtual intersection, because it does not really happen. This virtual intersection is the point where the image of the top of the object forms. The image is virtual and upright in this case.

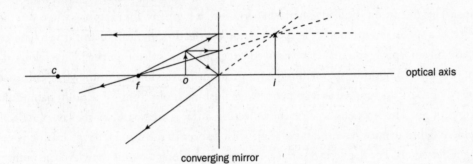

converging mirror

There is a simple equation that relates focal length to the object and image distances. Here we just state the formula, called the mirror equation when discussing mirrors or the thin lens equation when dealing with lenses, without proof:

$$\frac{1}{f} = \frac{1}{s_o} + \frac{1}{s_i}$$

This one equation applies to all of the curved mirrors and lenses used on the AP Physics exam.

DIVERGING SPHERICAL MIRRORS

A diverging spherical mirror is a convex mirror (i.e., it looks like part of the outside surface of a sphere). If parallel rays strike a diverging mirror, they reflect out as if they come from a single point; this point is behind the mirror and it is the focus. The focal length for a diverging mirror is negative. There is only one ray-tracing diagram for a diverging mirror because there is no way to place an object between the mirror and the focus, or even to place an object so that the focus is between it and the mirror. The image is always virtual.

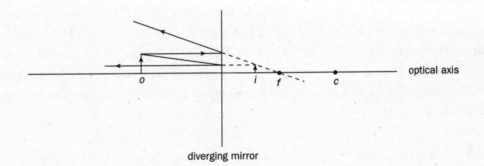

diverging mirror

Notice how the same three rays are drawn from the top of the object. The rays never really cross, so we must project the reflected rays backward to find the image. This means that a diverging mirror always produces a virtual, upright image. The mirror equation still applies to this system.

SPHERICAL ABERRATION

In a spherical mirror, there is noticeable distortion in the image if the incident light covers too much area. This distortion is called spherical aberration; it happens, for example, when you look at the back of a spoon, even though the spoon is not a perfectly spherical mirror. If you have ever walked around a hospital, you may have noticed large, spherical mirrors at the intersections of different hallways. In both the spoon and the hospital mirror, the image is somewhat "fish-eyed" (i.e., bowed outward and curved as if someone were pushing it out from some point behind the reflecting surface). This happens because the reflecting surfaces are bound by the law of reflection. Any incident light ray must have the same angle with respect to the normal to the mirror as the resulting reflected ray. For a spherical surface, the normal is along a radius drawn to the point the ray strikes the mirror. If you look at the following figure, you can see what happens when three parallel rays are reflected off the same surface.

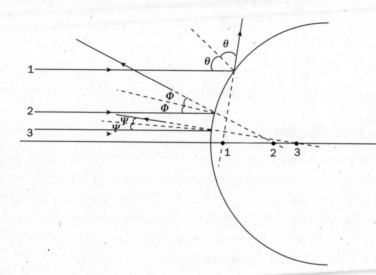

The intersection of the traced-back reflected ray and the optical axis defines the focal point for that ray. As the ray moves farther off the optical axis, the focal point moves toward the surface of the mirror. This means that each ray has its own focal length. Only when the rays are very close to the optical axis and nearly parallel with it do we get a reasonably well-defined focal length. This information should help drive home the paraxial assumption. Any deviation from this will result in a distortion of the image.

LENSES

We now turn our attention to lenses. A lens is an object that focuses parallel rays of light onto one point or defocuses parallel rays of light such that they appear to come from one point. The lenses that focus are called converging, or positive, lenses. The lenses that defocus are called diverging, or negative, lenses. The main difference between lenses and mirrors is that lenses exploit refraction to focus (or defocus) light, while mirrors exploit reflection. A second difference is that a lens gives us two surfaces (or faces) to contend with, each of which may be different, while a mirror has just one surface. We will narrow the scope of our discussion to lenses that can be treated as a single surface, or thin lenses.

CONVERGING LENSES

Converging, or convex, lenses have the property that they focus parallel light down to a single point. These lenses are usually thicker in the middle than at the edge. Like a converging mirror, a converging lens has a focal length. Unlike a converging mirror, a lens has two focal points. One focus is placed on either side of the lens on the optical axis. For simple lenses, the lens is halfway between the two focal points.

If you want to know the properties of an image produced by a converging lens, you can simply construct a ray-tracing diagram. The following figure shows one such diagram with the object placed beyond the focus.

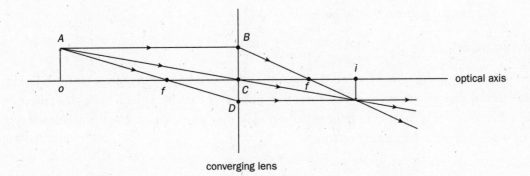

converging lens

We use the same three incident rays as we used for mirrors. This time, though, we must take care to note which ray goes with which focus. From the figure, ray *AB* passes through the lens and is refracted through the opposite focus. Ray *AC* passes through the center of the lens and continues on undeviated. Ray *AD* passes through the near focus and is refracted by the lens to parallel to the optical axis. The intersection of the three refracted rays is the location of the top of the image of the object. Note that in this case, the image is inverted and the refracted rays actually cross, which means that the image is real.

When constructing ray-tracing diagrams for lenses, remember that each focus gets exactly one ray through it. One focus gets an incident ray, and the other focus gets a refracted ray. For positive lenses, the near focus gets the incident ray, and the far, or opposite, focus gets the refracted ray. Recalling the method for a converging mirror with the object between the focus and the mirror, we can determine what happens with a converging lens under similar conditions. The following figure shows the result.

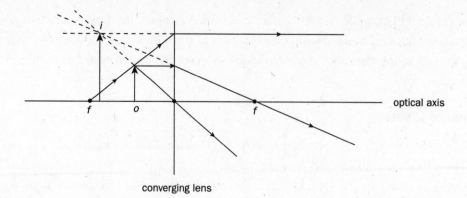

converging lens

The image is upright and virtual. These are the conditions for a simple magnifying glass; converging lenses are also commonly used to correct the visual defect known as farsightedness.

Whenever an image is real, it is a positive distance from the element; when an image is virtual, it is a negative distance from the element. For converging lenses, then, if the object distance is greater than the focal length, the image distance will be positive. If the object distance is less than the focal length, the image distance will be negative.

DIVERGING LENSES

A diverging lens is also sometimes called a negative lens. This is because the roles of the two foci in a ray-tracing diagram are reversed from those they play in the converging lens. For a glass lens in air, a diverging lens is concave, or thinner in the middle than on the edges. The figure below shows a ray-tracing diagram for a diverging lens.

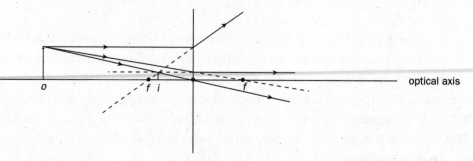

diverging lens

Again, we use the same three incident rays. This time the incident ray parallel to the optical axis refracts out as if it came from the near focus. The ray that goes through the center of the lens continues on undeviated, which leaves the far focus as the target for the third ray. When this ray passes through the lens, it refracts out parallel to the optical axis. Diverging lenses always produce virtual, upright images; they are commonly used to correct the visual defect known as nearsightedness.

PHYSICAL OPTICS

As stated earlier, physical optics treats light as a wave. Under this assumption, we could rederive all of geometric optics, but at this level that would not be especially productive. Instead, we will focus on basic phenomena that are not describable using geometric optics. We'll begin with interference and diffraction, a discussion that will mirror some of what is said in Chapter 10 but that will also note some peculiarities concerning light. We'll also cover the problem of multislit diffraction and the analysis and interpretation of interference patterns. Particular attention will be paid to the use of this machinery to make precise measurements of the track spacing on optical storage media. This section will end with a discussion of dispersion and the electromagnetic spectrum.

Because light is sometimes considered to be a wave, it must have wavelike properties. The most important are the wave speed and wavelength of visible light. As was stated earlier, the speed of light in a vacuum is approximately 3×10^8 m/s. In other media, it is slower by a factor of $\frac{1}{n}$, where n is the index of refraction. Visible light falls into a narrow band of the electromagnetic spectrum; the range for the wavelengths of visible light is about 400–700 nanometers (nm). Visible light's frequency of oscillation is on the order of 10^{14} Hz.

> **AP EXPERT TIP**
>
> Memorizing these wavelength and frequency values will help you to decide if a calculation has yielded a sensible result.

INTERFERENCE AND DIFFRACTION

We know that when two waves occupy the same place in a medium at the same time, they will interfere. This is true for light, with one caveat: light waves of different frequencies do not interact with each other. For example, red light and blue light will not interfere. This fact of nature is exploited in fiberoptic communications lines, where different data streams are sent via different frequencies of light and the signals are thus kept separate. A copper wire can carry only one data stream at a time.

We measure the effect of light interference by looking at the brightness of the light. A bright spot indicates constructive interference, and a dark spot indicates destructive interference. Constructive interference occurs when the waves are in phase (i.e., when the crests and troughs of the two waves line up crest-to-crest and trough-to-trough). Destructive interference occurs when the waves are out of phase, lining up crest-to-trough. We can also describe destructive interference by saying that the waves are shifted half a wavelength relative to each other. The following figure should help refresh your memory on the effects of constructive (a) and destructive (b) interference.

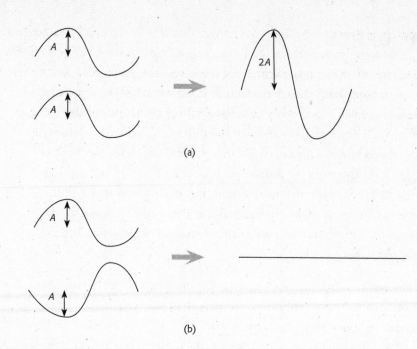

(a)

(b)

When light waves come to a sharp edge, they bend around that edge. You can see this if you stand a razor blade upright and shine a laser on its cutting edge. If you look at the shadow cast by the razor on a smooth surface, you will see the laser light spilling over. This is the diffraction, ultimately a result of Huygens's principle, which states that each point on a wave front acts as a source of spherical waves. The interference from this infinity of spherical waves creates the next wave front. If you use a thin enough edge to stop part of the wave, the end of the wave generates a diffracted wave. The following figure shows the net result of the diffraction of a light wave near a sharp edge.

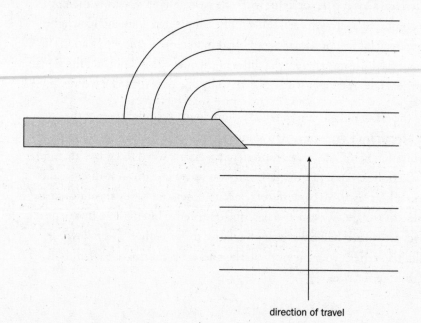

direction of travel

If you shine light on an opaque screen that has a small slit cut in it, there will be a diffraction pattern set up. If the light is a well-collimated, monochromatic beam (e.g., a laser), the beam will broaden perpendicular to the long axis of the slit. When we draw a picture of this, we typically do it as shown in the figure below, which depicts a diffraction pattern with an intensity plot. The light is brightest in the center and gets dimmer as you move away in either direction.

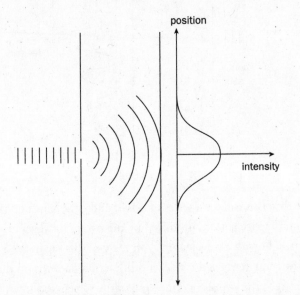

The light shines from the left onto the screen with the slit. A second screen, with no slit, is used to display the resulting pattern. Along the right edge of the second screen, we plot intensity (brightness) as a function of position; intensity increases toward the right. The diffracted light is brightest directly opposite the slit.

Single-slit diffraction is an interesting phenomenon, but the mathematics of determining the way that diffracted light intensity varies along the second screen is beyond the scope of this course. If we introduce more than one slit, though, we are able to determine a good deal of information via geometry and our rudimentary knowledge of how light interferes.

MULTISLIT DIFFRACTION

The following figure shows the setup for two-slit diffraction, originally performed by Thomas Young in the early 1800's. The intensity plot shows the bright and dark bands you would see. We will use this arrangement to determine all that we can about the phenomenon. We will then generalize our results to multiple slits that are evenly spaced, which is useful in the CD/DVD diffraction experiment mentioned in the introduction to this chapter.

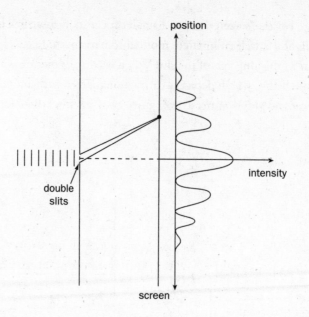

In an experiment like this, you will see a series of bright and dark bands on the screen. These bands are the result of the interference pattern produced by the two light waves originating from the two slits. Suppose we shine light with a wavelength λ on the two slits separated by a distance d. Assume that the slits are narrow enough to produce decent diffraction patterns on their own. If we imagine a line that is halfway between the two slits connecting to the screen, as shown, the intersection of this line with the screen will be $x = 0$. Where will the bright and dark bands occur? That is, we want to know the position of each bright and dark band on the screen at a distance L from the slits.

We can see what happens in the figure below.

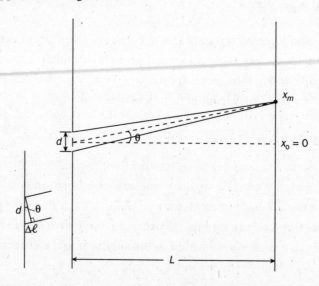

The point x_m is the location of the mth bright band, taking the central bright band to be at x_0. As drawn, the path that the upper wave must take is somewhat shorter than the path that the lower wave takes to reach x_m. For there to be a bright band, this path difference must be a whole number of wavelengths. Calling the path difference Δl, we have

$$\Delta l = m\lambda$$

The inset in the above figure shows that the path difference is also given by

$$\Delta l = d \sin \theta$$

Equating these two, we arrive at

$$m\lambda = d \sin \theta$$

This is an exact expression for the interference pattern set up by two slits.

If the apparatus is such that L is dramatically greater than d, then the angle θ is approximately the angle between the perpendicular to $x = 0$ and the line joining the point halfway between the slits and x_m. We can measure L and x_m, so we know

$$\tan \theta \approx \frac{x_m}{L}$$

Tangent and sine are approximately the same for small angles (which is what we assume), so after a tiny bit of algebra, we arrive at

$$x_m \approx \frac{m\lambda L}{d}$$

This formula can be rearranged to find the spacing between the slits, the wavelength of the light, or even the distance between the screens. All we have to supply is the value of m, which is called the order number. We determine m by starting with 0 at the central maximum. Each bright spot is numbered sequentially as we move away from the central maximum. If, for example, you are asked to find the location of the fourth-order maximum, you set $m = 4$ and plug in all the other required information.

These formulas apply even if there are many slits. The physical effect of additional slits is to increase the separation between adjacent bright bands, which is related to the constraints on placing many slits in the path of the light. The slits must be quite close together, and you can see that the location of the mth maximum is inversely proportional to the slit spacing. You should also note the dependence on wavelength. It is this wavelength dependence that accounts for the rainbow seen on the data sides of CDs and DVDs.

If you want to measure the track spacing on a CD, all you need to do is shine a laser of known wavelength perpendicular to the data side. The reflected light will give a diffraction pattern on the back wall. You can determine your angle by using the tangent function and then plug this information back into the equation:

$$m\lambda = d \sin \theta$$

What is interesting here is that the ratio of track spacing on a CD to that on a DVD is the square root of the ratio of the storage capacity of the DVD to that of the CD. This is related to the fact that the information is stored on the area of the disk.

DISPERSION AND THE ELECTROMAGNETIC SPECTRUM

When we stated that light travels more slowly in any medium than it does in a vacuum, we glossed over an important point: Light of different frequencies travels at different speeds in dense media. In other words, each color of light has its own index of refraction. In general, blue light travels more slowly in dense media than red light does. We will not attempt to explain why this is, nor will we try to determine the various indices of refraction for the various colors in a given material. We will simply avail ourselves of the rule as given and see what we predict.

The first thing we see is that this rule means that blue light must have a higher index of refraction than red light: $n_{blue} > n_{red}$. By Snell's law, this means that blue light tends to refract more than red light does. This explains why white light (a mixture of all of the colors of light) incident on a prism produces a rainbow, as shown in the following figure.

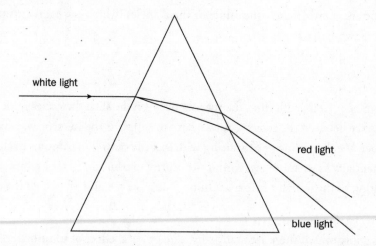

The light that we can see is only a narrow band of the entire electromagnetic spectrum. The sun emits light all across the electromagnetic spectrum, but we only see the small window of wavelengths between 400 and 700 nm. Maxwell showed light to be an electromagnetic wave. All objects that have a temperature greater than 0 K give off light in the sense that they emit radiation; the hotter the object, the "bluer" the light. This fact is exploited through the use of infrared cameras. These cameras are sensitive to light with wavelengths longer than we can see. Ultraviolet cameras detect light with wavelengths shorter than we can see.

REVIEW QUESTIONS

1. The frequency of a particular laser is 4.5×10^{14} Hz. If it enters glass with an index of refraction of $n = 2.5$, which of the following is the best estimate for the speed of the light in the glass?

   (A) 7.5×10^8 m/s

   (B) 3.0×10^8 m/s

   (C) 2.5×10^8 m/s

   (D) 1.2×10^8 m/s

   (E) 1.0×10^8 m/s

2. Light is incident on a water-air boundary from the water side. If the index of refraction for water is 1.33 and that of air is 1.00, which of the following is true?

   (A) $\theta_1 > \theta_2$

   (B) $\theta_1 < \theta_2$

   (C) $\theta_1 = \theta_2$

   (D) $\theta_1 \leq \theta_2$

   (E) $\theta_1 \geq \theta_2$

3. Determine the critical angle for light crossing the boundary from a block of transparent plastic ($n = 2$) into air ($n = 1$).

   (A) 0

   (B) $\dfrac{\pi}{2}$

   (C) $\dfrac{\pi}{4}$

   (D) $\dfrac{\pi}{6}$

   (E) $\dfrac{\pi}{8}$

4. In the following figure, the index of refraction for the block increases smoothly as you move from left to right. Which of the paths will the incident light ray take through the block?

   (A) A

   (B) B

   (C) C

   (D) D

   (E) E

Use the following figure to answer question 4.

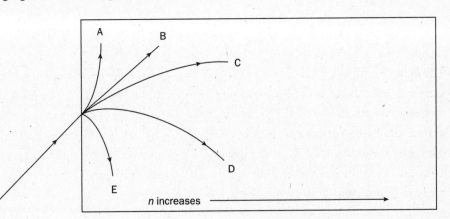

n increases

5. An object is placed near a thin converging lens such that $s_o < f$. Which of the following statements will always be true?

   I. $|s_i| < |s_o|$

   II. $|s_i| > |f|$

   III. $s_i < 0$

   (A) I only
   (B) III only
   (C) I and II only
   (D) I and III only
   (E) I, II, and III

6. A concave spherical mirror has a radius of curvature of 16 cm. What is the focal length of this mirror?

   (A) 64 cm
   (B) 32 cm
   (C) 16 cm
   (D) 8 cm
   (E) 4 cm

7. In each of the following figures, the arrow on the left is reflected in a plane mirror. Which of the following represents the actual reflection?

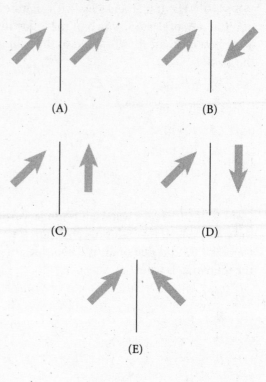

Use the following figure to answer question 8.

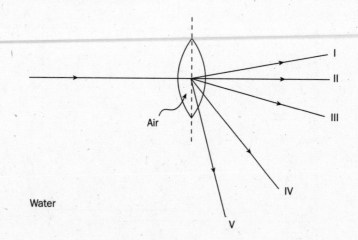

8. In the previous figure, which path represents a paraxial ray (a ray that is close to and nearly parallel with the optical axis) incident on a convex air lens under water?

(A) I

(B) II

(C) III

(D) IV

(E) V

9. An object is placed near a converging lens such that $s_o > f$. What can be said about the image?

(A) The image is real and upright.

(B) The image is virtual and upright.

(C) The image is real and inverted.

(D) The image is virtual and inverted.

(E) It is impossible to say anything important about the image based on the given information.

10. Which of the following is a possible expression for λ, the wavelength of the light used, in a two-slit diffraction experiment where x_m is the location of the mth maxima, d is the spacing between the slits, and L is the distance between the slits and the screen?

(A) $\lambda = mx_mLd$

(B) $\lambda = \dfrac{mLd}{x_m}$

(C) $\lambda = \dfrac{mx_m}{Ld}$

(D) $\lambda = \dfrac{x_md}{mL}$

(E) $\lambda = \dfrac{x_mL}{md}$

FREE-RESPONSE QUESTION

The figure below shows a two-slit diffraction experiment. Assume that you know d; your goal is to measure the wavelength of light used.

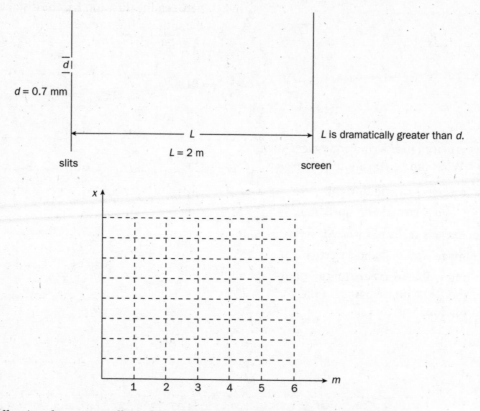

The following data were collected regarding the locations of the various maxima.

Order of maximum m	Position of maximum x_m in mm
1	1.7
2	3.4
3	5.1
4	6.9
5	8.6
6	10.3

(a) On the grid provided, plot the data given.

(b) Show how to extract the value of λ from the slope of the graph.

(c) Using the highest and lowest data points and the method shown in (b), calculate the value of λ.

(d) How would decreasing d affect the graph?

(e) Identify possible sources of error, and explain how you might minimize those errors.

ANSWERS AND EXPLANATIONS

1. D

The definition of the index of refraction is $n = \dfrac{c}{v}$, which means $v = \dfrac{c}{n}$. You should know that the speed must be less than 3.0×10^8 m/s. Writing n as a fraction $\left(\dfrac{5}{2}\right)$ makes performing the division quite easy. The result is a speed of 1.2×10^8 m/s. You should know the value for the speed of light in vacuum, but it is also on the constants sheet provided for the multiple-choice portion of the AP Physics exam.

2. B

By Snell's law, we have $n_1 \sin\theta_1 = n_2 \sin\theta_2$. Because $n_1 > n_2$, we know that $\sin\theta_1 < \sin\theta_2$. Since the angles for refraction are always less than or equal to $90°$, we can state with confidence that $\theta_1 < \theta_2$.

3. D

The critical angle is calculated when the refracted angle is $90°$, or $\dfrac{\pi}{2}$ rad. The sine of this angle is 1. Therefore, $n_1 \sin\theta_1 = n_2$. For this problem, $n_2 = 2$ and $n_1 = 1$. This means $\sin\theta_1 = \dfrac{1}{2}$. The angle that satisfies this is the critical angle. Numerically, it is 30 degrees, or $\dfrac{\pi}{6}$ radians.

4. C

Because the index of refraction is increasing, the light will bend toward the normal. This fact rules out choices (A) and (B). Once the light is going parallel to the long edges of the block, it is normally incident on each new index of refraction. With normal incidence, there is no refraction, which excludes choices (D) and (E). Choice (C), by process of elimination, is the correct answer.

5. B

When an object is between a converging lens and its focus, the image will be virtual. We know that statement III must always be true. We want to test statement I first, because it is paired with statement III in choices (D) and (E). If statement I is ever false, we have finished the problem. The easiest way to find this answer is to make a quick ray-tracing diagram. Place a small object near the focus and a larger object near the lens. Complete the diagram, and you will see that the small object produces an image well beyond the focus. The large object, though, produces an image very close to the lens. This makes statement I false, which leaves choice (B) as the correct answer.

6. D

The relationship between focal length and radius of curvature for a spherical mirror is $f = \dfrac{R}{2}$. Using this formula, we see that the focal length is 8 cm.

7. E

Remember that for a plane mirror, image distance equals object distance. Corners on the object near the surface of the mirror must lead to corners on the image that are near the surface of the mirror. You could also fold the paper along the mirror line and see which images overlap.

8. A

A convex lens is usually a converging lens. This is true only if the lens is made of a material that is more optically dense than its surroundings. A convex lens that is less dense than its surroundings will be diverging; a concave lens that is less dense than its surroundings will be converging. You should know that air is less optically dense than water.

9. C

By now you should know that if only one thin lens is used, real images are always inverted, and virtual images are always upright. All you have to recall, then, is what happens when an object is farther from a converging lens than the focal length: The image will be real and inverted.

10. D

The key here is first to eliminate choices based on the dimensions of each side of the equations. Wavelength is in meters. Anything on the right side that is not in meters can therefore be eliminated, which gets rid of choice (A), right side in cubic meters, and choice (C), right side in 1/meter. Now you need to rely on what you know about the physics of the situation. If the distance between the slits decreases, the separation between adjacent maxima increases. Therefore, x_m and d are inversely related. This means their product should show up in the equation. Because choice (D) is the only choice with this product, it must be the correct answer.

FREE-RESPONSE QUESTION

(a)

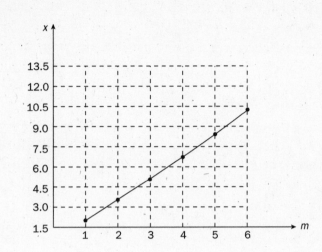

(b) Since the axes of the plot are (m, position) in (x,y), and the graph is linear, the wavelength must be related to the slope. From

$$x_m \approx \frac{m\lambda L}{d}$$

we can see that the y-axis is x_m and the x-axis is m. Thus, the slope of the graph is given by slope $= s = \dfrac{\lambda L}{d}$.

Solving for λ, we get

$$\lambda = \frac{sd}{L}$$

(c) The slope can be estimated by rise/run. Taking the two extreme points, we have $\dfrac{(10.3 - 1.7)}{(6 - 1)}$, which is about 1.72 mm. Converting to meters and placing this in the expression above gives us $\lambda = 602$ nm.

(d) If the distance between the slits is decreased, the slope of the graph will increase.

(e) The largest error comes from measuring the positions of the maxima. To decrease the effect of this error, we could increase L. Any increase in L will cause a corresponding increase in x_m. As x_m gets larger, the error introduced by the measuring device becomes less important.

CHAPTER 12: ATOMIC AND NUCLEAR PHYSICS

IF YOU MEMORIZE ONLY FOUR EQUATIONS
IN THIS CHAPTER . . .

$$\Delta E = (\Delta m)c^2$$

$$E = hf = pc$$

$$K_{max} = hf - \phi$$

$$\lambda = \frac{h}{p}$$

NOTE: This topic appears *only* on the B exam.

INTRODUCTION

Atomic and nuclear physics encompass a wide range of phenomena that were first observed only in the very late 19th and early 20th century. When they were first observed, these phenomena seemed so extraordinary that many physicists refused to believe that they were real. The work done by Planck, Einstein, Bohr, de Broglie, and others transformed our perceptions of the physical world.

Seemingly contradictory ideas were eventually accepted as the true nature of light. But when a similarly contradictory description of electrons (as both particle and wave) was offered, it was considered almost laughable by many physicists. The discovery that electrons do indeed "interfere" in the same way that waves do radically changed the way that physicists thought about electrons and provided a theoretical basis for the various types of electron microscopes currently used to "see" atoms in a solid.

This is merely one of the topics covered in this chapter. The physics presented here is the starting point for understanding devices such as lasers, GPS tracking, nuclear power, and nuclear weapons. The goal of this chapter is to provide a solid review of the atomic and nuclear physics needed to deal with the questions on the AP Physics B exam that cover these topics. Approximately 10 percent of the exam will be on these subjects; these questions may be in the multiple-choice part of the exam, the free-response portion, or both.

A FEW PRELIMINARIES

A good understanding of these topics requires that you build a unified picture of them in your mind. To help you achieve this goal, we will start here with a bit of background in modern physics, an area that typically encompasses special relativity and simple quantum mechanics.

A BIT OF SPECIAL RELATIVITY

You may have studied the law of conservation of mass in chemistry class. One of the rather odd results that arise in the field of special relativity is that the conservation of mass is not really true. It turns out that matter is merely a type of energy that should be included in the more accurately named conservation of energy law.

Einstein found that the total energy of a particle of mass m and momentum p is given by

$$E^2 = p^2c^2 + (mc^2)^2$$

where c is the **speed of light in vacuum** and has a value of approximately of $c = 3 \times 10^8$ m/s. This formula implies that even a stationary particle ($p = 0$) has some non-zero energy that is proportional to its mass. This concept is known as **mass-energy equivalence**.

Matter is a thus a form of energy, and like other types of energy, it can be converted into different forms. The amount of energy that is available from a given change in the amount of mass is given by Einstein's famous equation:

$$\Delta E = (\Delta m)c^2$$

This means that if 1 mg ($=10^{-6}$ kg) were converted to the equivalent amount of energy, it would release an energy of

$$\Delta E = 10^{-6}\,\text{kg} \times (3 \times 10^8\,\text{m/s})^2 = 3 \times 10^{10}\,\text{J}$$

You can see that this is an enormous amount; the productivity of this conversion is the reason that nuclear power is an appealing energy source.

QUANTIZATION OF ENERGY AND PHOTONS

One of the major forces driving the scientific method is the observation of unexpected phenomena and the desire to make sense of them. In the early part of the 20th century, the success of Maxwell's equations led most scientists to believe that light was a type of electromagnetic wave or radiation. In the wave description, the energy of the radiation is related to the square of the amplitude of the wave; it can take on any arbitrary value for any arbitrary frequency of light.

Researchers discovered that this theory of light had problematic implications, however. Specifically, it suggested that a heated body would give off infinite amounts of electromagnetic radiation (or E-M radiation) in the ultraviolet frequency range. Because this was certainly not occurring in real life, it became obvious that the wave description of light needed to be altered or possibly abandoned altogether.

Max Planck discovered that he could make this problem go away by assuming that the radiation could *not* be emitted in arbitrary amounts at any frequency. His idea was to restrict the emitted radiation for any given frequency to individual packets having identical energy, which would depend on the frequency. That is to say, the energy is **quantized**. He called each packet a **quantum** of energy (the plural is *quanta*) and found a relationship that led to results that were exactly in line with experimental data. This relationship is

$$E = hf$$

In this formula, E is the energy in a single packet or quantum of energy, f is the frequency of the radiation/wave, and h is an experimentally determined constant now known as **Planck's constant**. It has a value of $h = 6.6 \times 10^{-34}$ J·s.

The concept of quantization is also found in electrostatics. There is, after all, a well-defined bundle size of electric charge; charges are never observed to exist in any other amount. This charge quantum is called the elementary charge, e, and is the amount found on all electrons and protons.

As we said, if one makes the assumption that electromagnetic radiation is quantized such that the energy at a given frequency is found only in bundles, later to be known as **photons**, then the experimental data can be explained. However, if light is given off as a stream of individual photons and not as waves, how does interference occur? This apparent contradiction was not easily resolved.

PHOTONS AND THE PHOTOELECTRIC EFFECT

Initially, scientists tried to "explain" the contradiction by pretending that it didn't exist. They believed that the "patch" of assumed quantized photon energies was merely a temporary artifice that would eventually give way to a more satisfying (and wave-based) explanation of light. However, another set of experiments was causing additional concern about light's wave nature.

DESCRIPTION OF THE EFFECT

When light was shone onto certain metallic targets, researchers discovered that electrons (called **photoelectrons**) were ejected and could be used to create an electric current in a circuit. This phenomenon is called the **photoelectric effect**. The figure below shows a simplified version of a typical experimental setup.

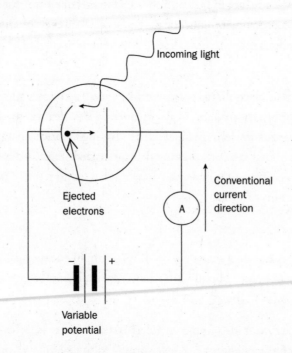

In the figure, light strikes a curved electrode, which then emits a series of electrons that have a range of kinetic energies. These kinetic energies might be thought to result from the transfer of the energy in the incident light to the electrons in the electrode. If this were the case, then as long as the variable potential has the polarity depicted in the figure, emitted electrons can cross the gap to the other electrode and create a current as measured by the ammeter. The current drops as the variable potential is reduced in size to zero. The current does not drop to zero, though, because the kinetic energy of the electrons is still adequate to allow some to cross the gap and create a current.

However, as the polarity of the variable potential is reversed and it becomes more and more negative, the potential difference between the electrodes eventually becomes great enough to stop all of the emitted electrons before they reach the other side of the gap. The current drops to zero as a result. The size of the potential difference that accomplishes this current cutoff is known as the **stopping potential**, V_s. At this point, the maximum kinetic energy of emitted electrons, K_{max}, is exactly offset by the potential energy difference eV_s between the electrodes.

PROBLEMS POSED BY THE PHOTOELECTRIC EFFECT

Our picture of this device up to this point could be explained by either a wave or a particle description of light. However, certain serious discrepancies emerge when one considers the details.

First, if light is understood as a wave, then the incoming energy involved in the photoelectric effect is related to the intensity of the light. Thus light of any frequency should be able to cause the emission of electrons, if only it is "bright" enough. What is experimentally observed, however, is that for light below a certain **cutoff frequency**, f_c, no photoelectrons are observed no matter how intense the light source is. Further, for light above the cutoff frequency, photoelectrons are seen even when using very feeble light sources.

Second, if light were only a wave, the maximum kinetic energy of the photoelectrons, as measured by the stopping potential, should be higher for higher intensities of incoming light. This is the expected result because the energy of a wave is related to the intensity (or the square of the amplitude of the wave). However, this is not the result that is observed. For higher intensities, higher currents did indeed occur, but the maximum kinetic energy of the photoelectrons was not impacted.

Third, for waves, the maximum kinetic energy should not have been related to the frequency of the incoming light. However, such a relationship was readily apparent in experiments.

Finally, classical physics would predict that some noninfinitesimal period of time should elapse between the incidence of the light and the emission of the photoelectrons. But experiments showed that as long as the light was above the cutoff frequency, the emission of the photoelectrons was essentially instantaneous.

EINSTEIN'S SOLUTION

In 1905, Einstein proposed a solution to these problems. Building on the quantization concept put forth earlier as a stopgap measure by Planck, Einstein proposed that light was indeed quantized into photons as a statement of material fact. He assumed, as Planck did, that light actually consisted of a stream of photons, each having an energy given by

$$E = hf$$

From this perspective, the data could be explained by assuming that this energy is completely absorbed by the electrons in the material and converted to the kinetic energy of the electrons.

Einstein further postulated that each material had some inherent property that resisted the emission of photoelectrons. This minimum amount of energy needed to pop an electron free of a material's surface was called its **work function,** ϕ, typically on the order of a few eV (recall that $1 \text{ eV} = 1.6 \times 10^{-19}$ J). One can then relate the maximum kinetic energy of the ejected electrons to the energy of the incoming photons as

$$K_{max} = hf - \phi$$

This formula shows that the relationship between the maximum kinetic energy of the photoelectrons and the frequency of the incident E-M radiation is linear with a slope equal to Planck's constant, h. That is, all such lines will be parallel; this is true for any material. The only difference between the curves for different materials is the y-intercept of the curve, given by the material-specific work function. Further, the x-intercept of the curve will be the cut-off frequency.

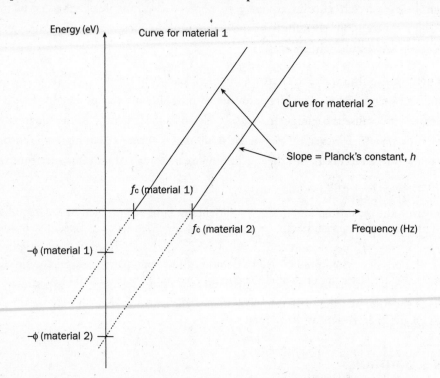

COMMON EXAM PROBLEMS DEALING WITH THE PHOTOELECTRIC EFFECT

There are several kinds of questions involving the photoelectric effect. For most of them, the overarching idea that should shape your answer is the conservation of energy.

The first type of problem involves the relationship between the stopping potential and the maximum kinetic energy of the photoelectrons. For example, suppose researchers discover that the electric potential needed to shut down a photoelectric current is 3 volts. What is the maximum kinetic energy of the photoelectrons?

The given potential is the stopping potential, V_s, and so the potential energy of the electrons when they get to the far electrode is given by:

$$U = e\Delta V_s = (1.6 \times 10^{-19}\ \text{C}) \times (3\ \text{V}) = 4.8 \times 10^{-19}\ \text{J}\,(= 3\ \text{eV})$$

But this must be exactly equal to the maximum kinetic energy of the photoelectrons.

Next, if the work function of the material is known to be 2.0 eV, what is the cutoff frequency of photons for this material?

The cutoff frequency is that frequency above which electrons can be freed from the material. That is to say, it is the frequency of radiation whose energy is equal to the work function. We use the relation for the energy of a photon to get

$$E = hf_c = \phi$$

$$f_c = \frac{\phi}{h} = \frac{2\ \text{eV}}{4.14 \times 10^{-15}\ \text{eV} \cdot \text{s}} = 4.83 \times 10^{14}\ \text{Hz}$$

What is the wavelength of light having this frequency?

The wavelength and frequency are related through

$$c = \lambda f$$

$$\lambda = \frac{c}{f} = \frac{3.00 \times 10^8\ \text{m/s}}{4.83 \times 10^{14}\ \text{Hz}} = 621\ \text{nm}$$

If we now shine a monochromatic source of light having a wavelength of $\lambda = 500$ nm onto this material, would any electrons be emitted, and if so, what would be their maximum kinetic energy?

To determine whether any electrons would be emitted, we need to see if the photon energy of the incident light is greater than the work function. Or equivalently, we could check to see if the frequency is greater than the cutoff frequency.

The photon energy is given by

$$E = hf = \frac{hc}{\lambda} = \frac{1.24 \times 10^3\ \text{eV} \cdot \text{nm}}{500\ \text{nm}} = 2.48\ \text{eV}$$

This energy is indeed greater than the 2.0 eV value for the work function, so electrons will be given off. The maximum kinetic energy will then be the difference between the photon energy and the work function, or $K_{\text{max}} = hf - \phi = 2.48\ \text{eV} - 2.00\ \text{eV} = 0.48\ \text{eV}$.

Notice that in all of these calculations, we have used the values of various constants (h, hc ($= h \times c$), etc.) in a variety of non-SI units. These alternate values are all tabulated in the Table of Information that will be available to you during the AP Physics exam. It is important for you to familiarize yourself with this table and to become comfortable using the different sets of units (mostly eV versus J). However, you can always resort to converting all quantities to SI units and then using the values of the various constants in SI units. This approach will always be correct, provided that you maintain consistency among your units.

WAVE-PARTICLE DUALITY

At this point, we have a dilemma. Light seems to exhibit the properties of both waves and particles (e.g., two-slit interference *and* the photoelectric effect). The simple fact of the matter is that light behaves both ways, depending on the situation. This apparently paradoxical behavior is known as **wave-particle duality**. While it was a bit controversial when it was first suggested, long years of experimental verification have established the reality of wave-particle duality.

Light's dual nature suggests a further question: "Which aspect shows itself in a given situation?" Fortunately, it turns out that no experiment can be performed in which light will exhibit both aspects of its nature at the same time. For example, imagine performing a two-slit interference experiment. You will see the interference fringes appear on a screen some distance from the slits. They appear because light is behaving as if it were a wave that can go through both slits at once and interfere with itself.

But if you now try to treat light as a stream of photons (each of which can go through only one slit or the other, after all), you find that anything that you do to determine which slit a given photon goes through immediately destroys the interference pattern. In fact, if you let only one photon go through the slits at a time, over time the interference pattern will build up as a series of dots on the screen. But, again, as soon as you do anything at all that would tell you which slit each photon actually goes through, this interference pattern stops.

The dual nature of light can be described by equating the photon energy given by Planck's formula to the total relativistic energy given by Einstein's formula. Because photons have no mass, we obtain

$$E^2 = (hf)^2 = p^2 c^2 + (mc^2)^2$$

or

$$E = hf = pc$$

If we now recall that $c = \lambda f$ for light, we get the relationship between the wave and particle aspects of light as

$$\lambda = \frac{h}{p}$$

In this formula, the wavelength, λ, represents the wave aspect of light while the momentum, p, represents the particle aspect. These formulas tell us that light transmits not only energy, as do other waves, but also linear momentum, as do classical particles. These two aspects are intimately connected through Planck's constant, h.

Once the concept of wave-particle duality was developed to explain the behavior of light, a French physicist named de Broglie asked why this paradoxical behavior should be limited to light. He postulated that moving particles, such as electrons, protons, or airplanes, might also be subject to the same wave-particle duality and obey the same formula. He predicted that electrons having a momentum p could be induced to exhibit wavelike behavior, such as interference, if only one could get a diffraction grating fine enough to resolve the wavelengths involved.

For example, the nearly infinitesimal magnitude of Planck's constant means that—for electrons with any appreciable momentum—one would need a diffraction grating with spacing on the order of the interatomic distances in crystals. Indeed, firing streams of electrons at crystals revealed interference patterns, just as if the electrons had a wavelength as given by the formula on the previous page. The wavelength associated with a moving particle as given by the above formula is therefore known as the **de Broglie wavelength** of the particle.

THE COMPTON EFFECT

Another phenomenon that was found to be explainable only in terms of quantization of E-M radiation was discovered as follows. In the early 1920s, Arthur Compton fired beams of very short wavelength E-M radiation (known as X-rays) at a target. When he investigated what came out the other side, he found that the wavelength of the radiation had changed, depending on the angle at which it came out. The deflected wavelength, λ, was longer than the incident wavelength, λ_0, if the wave had been deflected through some angle, θ, from its original path; the more it had been deflected, the greater this change in wavelength.

He reasoned that the increase in wavelength corresponded to a decrease in the energy of the incident photons as a result of particle-like collisions with loosely bound outer electrons. Recall that the energy is related to the wavelength as follows:

$$E = hf = \frac{hc}{\lambda}$$

Using the standard methodology for collisions between particles, he employed the relativistic conservation of energy and momentum equations for the photon-electron collision. He assumed that the electron was at rest and not attached to any atom; he also assumed that the photon had an energy as given in the previous equation, as well as a momentum given by

$$p = \frac{h}{\lambda}$$

as described in the previous section on wave-particle duality.

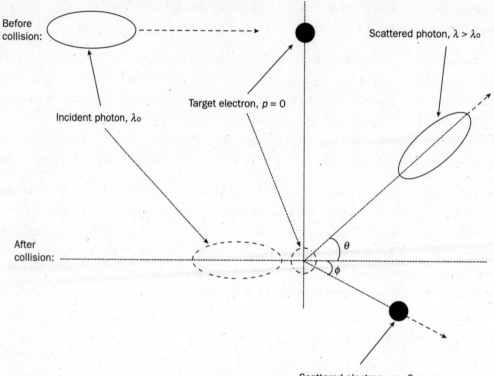

Before collision:

Incident photon, λ_0

Target electron, $p = 0$

Scattered photon, $\lambda > \lambda_0$

After collision:

θ

ϕ

Scattered electron, $p > 0$

He then solved these equations for the expected momentum of the photon as a function of scattering angle and converted the initial and scattered momenta into their respective wavelengths. The formula he derived agreed very closely with the experimental data he had collected:

$$\Delta\lambda = \lambda - \lambda_0 = \frac{h}{m_e c}(1 - \cos\theta)$$

This change in the wavelength of the incident photons is known as the **Compton shift** of the photon. It turns out that such a shift can be caused by a photon's collision with any particle. For such cases, a similar formula can be derived. The only difference would be that the mass term would need to be the mass of the particle in question. For this reason, the term $\dfrac{h}{m_e c}$ is known as the **Compton wavelength** of an electron, since that is the particle we are considering.

Compton's analysis provided important evidence for the idea that light can behave as a stream of particles. It lends credence to the idea that light carries linear momentum, as a particle would, and therefore can be considered to be a particle under these circumstances.

For example, suppose X-rays having an energy of 10.0 keV are incident on a target and are scattered by the electrons through an angle of 30°. We would like to know the wavelength of the incident photons, the Compton shift of the radiation at this angle, and the final wavelength and energy of the deflected photons.

The wavelength of the incoming photons is obtained from

$$\lambda_0 = \frac{hc}{E_0} = \frac{1.24 \times 10^3 \text{ eV} \cdot \text{nm}}{10.0 \times 10^3 \text{ eV}} = 0.124 \text{ nm}$$

The Compton shift is given by

$$\Delta\lambda = \frac{h}{m_e c}(1 - \cos\theta) = \frac{6.63 \times 10^{-34} \text{ J} \cdot \text{s}}{(9.11 \times 10^{-31} \text{ kg}) \times (3.00 \times 10^8 \text{ m/s})}(1 - \cos 30°)$$
$$= 3.25 \times 10^{-13} \text{ m} = 3.25 \times 10^{-4} \text{ nm}$$

Note that for this calculation, we reverted to SI units for all values. As we said previously, this is perfectly acceptable at any time as long as we are consistent within the same calculation. The final wavelength is then

$$\lambda = \lambda_0 + \Delta\lambda = 0.124 + 0.000325 = 0.124325 \text{ nm}$$

Finally, the energy of the deflected photons is

$$E = \frac{hc}{\lambda} = \frac{1.24 \times 10^3 \text{ eV} \cdot \text{nm}}{0.124325 \text{ nm}} = 9,974 \text{ eV} = 9.974 \text{ keV}$$

Notice the extremely small value of the Compton shift and the small loss of energy of the photons here. This is due to the size of the Compton wavelength of the electron.

$$\frac{h}{m_e c} = \frac{6.63 \times 10^{-34} \text{ J} \cdot \text{s}}{(9.11 \times 10^{-31} \text{ kg}) \times (3.00 \times 10^8 \text{ m/s})}$$
$$= 2.426 \times 10^{-12} \text{ m} = 2.426 \times 10^{-3} \text{ nm}$$

As a final note, we acknowledge a somewhat less than scrupulous use of significant figures in the above example. However, it was necessary in this problem to highlight the scale of the shifts that occur in this process.

ATOMIC ENERGY LEVELS

Another phenomenon that puzzled physicists during early part of the 20th century involved the highly specific electromagnetic radiation frequencies that are seen emanating from atoms. Tubes filled with various gases such as hydrogen were excited, and the light coming from the tube was passed through a prism. When this light underwent the same dispersion that occurs with normal "white" light, a strange effect was noted.

AP EXPERT TIP

The AP graders do not typically take off points for significant figure use as long as you choose to use a reasonable number in each case. It is generally safe to give answers to three significant figures.

Instead of a continuous band of colors, only very distinct lines of specific colors (or frequencies) were observed. Further, these patterns of lines, called **spectral lines** or **atomic spectra,** were characteristic of the elements in the discharge tube. For example, excited hydrogen only emitted visible light with wavelengths of approximately 410 nm, 434 nm, 486 nm, and 656 nm.

Instruments detected other wavelengths in the infrared and ultraviolet ranges, but always with unique and distinct values. It was as if hydrogen was only capable of emitting electromagnetic radiation with certain wavelengths or frequencies. Further, a type of pattern was discovered in these wavelengths that seemed completely arbitrary yet very real. This pattern was known as the Ritz combination principle, and though it seemed to be accurate, nobody could say why such a rule should work.

Near this time, the structure of matter was also undergoing something of an evolution. The best scientific guess at how things were constructed was called the "plum pudding" model. Neutrons had not yet been discovered at this time, and the positive charge in matter was assumed to be spread uniformly over the entire volume. The negative charge was thought to be carried by the newly discovered electrons, which were thought of as something like little "plums" embedded in a "pudding" of positive charge.

Ernest Rutherford discovered that this model was not accurate. He discovered that the positive charge was not a uniformly distributed material; instead, it was concentrated in a very small volume, which we now call the nucleus of an atom. The vast majority of the volume of any substance, it turned out, was empty space. He also reasoned that the electrons were meandering around in this vast space of the material.

Niels Bohr seized on this idea and used it to create a model of the atom in which the electrons actually orbited around the nucleus like little negatively charged planets moving around an attracting, positively charged sun. But classical physics required that these orbiting electrons radiate energy and very quickly "fall" into their "sun," the nucleus.

To get around this, Bohr proposed that as long as the orbits satisfied some quantization condition, they could remain in orbit without radiating. His quantization condition was that stable orbits had to be such that the product of the momentum of the electron and the circumference of the orbit (related to a quantity known as angular momentum) needed to be some multiple, n, of Planck's constant, h. This integer, n, by the way, is known as a quantum number; in this particular example, it is the number you may have called the energy level in chemistry class (i.e., n is the leading number in electron configurations such as $1s^2\, 2s^2\, 2p^3$).

> **AP EXPERT TIP**
>
> Rutherford's conclusion was reached in the famous gold-foil experiment, where he found that a thin foil of gold could change the path of fast-moving alpha particles.

Bohr's assumption for stable orbits required

$$mv\left(2\pi r\right) = nh$$

or

$$mvr = n\frac{h}{2\pi} = n\hbar$$

These stable orbits would then also have a specific characteristic total energy (kinetic plus electrostatic potential energy) associated with their radius. These energies associated with stable orbits due to quantized angular momentum came to be known as **atomic energy levels**. When the energies for these atomic energy levels were calculated, the frequencies of the spectral lines for hydrogen were seen to be exactly what would be expected by the change in energy of an electron that "fell" from a higher energy stable orbit into some lower energy stable orbit. This loss of energy would then be emitted as a photon with a frequency given by Planck's formula:

$$hf = E_i - E_f$$

or

$$f = \frac{E_i - E_f}{h}$$

which is exactly in line with the Ritz combination principle.

It was further noted that incoming photons could elevate an electron from a lower orbit to a higher orbit if its energy was exactly right. This **absorption** process was said to **excite** the electron or lift it into an excited state, from which it could then **decay** or fall back to its normal, lowest energy **ground state**, emitting an appropriate-frequency photon as described above. These excitations and decays are known collectively as **transitions**. It is important to keep in mind that these transitions do *not* need to be between adjacent energy levels; they can occur between any two energy levels in an atom. However, a transition that does not begin and end on one of the predefined energy levels cannot occur unless the electron is given sufficient energy to leave the atom entirely and carry off the excess energy as kinetic energy.

The de Broglie wavelengths described earlier in this chapter have also been used to provide an early explanation for the stability of electron energy levels in atoms. In this explanation, it was assumed that at certain orbital radii, centripetal force implied an electron momentum that corresponded to some de Broglie wavelength. If the circumference of this orbit was exactly some integral number of wavelengths for the electron, it would behave not as a particle but as a standing wave pattern that would not radiate energy. These standing waves thus provided another rationale for why the electron would not "fall" into the nucleus (i.e., why it would not behave as predicted by classical electromagnetic theory). If one then calculated where these standing waves could occur, the energy levels were in close agreement with the earliest experimental data for hydrogen as predicted by the Bohr model of the atom.

In practice, these atomic energy levels are usually depicted as a series of lines (known as an **energy-level diagram**) running up the page from the ground state ($n = 1$) through all the excited states ($n = 2, 3 \ldots \infty$). The following figure shows the energy-level diagram for hydrogen; notice that that as the value of n grows, the spacing between adjacent levels decreases.

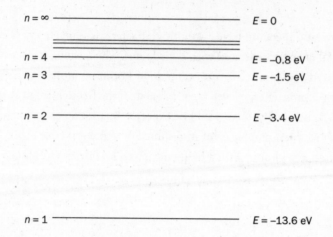

This is a common feature of the diagrams you will see on the AP Physics B exam; a variety of factors make real-world examples more complex, such as the subshells you may have seen in chemistry class. Transitions are typically shown as arrows moving from the initial state to the final state, and it is assumed that the energy is always zero at the $n = \infty$ level, making all other n levels equivalent to negative values of energy.

When an electron is lifted to the highest ($n = \infty$) energy level or state, it has been set free from the atom altogether, and the atom is said to be **ionized**. Further, the energy necessary to free an electron from state n is simply $-E_n$. This energy is known as the **binding energy** of the state n. For hydrogen, the energy levels are given by the following formula:

$$E_n(\text{eV}) = \frac{-13.6}{n^2}$$

These concepts are used in the following way. Imagine that we are considering a hypothetical element with the energy-level diagram given in the following figure. (Also shown are a number of possible transitions labeled **a** through **j**.)

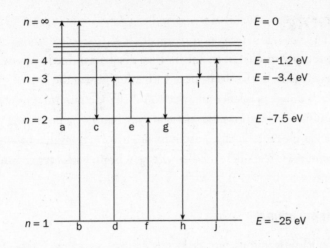

Notice that all transitions with an arrow pointing up correspond to the absorption of a photon, while those with an arrow pointing down correspond to the emission of a photon. Note further that this is not an exhaustive listing of all possible transitions and that transitions may occur in multiple steps—for example, an absorption **j**, followed by the two emissions **i** and **h**, returning to the ground state.

An electron may stay in an excited state for some period of time. For example, the absorption **j** could just as well be followed the two emissions **i** and **g**, with the electron then simply remaining in the $n = 2$ excited state. It could even absorb another photon and leave the atom altogether via transition **a**.

Let's determine which, if any, of the transitions in the figure correspond to photons of visible light. The range of wavelengths in the visible spectrum runs from about 400 nm to about 700 nm. If we convert these two ends of the spectrum into their equivalent energies, we can compare the possible transition energies to this range and pick out the ones we desire.

400 nm:

$$E = hf = \frac{hc}{\lambda} = \frac{1.24 \times 10^3}{400} = 3.1 \text{ eV}$$

700 nm:

$$E = hf = \frac{hc}{\lambda} = \frac{1.24 \times 10^3}{700} = 1.77 \text{ eV}$$

Of all the transitions, only **i** (3.4 − 1.2 = 2.2 eV) has an energy in the visible range. We see that transitions **a** (7.5 eV), **b** (25 eV), **c** (6.3 eV), **d** and **h** (both 21.6 eV), **e** and **g** (both 4.1 eV), **f** (17.5 eV), and **j** (23.8 eV) are outside of the visual range of energies.

AP EXPERT TIP

Remember that your answer for *E* is the **difference** between the energy level values in the transition, not one of the values itself.

NUCLEAR REACTIONS

The type of reactions you probably studied in chemistry class involved different atoms interacting to form compounds. In these reactions, the electrons of each atom played the crucial role, and you were concerned with such things as valence and electron shells. Certain rules were applied to balance the reactants, but in these reactions, the elements you had at the beginning were the exact same elements you had at the end. Nuclear reactions also involve rules that balance certain aspects of the reactants. But in these reactions, the elements you begin with may transmute into completely different elements.

DEFINITIONS AND NOMENCLATURE

To begin our discussion of nuclear reactions, we need a few definitions. In nuclear reactions, the components of the nucleus are the important thing. All nuclei are composed of protons and/or neutrons. **Nucleon** is the generic term given to particles that can be in a nucleus; both neutrons and protons are considered nucleons. The **atomic number**, Z, corresponds to the number of protons in a given nucleus. The atomic number is the quantity that actually defines which element you are dealing with. The **neutron number**, N, corresponds to the number of neutrons in a nucleus. The total number of protons and neutrons in a nucleus is conveyed by the **mass number**, A, which is equal to $Z + N$. Atoms of the same element (same value of Z) but having different number of neutrons, N, are called **isotopes** of that element.

We use a type of shorthand to identify the elements in a nuclear reaction. However, we will need to modify the common abbreviations from chemistry because nuclear reactions can cause changes in elements and the isotopes of an element. For example, it is not enough just to say that our reaction involves hydrogen, H, because three isotopes of hydrogen are common in nuclear reactions. We put the mass number as a superscript in *front* of the symbol to distinguish among these various isotopes. The most common isotope of hydrogen has one proton and no neutrons and, therefore, has a mass number of 1; its symbol would be ^{1}H. Another common isotope, known as deuterium, has one proton (as all hydrogen isotopes do) and one neutron; its mass number is 2 and its symbol is ^{2}H. Note that this is (importantly) different from the symbol for the diatomic hydrogen gas molecule, H_2. The final common isotope of hydrogen, called tritium, combines a single proton with two neutrons for a mass number of 3 and the symbol ^{3}H.

You will also sometimes see the atomic number in a symbol in addition to the mass number. This allows you to know Z without needing to look up the element that corresponds to a specific abbreviation. For example, if you see the symbol $^{235}_{92}$U (read as "uranium 235"), this indicates that the element is uranium, its atomic number is 92, and its mass number is 235. Subtraction reveals the number of neutrons in this isotope of uranium: $235 - 92 = 143$.

Two other types of particles that are often seen in nuclear reactions are neutrons (symbol ^{1}n) and alpha particles, which are actually helium nuclei (symbol ^{4}He, or often α). Other reactants are treated the same way. For example, the atomic number of lithium is 3 ($Z = 3$), meaning that there three protons in a lithium nucleus. If we want to know how many neutrons are in the isotope ^{7}Li, we can determine that lithium's mass number, A, is 7, which tells us that there must be four neutrons in this isotope.

MASS-ENERGY EQUIVALENCE

We mentioned mass-energy equivalence near the start of the chapter, and it returns here at the end because it is an important concept in nuclear reactions. In these reactions, it often turns out that the total mass of the initial reactants does not equal the total mass of the final products. This discrepancy must be accounted for by calculating the energy that corresponds to the missing mass.

The **unified atomic mass unit** is one final piece of information that can be useful in analyzing nuclear reactions. This is a unit of mass, given the abbreviation u, that simplifies the arithmetic in reaction calculations. It is approximately, but not exactly, the mass of proton, and the conversion is

$$1 \text{ u} = 1.66 \times 10^{-27} \text{ kg}$$

The units are typically used for atomic masses, allowing you to identify atomic masses in whole numbers that are approximately equal to the mass number of a given isotope.

RULES FOR ANALYZING NUCLEAR REACTIONS

In any nuclear reaction, elements are changing into other elements, into other isotopes of the same elements, or into a combination of both. The first rule of nuclear reactions is that the total charge must remain the same before and after the reaction. The second rule of nuclear reactions is that the total number of nucleons must remain the same before and after the reaction. Note that the number of protons does not have to remain the same, nor does the total number of neutrons—only the sum of both.

For example, one type of nuclear reaction that occurs in heavy nuclei is known as **alpha decay**. In this type of reaction, a heavy nucleus spontaneously emits an alpha particle and transmutes into another element. One type of alpha emitter is found in many home smoke alarms. It is an element known as americium, and the isotope used is typically $^{241}_{95}$Am, which has an atomic number of 95 and a mass number of 241. If it undergoes alpha decay, the reaction is written as $^{241}_{95}$Am $\longrightarrow$ $^{237}_{93}$Np + $^4_2$He.

You can see that the nuclear reaction rules are followed here, because the total number of nucleons in the americium nucleus ($A = 241$) is identical to the total number of nucleons in the product nuclei (237 in neptunium + 4 in the alpha particle). Further, we see that the total charge in this reaction stays the same because no electrons are explicitly called out on either side and the total number of protons on each side of the equation is the same (95 in americium and a total of $93 + 2 = 95$ in the neptunium and the alpha particle).

In fact, a common exam problem asks students to determine the atomic number and mass number in the initial nucleus if given the product nucleus (i.e., daughter nucleus) or in the product nucleus if given the initial nucleus. As we see, this is accomplished through the use of the two conservation rules for nuclear reactions.

Imagine we are told that ^{241}Am has an atomic mass of 241.056885 u, ^{237}Np has an atomic mass of 237.048217 u, and ^{4}He has an atomic mass of 4.002604 u. We can then determine how much energy is released in this reaction, energy that will be predominantly contained in either photons or the kinetic energy of the products. To accomplish this, we find the mass discrepancy between the two sides and convert to energy units:

$$m_{in} = m_{Am} = 241.056885 \text{ u}$$
$$m_{out} = m_{Np} + m_{He} = 237.048217 \text{ u} + 4.002604 \text{ u} = 241.050821 \text{ u}$$

So

$$\Delta m = m_{in} - m_{out} = 241.056885 - 241.050821 = 0.006064 \text{ u}$$

This mass discrepancy must then be converted into an equivalent energy using

$$\Delta E = \Delta m c^2$$

Your equation sheet will contain a conversion for the atomic mass units, u, into units of MeV/c^2. If we convert into these new units and multiply by c^2, we get energy units directly:

$$\Delta E = 0.006064 \text{ u} \times \left(\frac{931 \text{ MeV}/c^2}{1 \text{ u}} \right) \times \frac{c^2}{1} = 5.646 \text{ MeV}$$

This conversion of mass into energy is one of the hallmarks of nuclear reactions. When multiplied by the vast number of reactions that take place in a macroscopic-sized sample, the energies involved are the primary reason that nuclear power is seen as a viable alternative to energy derived from fossil fuels.

REVIEW QUESTIONS

1. Two photons are completely absorbed by a target. The first has a wavelength of $\lambda_1 = 450$ nm, and the second has a wavelength of $\lambda_2 = 700$ nm. Which one transfers the most energy to the target?

   (A) The first photon

   (B) The second photon

   (C) They both transfer the same energy.

   (D) It depends on the target.

   (E) It cannot be determined from the information given.

2. A photoelectric tube is illuminated by electromagnetic radiation having a frequency of 5 GHz. The tube is found to have a stopping potential of 2.5 V. What is the maximum kinetic energy of the electrons released from the photo-electrode?

   (A) 7.5 GeV

   (B) 2.0 GeV

   (C) 2.5 eV

   (D) 0.5 eV

   (E) 2.5 GeV

3. A photon strikes a free electron that is initially at rest. The electron and photon go off at an acute angle to each other. As a result of the collision, the wavelength of the photon

   (A) decreases.

   (B) increases.

   (C) remains the same, but the frequency decreases.

   (D) remains the same, but the frequency increases.

   (E) increases, but the frequency remains the same.

Use the following figure to answer questions 4 and 5.

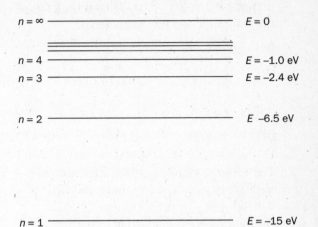

4. The figure shows an atomic energy-level diagram for a hypothetical atom. The frequency of light that is emitted when an electron makes a transition from the $n = 3$ state to the ground state is most nearly

   (A) 5.8×10^{15} Hz.

   (B) 4.2×10^{15} Hz.

   (C) 3.6×10^{15} Hz.

   (D) 3.0×10^{15} Hz.

   (E) 2.4×10^{14} Hz.

5. What range of photon wavelengths will be certain to ionize the atom from the ground state?

   (A) λ greater than 82.7 nm

   (B) λ less than 82.7 nm

   (C) λ greater than 1,240 nm

   (D) λ less than 1,240 nm

   (E) λ greater than 100 nm

6. The cutoff frequency for a photo-electrode material is 2.00×10^{15} Hz. What is the value of the work function for this material?

   (A) 2.00 eV

   (B) 1.33×10^{-18} eV

   (C) 8.28 keV

   (D) 8.28 eV

   (E) 14.3 eV

7. Light from a monochromatic light source with a wavelength of 400 nm is absorbed by a metal with a work function of 1.5 eV. If the light source is doubled in intensity, what effect will this have on the maximum kinetic energy, K_{max}, of the photoelectrons and on the current, I, that is produced?

   (A) K_{max} and I will both increase.

   (B) K_{max} will increase and I will remain constant.

   (C) K_{max} and I will both remain constant.

   (D) K_{max} will remain constant and I will increase.

   (E) K_{max} and I will both decrease.

8. Gamma radiation is the name given to the photons created during a nuclear reaction. One such photon has a wavelength of 2.00×10^{-15} m. How much mass is lost in this reaction?

   (A) 620 MeV/c^2

   (B) 6.20×10^8 MeV/c^2

   (C) 6.20×10^{11} Mev/c^2

   (D) 2.48×10^{-18} MeV/c^2

   (E) 2.48×10^{-6} MeV/c^2

9. $^{235}_{92}U + {}^1n \rightarrow {}^A_Z X + {}^{140}_{55}Cs + 3 {}^1n$

   Given this equation for a nuclear reaction, determine the values of Z and A for the unknown daughter product X.

   (A) $Z = 147, A = 378$

   (B) $Z = 92, A = 235$

   (C) $Z = 140, A = 92$

   (D) $Z = 37, A = 93$

   (E) $Z = 93, A = 37$

10. If the frequency of the light shining on a photoelectric tube is doubled, what is the effect on the maximum kinetic energy of the photoelectrons?

    (A) The maximum kinetic energy doubles.

    (B) The maximum kinetic energy is not affected.

    (C) The maximum kinetic energy increases, but it is less than twice the original value.

    (D) The maximum kinetic energy increases, but it is more than twice the original value.

    (E) The maximum kinetic energy decreases.

FREE-RESPONSE QUESTION

An inventor wants to build an instrument that will operate only when a specific wavelength of light is shone on it. She chooses a wavelength identical to that which will excite a certain atom from its ground state to its first excited state ($n = 2$). When the excited electrons decay back to their ground state, the emitted photons are of an energy that is greater than the work function associated with a given photoelectric tube. (See the figure below.)

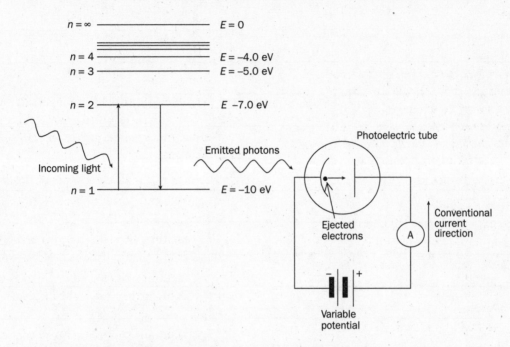

(a) What is the wavelength of light that will cause the required excitation?

(b) What is the energy of the photon that is emitted from the excited atom and absorbed by the photoelectric tube?

(c) If the inventor finds that the stopping potential for this device is −1.0 V, what is the value of the work function for the photoelectric tube?

(d) If the variable potential in the figure is set to a value of 2.0 V with a polarity as shown in the figure, what is the maximum kinetic energy of the electrons striking the flat electrode in the photoelectric tube?

ANSWERS AND EXPLANATIONS

1. A

If both photons are completely absorbed, all of their energy is transferred to the target, regardless of the nature of the target. The energy of a photon is given by

$$E = hf = \frac{hc}{\lambda}$$

Therefore, a smaller wavelength corresponds to a higher-energy photon.

2. C

The stopping potential is exactly the potential that creates just the electrostatic potential energy to offset the maximum kinetic energy of the photoelectrons. This is given by

$$K_{max} = U_e = eV_S = e \times 2.5\,\text{V} = 2.5\,\text{eV}$$

3. B

Photons carry and transmit momentum. This interaction should be treated as if the photon is merely a particle that collides with the electron and transfers some of its momentum to the electron. If the momentum of the photon decreases, we must use this information to choose the correct answer. The momentum of the photon is given by

$$p = \frac{h}{\lambda}$$

so a loss of momentum must be accompanied by an increase in wavelength. This eliminates choices (A), (C), and (D). To choose between the remaining responses, we recall that for photons

$$c = \lambda f$$

Because the speed of the photon does not change, the frequency must change. If the wavelength increases, then the frequency must decrease, so choice (E) is incorrect, leaving choice (B) as the correct answer.

4. D

The difference in energy between the $n = 3$ state and the ground state ($n = 1$) is

$$E = -2.4 - (-15) = 12.6\,\text{eV}$$

This corresponds to a frequency of

$$f = \frac{E}{h} = \frac{12.6\,\text{eV}}{4.14 \times 10^{-15}\,\text{eV} \cdot \text{s}} \approx 3.0 \times 10^{15}\,\text{Hz}$$

5. B

A photon of energy greater than 15 eV is required to ensure that any electron can be removed from the ground state. This corresponds to a wavelength given by

$$\lambda = \frac{hc}{E} = \frac{(4.14 \times 10^{-5}\,\text{eV} \cdot \text{s})(3.00 \times 10^8\,\text{m/s})}{15\,\text{eV}}$$
$$= 82.7\,\text{nm}$$

If we require that the energy be greater than the cutoff, then the wavelength needs to be less than the corresponding cutoff.

6. D

The cutoff frequency corresponds to the energy of a photon that is just sufficient to free an electron from the electrode. This is, by definition, the work function, ϕ.

$$\phi = E = hf$$
$$= (4.14 \times 10^{-15}\,\text{eV} \cdot \text{s})(2.00 \times 10^{15}\,\text{Hz})$$
$$= 8.28\,\text{eV}$$

7. D

The maximum kinetic energy depends on the incident photon energy and the work function of the material, neither of which changes in this problem. The current is a measure of how many electrons are emitted per unit time from the photoelectrode. More intense light corresponds to more photons per unit time, which in turn leads to more electrons being emitted per unit time.

8. A

This problem is almost as much about scientific notation and conversion of units as it is about nuclear physics. To find the mass that has been converted into the given photon energy, we equate the photon energy with the Einstein mass energy to get

$$E = hf = \frac{hc}{\lambda} = mc^2$$

or, solving for the mass in the units given in the responses,

$$m = \frac{hc}{\lambda} \times \frac{1}{c^2} = \left(\frac{1.24 \times 10^3 \text{ eV} \cdot \text{nm}}{2.0 \times 10^{-15} \text{ m}} \right)$$

$$\times \left(\frac{10^{-9} \text{ m}}{1 \text{ nm}} \right) \times \left(\frac{1 \text{ MeV}}{10^6 \text{ eV}} \right)$$

$$\times \frac{1}{c^2} = 0.62 \times 10^3 = 620 \, \frac{\text{MeV}}{c^2}$$

9. D

The rules for nuclear reactions say that the total amount of charge must be the same on both sides of the equation. A uranium atom is neutral, as is a neutron; thus we start with 0 net charge. Because there is no indication of ionization or free charged particles on the right side, there is 0 net charge there also, as long as all the protons are accounted for in the two nuclei. That means that the number of protons in element X must be $92 - 55 = 37$. Next, the total number of nucleons must be conserved. Before the reaction, we have a total of $235 + 1 = 236$ nucleons. That means that the element X must have $236 - 140 - 3 = 93$ total nucleons. The element X is actually $^{93}_{37}\text{Rb}$.

10. D

The easiest way to see this is to consider an example. Say we have a material with a work function of 5 eV and photons with an original energy of 10 eV. Then the maximum kinetic energy of the photoelectrons is given by

$$K_{\max} = hf - \phi = 10 - 5 = 5 \text{ eV}$$

Doubling the frequency of the photons doubles the energy of the photons, but it doesn't change the work function at all. The new energy of the photons is 20 eV, so the maximum kinetic energy of the photoelectrons is:

$$K_{\max} = hf - \phi = 20 - 5 = 15 \text{ eV}$$

Obviously, the maximum kinetic energy is more than double the original.

FREE-RESPONSE QUESTION

(a) The required wavelength of the incoming light is obtained from the required excitation energy.

$$E = E_2 - E_1 = hf = \frac{hc}{\lambda}$$

or

$$\lambda = \frac{hc}{E_2 - E_1} = \frac{1.24 \times 10^3 \text{ eV} \cdot \text{nm}}{(-7-(-10)) \text{ eV}} = 413 \text{ nm}$$

(b) The energy of the emitted photon is merely the difference in energy between the two states:

$$E = E_2 - E_1 = -7 - (-10) = 3.0 \text{ eV}$$

(c) The value of the stopping potential corresponds to the maximum kinetic energy of the photoelectrons. In this case

$$K_{max} = -e\Delta V_S = -e \times 1.0 \text{ V} = 1.0 \text{ eV}$$

We can then use this information in the photoelectric equation, together with the photon energy:

$$K_{max} = hf - \phi$$

or

$$\phi = hf - K_{max} = 3.0 \text{ eV} - 1.0 \text{ eV} = 2.0 \text{ eV}$$

(d) To find the maximum kinetic energy of the electrons striking the electrode, we start with the maximum kinetic energy of the electrons as they are emitted from the photoelectrode, 1.0 eV. These electrons are then accelerated due to the potential difference set up between the two electrodes by the variable potential. The kinetic energy as they reach the flat electrode is therefore the sum of these two components:

$$K_{total} = K_{max} + e\Delta V = 1.0 \text{ eV} + e \times 2.0 \text{ V} = 3.0 \text{ eV}$$

| Part Four |

PRACTICE TESTS

HOW TO TAKE THE PRACTICE TESTS

The next section of this book consists of two practice tests: one Physics B test and one Physics C test. Take both practice tests if you are preparing for both AP Physics exams, or just take the one for the AP exam you plan to take.

Taking a practice AP exam gives you an idea of what it's like to answer test questions for a longer period of time. You'll find out which areas you're strong in and where additional review may be required. Any mistakes you make now are ones you won't make on the actual exam, as long as you take the time to learn where you went wrong.

Before taking a practice test, find a quiet place where you can work uninterrupted for two and a half hours. Time yourself according to the time limit at the beginning of each section. It's okay to take a short break between sections, but for the most accurate results, you should approximate real test conditions as closely as possible.

As you take the practice tests, remember to pace yourself. Train yourself to be aware of the time you are spending on each problem. Try to be aware of the general types of questions you encounter, as well as being alert to certain strategies or approaches that help you handle the various question types more effectively.

After taking a practice exam, be sure to read the detailed answer explanations that follow. These will help you identify areas that could use additional review. Even when you answered a question correctly, you can learn additional information by looking at the answer explanation.

Finally, it's important to approach the test with the right attitude. You're going to get a great score because you've reviewed the material and learned the strategies in this book.

Good luck!

HOW TO COMPUTE YOUR SCORE

SCORING THE MULTIPLE-CHOICE QUESTIONS

To compute your score on the multiple-choice portions of the two Practice Tests, calculate the number of questions you got right on each test, then divide by 70 to get the percentage score for the multiple-choice portion of that test.

SCORING THE FREE-RESPONSE QUESTIONS

The reviewers have specific points that they will be looking for in each essay. In addition to spelling, grammar, and overall structure of the essay, they look for key bits of information. Each piece of information that they are able to check off in your essay is a point toward a better score.

To figure out your approximate score for the free-response questions, look at the key points found in the sample response for each question. For each key point you included, add a point. Add together all the points you received for each question, and add together all the possible points available for each question. Divide the total number of points you received by the total number of possible points to get the percentage score for the free-response portion of that test.

CALCULATING YOUR COMPOSITE SCORE

Your score on each AP Physics exam is a combination of your score on the multiple-choice portion and the free-response section of the exam.

Add together your score on the multiple-choice portion of the exam and your approximate score on the free-response section of the exam. Divide this sum by two and multiply by 100 to obtain your approximate score for each full-length exam. Round up to a whole number if your score is a decimal.

Remember, however, that much of this depends on how well all of those taking the AP test do. If you do better than average, your score would be higher. The numbers here are just approximations.

The approximate score range is as follows:

Physics B:
5 = 65–100 (extremely well qualified)
4 = 50–64 (well qualified)
3 = 35–49 (qualified)
2 = 30–34 (possibly qualified)
1 = 0–29 (no recommendation)

Physics C:
5 = 57–100 (extremely well qualified)
4 = 45–56 (well qualified)
3 = 35–44 (qualified)
2 = 25–34 (possibly qualified)
1 = 0–24 (no recommendation)

If your score falls between 50 and 100, you're doing great; keep up the good work! If your score is lower than 49, there's still hope—keep studying, and you will be able to obtain a much better score on the exam before you know it.

AP Physics B Practice Test
Answer Grid

1. Ⓐ Ⓑ Ⓒ Ⓓ Ⓔ
2. Ⓐ Ⓑ Ⓒ Ⓓ Ⓔ
3. Ⓐ Ⓑ Ⓒ Ⓓ Ⓔ
4. Ⓐ Ⓑ Ⓒ Ⓓ Ⓔ
5. Ⓐ Ⓑ Ⓒ Ⓓ Ⓔ
6. Ⓐ Ⓑ Ⓒ Ⓓ Ⓔ
7. Ⓐ Ⓑ Ⓒ Ⓓ Ⓔ
8. Ⓐ Ⓑ Ⓒ Ⓓ Ⓔ
9. Ⓐ Ⓑ Ⓒ Ⓓ Ⓔ
10. Ⓐ Ⓑ Ⓒ Ⓓ Ⓔ
11. Ⓐ Ⓑ Ⓒ Ⓓ Ⓔ
12. Ⓐ Ⓑ Ⓒ Ⓓ Ⓔ
13. Ⓐ Ⓑ Ⓒ Ⓓ Ⓔ
14. Ⓐ Ⓑ Ⓒ Ⓓ Ⓔ
15. Ⓐ Ⓑ Ⓒ Ⓓ Ⓔ
16. Ⓐ Ⓑ Ⓒ Ⓓ Ⓔ
17. Ⓐ Ⓑ Ⓒ Ⓓ Ⓔ
18. Ⓐ Ⓑ Ⓒ Ⓓ Ⓔ
19. Ⓐ Ⓑ Ⓒ Ⓓ Ⓔ
20. Ⓐ Ⓑ Ⓒ Ⓓ Ⓔ
21. Ⓐ Ⓑ Ⓒ Ⓓ Ⓔ
22. Ⓐ Ⓑ Ⓒ Ⓓ Ⓔ
23. Ⓐ Ⓑ Ⓒ Ⓓ Ⓔ
24. Ⓐ Ⓑ Ⓒ Ⓓ Ⓔ

25. Ⓐ Ⓑ Ⓒ Ⓓ Ⓔ
26. Ⓐ Ⓑ Ⓒ Ⓓ Ⓔ
27. Ⓐ Ⓑ Ⓒ Ⓓ Ⓔ
28. Ⓐ Ⓑ Ⓒ Ⓓ Ⓔ
29. Ⓐ Ⓑ Ⓒ Ⓓ Ⓔ
30. Ⓐ Ⓑ Ⓒ Ⓓ Ⓔ
31. Ⓐ Ⓑ Ⓒ Ⓓ Ⓔ
32. Ⓐ Ⓑ Ⓒ Ⓓ Ⓔ
33. Ⓐ Ⓑ Ⓒ Ⓓ Ⓔ
34. Ⓐ Ⓑ Ⓒ Ⓓ Ⓔ
35. Ⓐ Ⓑ Ⓒ Ⓓ Ⓔ
36. Ⓐ Ⓑ Ⓒ Ⓓ Ⓔ
37. Ⓐ Ⓑ Ⓒ Ⓓ Ⓔ
38. Ⓐ Ⓑ Ⓒ Ⓓ Ⓔ
39. Ⓐ Ⓑ Ⓒ Ⓓ Ⓔ
40. Ⓐ Ⓑ Ⓒ Ⓓ Ⓔ
41. Ⓐ Ⓑ Ⓒ Ⓓ Ⓔ
42. Ⓐ Ⓑ Ⓒ Ⓓ Ⓔ
43. Ⓐ Ⓑ Ⓒ Ⓓ Ⓔ
44. Ⓐ Ⓑ Ⓒ Ⓓ Ⓔ
45. Ⓐ Ⓑ Ⓒ Ⓓ Ⓔ
46. Ⓐ Ⓑ Ⓒ Ⓓ Ⓔ
47. Ⓐ Ⓑ Ⓒ Ⓓ Ⓔ
48. Ⓐ Ⓑ Ⓒ Ⓓ Ⓔ

49. Ⓐ Ⓑ Ⓒ Ⓓ Ⓔ
50. Ⓐ Ⓑ Ⓒ Ⓓ Ⓔ
51. Ⓐ Ⓑ Ⓒ Ⓓ Ⓔ
52. Ⓐ Ⓑ Ⓒ Ⓓ Ⓔ
53. Ⓐ Ⓑ Ⓒ Ⓓ Ⓔ
54. Ⓐ Ⓑ Ⓒ Ⓓ Ⓔ
55. Ⓐ Ⓑ Ⓒ Ⓓ Ⓔ
56. Ⓐ Ⓑ Ⓒ Ⓓ Ⓔ
57. Ⓐ Ⓑ Ⓒ Ⓓ Ⓔ
58. Ⓐ Ⓑ Ⓒ Ⓓ Ⓔ
59. Ⓐ Ⓑ Ⓒ Ⓓ Ⓔ
60. Ⓐ Ⓑ Ⓒ Ⓓ Ⓔ
61. Ⓐ Ⓑ Ⓒ Ⓓ Ⓔ
62. Ⓐ Ⓑ Ⓒ Ⓓ Ⓔ
63. Ⓐ Ⓑ Ⓒ Ⓓ Ⓔ
64. Ⓐ Ⓑ Ⓒ Ⓓ Ⓔ
65. Ⓐ Ⓑ Ⓒ Ⓓ Ⓔ
66. Ⓐ Ⓑ Ⓒ Ⓓ Ⓔ
67. Ⓐ Ⓑ Ⓒ Ⓓ Ⓔ
68. Ⓐ Ⓑ Ⓒ Ⓓ Ⓔ
69. Ⓐ Ⓑ Ⓒ Ⓓ Ⓔ
70. Ⓐ Ⓑ Ⓒ Ⓓ Ⓔ

AP PHYSICS B: **PRACTICE TEST**

SECTION I
Time—90 minutes 70 Questions

Note: Units associated with numerical quantities are abbreviated using the abbreviations listed in the table of information included with the exams. To simplify calculations, you may use $g = 10$ m/s^2 in all problems.

Directions: Each of the questions or incomplete statements below is followed by five suggested answers or completions. Select the best choice in each case.

1. Two objects collide, making a sound. Is it possible for this collision to be completely elastic?

    (A) No, because the sound comes from the momentum, which is therefore not conserved.

    (B) Yes, because the elasticity of a collision does not depend the sound it makes.

    (C) No, because the sound means it must be an explosive collision.

    (D) Yes, because it is still possible for momentum and kinetic energy to be conserved.

    (E) No, because the sound comes from the kinetic energy, which is therefore not conserved.

2. A 50-pound child and a 200-pound adult are at a park swinging on two swings of identical length. In the time it takes the adult to make a complete oscillation, the child can make

    (A) one-quarter of an oscillation.

    (B) half of an oscillation.

    (C) one oscillation.

    (D) two oscillations.

    (E) four oscillations.

GO ON TO THE NEXT PAGE

3. Which of the following best describes the size and placement of a full-length mirror that will allow viewing of one's entire body from head to foot? (There are no obstructions in front of the mirror.)

 (A) Half a person's height with the top at head level

 (B) Half a person's height with the bottom at foot level

 (C) Equal to a person's height with the top at head level

 (D) One-third a person's height with the top at head level

 (E) It depends on how far one stands from the mirror.

4. We are given a fixed length of a heating wire with a known resistance, which we connect across a 120 V line. If we cut the wire in half and connect each half across the 120 V line, how will the total power output change?

 (A) It will double.

 (B) It will be half as large.

 (C) Total power output will not change.

 (D) It will increase by a factor of four.

 (E) Not enough information is provided.

5. Three capacitors having capacitances C_1, C_2, and C_3 are connected in series. Which formula gives the equivalent capacitance, C_{eq}, of the three capacitors?

 (A) $C_{eq} = C_1 + C_2 + C_3$

 (B) $\dfrac{1}{C_{eq}} = \dfrac{1}{C_1} + \dfrac{1}{C_2} + \dfrac{1}{C_3}$

 (C) $C_{eq} = \dfrac{C_1 + C_2 + C_3}{C_1 C_2 C_3}$

 (D) $\dfrac{1}{C_{eq}} = C_1 + C_2 + C_3$

 (E) $C_{eq} = \dfrac{1}{C_1} + \dfrac{1}{C_2} + \dfrac{1}{C_3}$

6. Newton's second law tells us that

 (A) the force acting on a body is the product of its mass and its acceleration.

 (B) the force acting on a body is the rate of change of its momentum.

 (C) the force acting on a body is the product of its mass and the rate of change in velocity.

 (D) the weight of an object on earth is the product of its mass and g.

 (E) All of the above

7. A race car circles once around an oval track, covering a distance of 1 mile (1.61 km). It starts at rest and steadily increases its speed to 80.5 m/s as it completes the lap in 40 seconds. From start to finish at the same location, what is its average acceleration?

 (A) 0 m/s^2

 (B) 2.01 m/s^2

 (C) 1.01 m/s^2

 (D) 20 s

 (E) 129.6 km · m/s

8. A thin convex lens has a focal length of 4 cm. An object is 8 cm from the lens. Which of the following best describes the image that this arrangement will produce?

 (A) The image will be 4 cm on the other side of the lens from the object, real, inverted

 (B) The image will be 8 cm on the other side of the lens from the object, real, inverted

 (C) The image will be 8 cm on the other side of the lens from the object, virtual, erect

 (D) The image will be 8 cm on the same side of the lens as the object, virtual, inverted

 (E) The image will be 4 cm on the same side of the lens as the object, virtual, erect

GO ON TO THE NEXT PAGE ⟹

9. You are using a convex magnifying lens to burn a hole in a small piece of paper. When you hold the lens 12 cm from the paper, sunlight will focus onto the smallest possible point and provide maximum potential for burning the paper. Which of the following is closest to the focal length of the lens?

(A) 24 cm

(B) 12 cm

(C) 6 cm

(D) −6 cm

(E) −12 cm

10. The magnitude of the electromagnetic force acting on a charge depends on

(A) the strength and direction of the magnetic field.

(B) the magnitude of the charge.

(C) the velocity of the charge.

(D) the strength and direction of the electric field.

(E) All of the above

11. A ball is thrown straight down from the roof of a building with an initial downward speed of 5 m/s. It is thrown from a height of 10 m. Which of the following is closest to the speed at which the ball strikes the ground?

(A) 15 m/s

(B) 225 m/s

(C) 50 m/s

(D) 500 m/s

(E) 5 m/s

12. A person with a weight of 800 N is standing on the ground. What is the reaction, as required by Newton's third law, to the Earth's gravity pulling the person down with a force of 800 N?

(A) The ground pushing the person up with a force of 800 N

(B) The person compressing the ground with a force of 800 N

(C) The person's shoes being compressed with a force of 800 N

(D) The person's gravitational force pulling up on the Earth with a force of 800 N

(E) Intermolecular forces pushing against the person's muscles and bones with a force of 800 N

13. A weight is hanging on the end of a vertical spring hanging from a hook so that the spring is stretched slightly. Which of the following best describes the force or forces acting on the weight?

(A) Gravity acting downward and the spring acting upward

(B) Gravity and the spring acting downward

(C) Gravity acting downward and the spring and the hook both acting upward

(D) Gravity and the spring acting downward and the hook acting upward

(E) Gravity acting downward and the hook acting upward

14. A boat displaces 5 m^3 of water. If the density of water is 1,000 kg/m^3, which of the following is closest to the buoyant force acting on the boat?

(A) 5 N

(B) 50 N

(C) 500 N

(D) 5,000 N

(E) 50,000 N

GO ON TO THE NEXT PAGE

15. An 80 kg driver sits in a car and causes the springs to compress by 0.02 m from its position without the driver. Treat the car's entire suspension as if it were a single spring. Which of the following would be closest to the effective spring constant of the spring system?

    (A) 40,000 N/m

    (B) 4,000 N/m

    (C) 16 N/m

    (D) −4,000 N/m

    (E) −40,000 N/m

16. Consider a 5 m bar made of a material with a coefficient of thermal expansion of 4×10^{-5} $(C°)^{-1}$. After being heated from 20° C to 120° C, how long is the bar?

    (A) 0.02 m

    (B) 1.15 m

    (C) 4.98 m

    (D) 5.00 m

    (E) 5.02 m

17. Which of the following best describes the efficiency of a real heat engine?

    (A) Always greater than 1

    (B) Always equal to 1

    (C) Always less than 1

    (D) Usually less than 1, but with good engineering, occasionally equal to 1

    (E) Usually less than or equal to 1, but with good engineering occasionally greater than 1

18. Which of the following particles will feel the greatest force due to a magnetic field?

    (A) An electron moving perpendicular to the magnetic field lines

    (B) A neutron moving perpendicular to the magnetic field lines

    (C) An electron moving parallel to the magnetic field lines

    (D) A neutron moving parallel to the magnetic field lines

    (E) An electron at rest

19. A light wave traveling in air strikes a flat glass surface having an index of refraction of n_g with a incident angle of θ_i. Which of the following expressions gives the angle of refraction, θ_r?

    (A) $\sin\theta_r = n_g \sin\theta_i$

    (B) $\sin\theta_r = \sin\theta_i$

    (C) $\sin\theta_r = n_g$

    (D) $\sin\theta_r = \dfrac{\sin\theta_i}{n_g}$

    (E) $\sin\theta_r = \dfrac{n_g}{\sin\theta_i}$

20. Which of the following physical principles is commonly applied in solar power cells?

    (A) Brownian motion

    (B) The photoelectric effect

    (C) Compton scattering

    (D) Electron diffraction

    (E) The Doppler effect

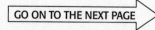
GO ON TO THE NEXT PAGE

21. An 8×10^{-4} C charge and a 2×10^{-4} C charge are 2 m apart. Which of the following is closest to the force between them?

(A) 360 N

(B) 720 N

(C) 3.6×10^6 N

(D) 7.2×10^6 N

(E) 3.6×10^{10} N

22. A water wave of amplitude A is traveling across the ocean. The wave has a speed v, a wavelength λ, and a frequency f. Which of the following equations describes the wave?

(A) $v = Af$

(B) $v = f\lambda$

(C) $f = v\lambda$

(D) $f = A\lambda$

(E) $v = \dfrac{\lambda}{f}$

23. After passing through a double slit, two light waves have a path length that differs by one half of a wavelength. If these two waves are superimposed, which of the following best describes what will happen?

(A) Destructive interference, resulting in a bright band

(B) Constructive interference, resulting in a bright band

(C) Destructive interference, resulting in a dark band

(D) Constructive interference, resulting in a dark band

(E) Partial interference, resulting in a gray band

24. Two parallel plate capacitors have an identical design, with one exception: The distance between the plates is twice as far for one capacitor as the other. The capacitor that has plates twice as far apart will have a capacitance that is

(A) one-fourth as much as the other.

(B) half as much as the other.

(C) the same value as the other.

(D) twice as much as the other.

(E) four times as much as the other.

25. You wish to build an air-filled parallel plate capacitor that carries a capacitance of 0.5 µF. You can bring the plates as close as 0.177 mm apart. What would the area of the plates need to be in order to get this capacitance?

(A) 1 m^2

(B) 1,000 m^2

(C) 0.1 m^2

(D) 100 m^2

(E) 10 m^2

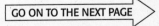
GO ON TO THE NEXT PAGE

26. In a laboratory, you connect two identical small light bulbs in a series circuit with a battery. Which of the following best describes what happens and gives the correct explanation?

(A) The brightness of both bulbs is the same because the bulbs are the same size.

(B) The first bulb in the circuit is brighter because it uses the current first.

(C) The brightness of both bulbs is the same because current is the same throughout a series circuit and there is the same potential difference across the bulbs.

(D) The first bulb in the circuit is brighter because the current and potential difference are both higher in this bulb.

(E) The second bulb in the circuit is brighter because the current and potential difference are both higher in this bulb.

27. A laboratory sample spinning at a constant speed in a centrifuge is 0.1 m from the axis of rotation and experiences a centripetal acceleration of 9,000 m/s^2. Which of the following is closest to its linear velocity?

(A) 30 m/s

(B) 90 m/s

(C) 300 m/s

(D) 900 m/s

(E) 3,000 m/s

28. There are two photons with energies E_1 and E_2. If the photon with energy E_1 has half the wavelength of the other photon, which of the following equations describes the correct relationship between the two energies?

(A) $E_1 = E_2$

(B) $E_1 = \frac{1}{2}E_2$

(C) $E_1 = 2E_2$

(D) $E_1 = \frac{1}{4}E_2$

(E) $E_1 = 4E_2$

29. The sun produces energy by a nuclear reaction that converts hydrogen into helium. Which of the following statements best describes why the reaction releases energy?

(A) The reaction is endothermic.

(B) The total mass before the reaction is slightly less than the total mass after the reaction.

(C) The total mass before the reaction is slightly more than the total mass after the reaction.

(D) The total momentum before the reaction is slightly less than the total momentum after the reaction.

(E) The total momentum before the reaction is slightly more than the total momentum after the reaction.

30. Which of the following best describes what happens when an electron in an oxygen atom jumps from a higher energy level to a lower energy level?

   (A) A photon is emitted with its energy determined by the higher level.

   (B) A photon is emitted with its energy determined by the difference between the levels.

   (C) A photon is absorbed with its energy determined by the higher level.

   (D) A photon is absorbed with its energy determined by the difference between the levels.

   (E) A photon is absorbed with its energy determined by the lower level.

Use the following figure to answer questions 31–33.

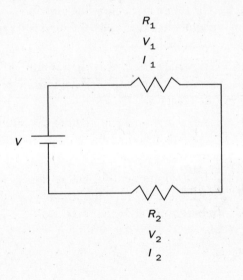

31. In the circuit shown above, the battery has a voltage V, and the resistors have resistances R_1 and R_2. What is the current I_1 in resistor R_1?

   (A) $I_1 = \dfrac{V}{(R_1 + R_2)}$

   (B) $I_1 = \dfrac{V}{R_1}$

   (C) $I_1 = \dfrac{V}{R_2}$

   (D) $I_1 = V(R_1 + R_2)$

   (E) $I_1 = VR_1$

32. In the same circuit, what is the potential difference V_1 across resistor R_1?

   (A) $V_1 = V$

   (B) $V_1 = \dfrac{V(R_1 + R_2)}{R_2}$

   (C) $V_1 = \dfrac{VR_2}{(R_1 + R_2)}$

   (D) $V_1 = \dfrac{VR_1}{(R_1 + R_2)}$

   (E) $V_1 = \dfrac{V(R_1 + R_2)}{R_1}$

33. In the same circuit, which of the following statements is correct about the currents I (the total current in the circuit), I_1, and I_2 and the potential differences V, V_1, and V_2?

   (A) $I = I_1 + I_2$ and $V = V_1 + V_2$

   (B) $I = I_1 = I_2$ and $V = V_1 + V_2$

   (C) $I = I_1 + I_2$ and $V = V_1 = V_2$

   (D) $I = I_1 = I_2$ and $V = V_1 = V_2$

   (E) $I_1 = I + I_2$ and $V_1 = V + V_2$

GO ON TO THE NEXT PAGE
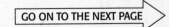

34. A bullet is shot straight into the air with an upward speed of v_0. It reaches its maximum height, h, and falls back down. Neglecting the effect of air resistance, what is its downward speed when it returns to its original level?

    (A) $< v_0$
    (B) $> v_0$
    (C) v_0
    (D) $\sqrt{gh}$
    (E) g

35. Identical small, strong magnets are dropped down two pipes. The pipes are the same size and shape but made of different materials. One pipe is made of a PVC-type material that is neither magnetic nor an electrical conductor. The other pipe is copper; it is non-magnetic but conducts electricity. Another identical magnet is dropped at a significant distance from both pipes; it accelerates downward with the standard acceleration of gravity, g. Which of the following statements is correct?

    (A) All three magnets fall at the same rate.
    (B) The magnet in the PVC pipe and the magnet outside the pipes fall at the same rate, while the magnet in the copper pipe falls more rapidly.
    (C) The magnet in the copper pipe and the magnet outside the pipes fall at the same rate, while the magnet in the PVC pipe falls more rapidly.
    (D) The magnet in the copper pipe and the magnet outside the pipes fall at the same rate, while the magnet in the PVC pipe falls more slowly.
    (E) The magnet in the PVC pipe and the magnet outside the pipes fall at the same rate, while the magnet in the copper pipe falls more slowly.

36. A weight lifter lifts a 400 N weight a distance of 0.3 m. The weight is completely at rest at the beginning of the lift and at the end of the lift. What is the *net* work done on the weight during this lift?

    (A) 1,330 J
    (B) 120 J
    (C) 0 J
    (D) −120 J
    (E) −1,330 J

37. As shown in the figure below, a force, F, acting on a block at the angle shown, displaces the block a distance, s. What is the work that the force does on the block?

    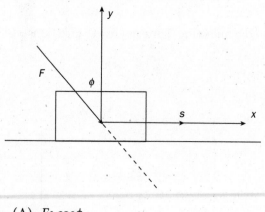

    (A) $Fs \cos\phi$
    (B) $Fs \sin\phi$
    (C) $Fs \tan\phi$
    (D) Fs
    (E) 0

GO ON TO THE NEXT PAGE

38. An electron with charge −e and mass m is traveling with a speed v, perpendicular to a uniform magnetic field, B. Which of the following best describes the path the electron is following?

    (A) A straight line perpendicular to B

    (B) A circle of radius $\dfrac{eB}{mv}$

    (C) A circle of radius $\dfrac{mv}{eB}$

    (D) A straight line parallel to B

    (E) A circle of radius $\dfrac{em}{Bv}$

39. A stringed instrument has a string with a distance of 0.5 m between the ends. Which of the following is a possible wavelength of the standing wave that is set up in the string after it is plucked?

    (A) 0.75 m

    (B) 1 m

    (C) 2 m

    (D) 4 m

    (E) 5 m

40. What will be the frequency of radio waves with a wavelength of 1 cm?

    (A) 3×10^2 Hz

    (B) 3×10^4 Hz

    (C) 3×10^6 Hz

    (D) 3×10^8 Hz

    (E) 3×10^{10} Hz

41. Laboratory experiments that show a diffraction pattern produced by a beam of electrons transmitted through aluminum foil confirm that

    (A) aluminum atoms have wavelike as well as particle-like properties.

    (B) electrons have particle-like but not wavelike properties.

    (C) electrons have wavelike but not particle-like properties.

    (D) electrons have wavelike as well as particle-like properties.

    (E) aluminum atoms have particle-like but not wavelike properties.

GO ON TO THE NEXT PAGE

42. The temperature of a tank of pure nitrogen gas increases from 200 K to 400 K. What happens to the root mean square speed of the nitrogen molecules?

(A) It doubles.

(B) It increases by four times.

(C) It increases by a factor of $\sqrt{2}$.

(D) It decreases by a factor of 2.

(E) It remains the same.

43. Which of the following best describes the physical consequences of Bernoulli's equation?

(A) The pressure of a gas is lower when the gas is flowing at a higher speed.

(B) The pressure of a gas is higher when the gas is flowing at a higher speed.

(C) The pressure of a gas is independent of the speed.

(D) The pressure of a gas is lower at a lower temperature.

(E) The pressure of a gas is lower at a higher temperature.

Use the following diagram to answer questions 44–45.

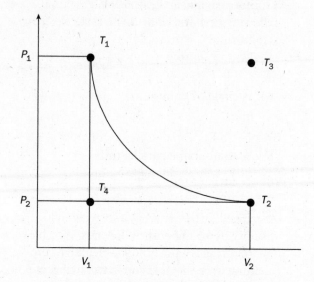

44. For the P-V diagram shown above, the path from T_1 to T_2 represents an isothermal process. Which of the following statements about the temperatures is correct?

(A) $T_3 = T_4$

(B) $T_3 = T_2$

(C) $T_3 > T_4$

(D) $T_3 < T_4$

(E) $T_3 < T_2$

45. Which of the following statements about the pressures on the same diagram is correct?

(A) $P_3 = P_4$

(B) $P_3 < P_4$

(C) $P_3 = P_2$

(D) $P_3 > P_2$

(E) $P_3 > P_1$

46. A cart is rolling along a smooth, level sidewalk with no friction. A child jumps into the cart, striking the cart in a perfectly vertical direction when she lands in the cart, and the cart continues to move along the sidewalk with the child inside. Which of the following statements about the cart's speed is correct?

   (A) The cart's speed stays the same because momentum is conserved.

   (B) The cart's speed decreases because momentum is conserved.

   (C) The cart's speed increases because momentum is conserved.

   (D) The cart's speed increases because kinetic energy is conserved.

   (E) The cart's speed decreases because kinetic energy is conserved.

47. In terms of the principles of thermodynamics, which of the following statements best describes the reason that a person will feel warm after exercising?

   (A) According to the third law of thermodynamics, work generates heat.

   (B) Because of the second law of thermodynamics, entropy is decreasing, which increases the temperature.

   (C) The temperature increases as the speed increases and the exercise moves the muscles, giving them a higher temperature.

   (D) Working muscles must absorb more heat according to the first law of thermodynamics.

   (E) Muscles are not 100 percent efficient and must therefore generate waste heat according to the second law of thermodynamics.

48. A string of length L has a rock of mass m tied to the end. An astronaut is holding the other end of the string, twirling the rock and rope around in a horizontal circle above his head in a weightless environment. It takes a time t for the rock to make a complete circle around his head. What is the expression for the tension T in the rope?

   (A) $T = \dfrac{4\pi^2 mL}{t^2}$

   (B) $T = \dfrac{4\pi^2 m}{t^2}$

   (C) $T = \dfrac{mL}{t^2}$

   (D) $T = \dfrac{m}{Lt^2}$

   (E) $T = \dfrac{4\pi^2 mt^2}{L}$

49. A sealed door has an area of 6,500 cm^2. On one side of the door, the atmospheric pressure is 10,000 N/m^2; on the other side, the pressure is 9,000 N/m^2. What is the minimum force needed to push the door open into the side with higher pressure?

   (A) 650 N/m^2

   (B) 7.2 atm

   (C) 650 N

   (D) 72.2 atm

   (E) 6.5 N

50. A fully loaded boat displaces 20 m^3 of water that has a density of 1,000 kg/m^3. What is the weight of the loaded boat?

   (A) 20,000 N

   (B) 200,000 N

   (C) 20 N

   (D) 200,000 kg

   (E) 20,000 kg

GO ON TO THE NEXT PAGE

51. When light photons strike metal atoms, the photoelectric effect can cause electrons to escape from the atoms. Which of the following will cause the electrons to escape the atoms at a higher speed?

(A) Increasing the light intensity

(B) Decreasing the wavelength of the light

(C) Flashing the light on and off very rapidly

(D) Decreasing the light intensity

(E) Increasing the wavelength of the light

Use the following information to answer questions 52–53.

A car is initially traveling at a speed of 8 m/s. Reaching the open highway, it begins accelerating at 4 m/s^2 for a period of 5 s.

52. What is the car's speed after this time period?

(A) 12 m/s

(B) 20 m/s

(C) 28 m/s

(D) 100 m/s

(E) 108 m/s

53. What is the total distance the car travels during the 5 s that it is accelerating?

(A) 10 m

(B) 40 m

(C) 50 m

(D) 90 m

(E) 250 m

54. A parallel plate capacitor is charged and then disconnected from a circuit. While keeping the capacitor fully charged, the distance between the plates is doubled. What happens to the amount of energy stored in the capacitor?

(A) It remains unchanged.

(B) It is cut in half.

(C) It increases by a factor of $\sqrt{2}$.

(D) It doubles.

(E) It increases by four times.

55. An isolated sphere has a uniform positive charge distributed over its surface. Which of the following best describes the electric field lines outside the sphere?

(A) They are directed radially inward.

(B) They are directed radially outward.

(C) They are a series of circles concentric with the sphere, pointing clockwise.

(D) They are a series of circles concentric with the sphere, pointing counterclockwise.

(E) The electric field is zero outside the sphere, so there are no field lines.

56. A 1,500 kg car is accelerating forward at a rate of 5 m/s^2. There is a 2,500 N frictional force acting on the car. How much force is acting on the car in the direction of the acceleration?

(A) 10,000 N

(B) 7,500 N

(C) 5,000 N

(D) 2,500 N

(E) 0 N

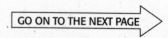

GO ON TO THE NEXT PAGE

Use the following diagram to answer question 57.

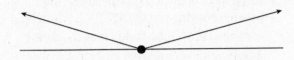

57. A person having a mass m is standing on the middle of a rope bridge as shown in the figure. The rope supporting the person makes an angle θ with the horizontal on both sides of the person. Which of the following expressions correctly describes the tension, T, in the rope?

   (A) $T = mg$

   (B) $T = \dfrac{mg}{2\cos\theta}$

   (C) $T = \dfrac{mg}{\cos\theta}$

   (D) $T = \dfrac{mg}{\sin\theta}$

   (E) $T = \dfrac{mg}{2\sin\theta}$

58. If Earth's mass is m and the sun's mass is M, which of the following expressions correctly describes the Earth's orbital speed around the sun? Make the simplifying assumption that Earth orbits the sun in a circular orbit of radius r.

   (A) $v^2 = \dfrac{GM}{r}$

   (B) $v^2 = \dfrac{Gm}{r}$

   (C) $v^2 = \dfrac{GMm}{r}$

   (D) $v^2 = \dfrac{gM}{r}$

   (E) $v^2 = \dfrac{gm}{r}$

59. When monochromatic light passes through a double slit (as in Young's double-slit experiment), it produces a pattern of bright and dark fringes. Which of the following effects causes this bright-dark pattern?

   (A) Refraction

   (B) Polarization

   (C) Dispersion

   (D) Interference

   (E) Diffraction

60. The acceleration of gravity near the Earth's surface is represented by g. What is the gravitational acceleration on a satellite with an orbit that is twice the Earth's radius from the center of the Earth?

   (A) $\dfrac{g}{2}$

   (B) $\dfrac{g}{4}$

   (C) $\dfrac{g}{\sqrt{2}}$

   (D) g

   (E) 0

Use the following information to answer questions 61–62.

In a pairs figure skating competition, a 100 kg man and 50 kg woman begin their routine at rest, standing close to each other. They then push away from each other. Assume that they both slide across the ice with no friction.

61. When the skaters push apart, the man exerts a 200 N force on his partner. What is the magnitude of his acceleration?

   (A) 10 m/s^2

   (B) 5 m/s^2

   (C) 4 m/s^2

   (D) 2 m/s^2

   (E) 0 m/s^2

GO ON TO THE NEXT PAGE

62. After they break contact with each other, the woman is moving with a speed of 6 m/s. What is the man's speed?

(A) 0 m/s

(B) 3 m/s

(C) 6 m/s

(D) 9 m/s

(E) 12 m/s

63. A child continuously blows a whistle at a constant intensity and pitch while she is swinging back and forth on a swing. A musician with perfect pitch is standing in front of the swing listening to her blow the whistle. Which of the following best describes how he hears the sound?

(A) The sound is louder when she is swinging toward him.

(B) The sound is louder when she is swinging away from him.

(C) The sound has a higher pitch when she is swinging toward him.

(D) The sound has a higher pitch when she is swinging away from him.

(E) The loudness and pitch remain constant.

64. Which of the following is an illustration of the Doppler effect?

(A) Different perceptions of ocean waves depending on whether the wave moves toward or away from you

(B) The drop in pitch of an ambulance siren as it drives past you

(C) Decreasing wavelengths from a galaxy moving away from us

(D) Observations from Young's double slit experiment

(E) Observations from Millikan's oil drop experiment

65. A child is pushing horizontally on a wagon with a force of 50 N. With this force, the wagon is moving at a speed of 3 m/s on a level sidewalk. What is the child's power output?

(A) 0 W

(B) 3 W

(C) 17 W

(D) 50 W

(E) 150 W

66. An electron is moving in a straight line through a region of space in which no forces act on the electron. A uniform magnetic field is suddenly turned on in this region of space, such that the magnetic field lines are not parallel (or antiparallel) or perpendicular to the electron's velocity. Which of the following best describes the path the electron will take after the magnetic field is turned on?

(A) It will continue to move at the same speed and direction.

(B) It will move in a circular path, with the plane of the circle perpendicular to the magnetic field lines.

(C) It will move in a circular path, with the plane of the circle parallel to the magnetic field lines.

(D) It will move in a helical (spiral) path around the magnetic field lines.

(E) It will come to a complete stop.

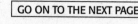GO ON TO THE NEXT PAGE

67. In a certain region of space, a fictional conservative force has the form $F = bpi$. Without knowing the physical meaning of these quantities, we know that the product has the correct units for force and does not depend on the position, x. Which of the following expressions could give the potential energy associated with this force?

    (A) $\dfrac{x}{bpi}$

    (B) $\dfrac{1}{2} bpix^2$

    (C) $\dfrac{bpi}{x}$

    (D) $bpix$

    (E) There will be no potential energy associated with this force.

68. There is a point charge, Q, in a region of space that is isolated from any other influences. In another region of space, also isolated from any other influences, there is a spherical charge distribution of radius R having the same total charge Q. At a distance r from each of these charges, where $r > R$, which of the following statements best describes the electric fields from the point charge and the spherical charge distribution?

    (A) The electric fields are the same.

    (B) The electric field from the point charge is larger.

    (C) The electric field from the point charge is smaller.

    (D) The electric field from the point charge is outward, while the electric field from the spherical charge is inward.

    (E) The electric field from the point charge is inward, while the electric field from the spherical charge is outward.

69. An object is held in uniform circular motion in a vertical circle by a length of taut rope. Which of the following best describes the forces acting on the object at the top of the circle?

    (A) The tension in the rope upward and gravity downward

    (B) The tension in the rope and gravity both downward

    (C) The tension in the rope and gravity downward and the centripetal force upward

    (D) The tension in the rope and gravity downward and the centrifugal force upward

    (E) The tension in the rope and the centrifugal force upward and gravity downward

70. An object is moving at some velocity when a force is applied perpendicular to the velocity. Which of the following best describes what will happen?

    (A) The velocity will not change.

    (B) The speed will increase, but the direction will not change.

    (C) The speed will decrease, but the direction will not change.

    (D) The speed will not change, but the direction will change.

    (E) Both the speed and direction will change.

IF YOU FINISH BEFORE TIME IS CALLED, YOU MAY CHECK YOUR WORK ON THIS SECTION ONLY. DO NOT TURN TO ANY OTHER SECTION IN THE TEST. **STOP**

SECTION II

Time—90 Minutes 8 Questions

> **Note:** Calculators are permitted, except for those with typewriter (QWERTY) keyboards.
>
> **Directions:** A table of information and lists of equations that may be helpful are on the green insert in the AP exam. Write out the answers to the following questions. Clearly show the method used and steps involved in arriving at your answers. Partial credit can be given only if your work is clear and demonstrates an understanding of the problem.

1. A 0.65 kg glass vase falls off a 1.3 m-high table onto the floor.

   (a) The vase's owner, on the other side of the room, wants to catch the vase before it hits the floor. How long does it take the vase to fall from the table to the floor?

   (b) At what speed does the vase hit the floor?

   (c) Fortunately the floor is carpeted, so the vase does not break. It compresses the carpet a distance of 0.005 m when it hits the carpet. What was the average stopping acceleration on the vase?

   (d) What average force was applied to the vase to stop it?

   (e) What was the impulse acting on the vase to stop it?

2. Imagine the following physically impossible situation: A planet is orbiting a star in a region of space where there is no gravitational force between the star and planet. Instead, an amount of charge has been transferred from the star to the planet so that each has a net charge of 5×10^{15} C. This charge is negative on the star and positive on the planet. Assume the planet is a distance of 2×10^{11} m from the star.

   (a) Ignoring the star, what is the value and direction of the electric field at the position of the planet?

   (b) Now consider both the planet and the star. Draw a sketch of what the electric field lines would look like if no other charges are nearby.

   (c) What is the amount of electrical force between the planet and the star? Is it attractive or repulsive?

   (d) If the star has the same mass as the Sun, $M = 2 \times 10^{30}$ kg, and the planet has the same mass as the Earth, $m = 6 \times 10^{24}$ kg, calculate both the planet's orbital speed in m/s and the planet's orbital period in years around the star.

GO ON TO THE NEXT PAGE ⟶

3. A power supply is placed in series with a 50-ohm resistor and a Slinky™ (a child's toy made out of a flexible coil of wire) as shown in the diagram. The Slinky has a total of 58 loops and is 35 cm long. A voltmeter is used to measure the voltage across the 50-ohm resistor. The power supply is adjusted to change the current flowing in the circuit, and a magnetic field sensor is used to measure the magnetic field in the central region of the Slinky. The following data are gathered:

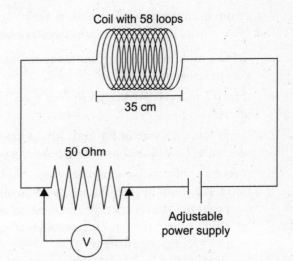

Coil with 58 loops

35 cm

50 Ohm

Adjustable power supply

V

Voltmeter

Voltage Across 50-Ohm Resistor (V)	Current (A)	Magnetic Field (T)
0		0
4.0		1.5×10^{-5}
7.9		3.0×10^{-5}
12.3		4.8×10^{-5}
16.1		6.3×10^{-5}
20.0		7.9×10^{-5}

(a) Fill out the values of the currents in the data table above and show a sample calculation for current in the 4.0 V case.

(b) Graph the magnetic field versus current.

(c) Calculate the slope of the graph and identify its units.

(d) Write an equation relating the magnetic field, B, and the current, I.

(e) The coil density, n, is defined as the ratio of the total number of loops to the meters of length. Calculate the coil density for this Slinky.

(f) Theoretically, the magnetic field in the central region of a long, thin coil of wire is given by $B = \mu_0 n I$, where μ_0 is the vacuum permeability constant. Calculate the experimental value for the vacuum permeability constant and show how you obtained its units.

(g) Find the percentage error between the theoretical value and your experimental value of the permeability constant. Discuss possible sources of error.

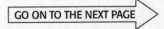

GO ON TO THE NEXT PAGE

4. You are doing an experiment to measure the resistivity of a piece of wire of an unknown composition. This piece of wire has a length l and a circular cross section with a radius r.

   (a) Using a multimeter, you measure the resistance, R, of this wire. What is the resistivity, ρ, of this wire in terms of l, r, and R?

   (b) If $l = 0.5$ m, $r = 0.002$ m, and $R = 6.8 \times 10^{-4}\ \Omega$, what is the resistivity of the material that makes up the wire?

   (c) If you wanted a wire of the same length, made of the same type of material, but with a lower resistance, how should you change the cross-sectional radius of the wire? Explain your answer.

   (d) A member of your lab group who tends to do foolish things wants to stick each end of the wire into a 120 V wall socket. You wisely try to talk the person out of this foolish action by explaining the probable consequences. If someone actually tried this, what would the current in the wire be? Explain your answer. What would be the likely result of trying to induce this much current? Again, explain your answer.

5. An electron in a hypothetical atom jumps from an energy level at 7 eV above the ground state to another level that is 18 eV above the ground state.

   (a) When making this jump, does the electron absorb or emit a photon? Explain why, in terms of conservation of energy.

   (b) What is the energy of this photon?

   (c) What is the frequency of this photon?

   (d) What is the wavelength of this photon?

6. A 90 cm guitar string is plucked such that a standing wave pattern is formed as shown below. The string vibrates at a frequency of 1,200 Hz.

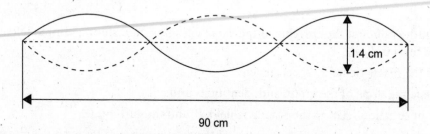

90 cm

   (a) What are the wavelength and amplitude of the standing wave?

   (b) Determine the speed of the wave on the string in m/s.

   (c) How could you alter the speed of the wave on the guitar string?

   (d) Is the wave on the string transverse or longitudinal? Explain.

   (e) What is the frequency of the sound wave produced by the vibrating string?

   (f) Explain whether this sound wave is transverse or longitudinal.

   (g) Is there a way to change the speed of the sound wave? If so, what would you do?

GO ON TO THE NEXT PAGE ▷

7. A spherical, shiny holiday decoration ball is acting as a convex mirror. The sphere has a radius of 4 cm.

  (a) What is the focal length of this mirror?

  (b) You are looking at your image in this mirror. Your face is 10 cm from the mirror. Where do you see the image of your face?

  (c) How much bigger or smaller is the image of your face than the actual size of your face?

  (d) Is this image real or virtual? Is it upright or inverted?

8. A portion of an electric circuit is connected to a 60-ohm resistor and is embedded in 0.50 kg of a solid substance in a calorimeter. The external portion of the circuit is connected to a 120 V DC power supply.

  (a) Draw a schematic diagram of the situation.

  (b) Calculate the current in the resistor.

  (c) Calculate the rate at which heat is generated in the resistor.

  (d) Assuming that all of the heat generated by the resistor is absorbed by the solid substance and that it takes 5 minutes to raise the temperature of the substance from 25°C to 85°C, calculate the specific heat of the substance.

  (e) At 85°C the substance begins to melt. The heat of fusion of the substance is 1.35×10^5 J/kg. How long after the temperature reaches 85°C will it take to melt all of the substance?

AP Physics B Practice Test: **Answer Key**

1. E	19. D	37. B	55. B
2. C	20. B	38. C	56. A
3. A	21. A	39. B	57. E
4. D	22. B	40. E	58. A
5. B	23. C	41. D	59. D
6. E	24. B	42. C	60. B
7. A	25. E	43. A	61. D
8. B	26. C	44. C	62. B
9. B	27. A	45. D	63. C
10. E	28. C	46. B	64. B
11. A	29. C	47. E	65. E
12. D	30. B	48. A	66. D
13. A	31. A	49. C	67. D
14. E	32. D	50. B	68. A
15. A	33. B	51. B	69. B
16. E	34. C	52. C	70. E
17. C	35. E	53. D	
18. A	36. C	54. D	

ANSWERS AND EXPLANATIONS

SECTION I

1. E

In a completely elastic collision, kinetic energy is conserved. The energy that produces the sound must come from somewhere, according to conservation of energy. The only energy source available is the kinetic energy of the objects in the collision. Therefore, the fact that a sound is produced by the collision means that the kinetic energy cannot be conserved and the collision cannot be completely elastic.

2. C

The swings act as pendulums. The period of a pendulum depends on the length of the pendulum and the local acceleration of gravity, not the mass on the end of the pendulum. If the swings both have the same length, they will swing back and forth in the same amount of time (i.e., they will have the same period, regardless of the weight of the person on the swing). Both the child and the adult take the same amount of time to make a complete oscillation.

3. A

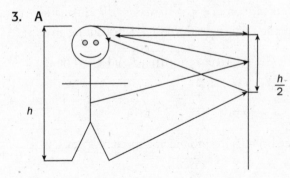

The ray diagram illustrates the light path from the feet and other parts of the body to the eyes. If we make the approximation that our eyes are essentially at the top of our heads, then the ray from the feet to the eyes reflects off of a point on the mirror at half the body height. Rays from the knees, waist, and other parts of the body will reflect off points on the mirror higher than where the feet reflect. Hence, the mirror needs to extend from the level of the body midpoint to the top of the head level. It must be half a person's height with the top at head level.

4. D

The total power for the wire(s) is given by $P = \dfrac{V^2}{R}$, where $V = 120$ for both situations. When we cut the wire in two, the resistance of each new wire is half the original. Therefore, the power output of each wire is double the original. Adding the effects of both wires, the power output quadruples.

5. B

For capacitors in series, add the reciprocals of the individual capacitances to get the reciprocal of the equivalent capacitance: $\dfrac{1}{C_{eq}} = \dfrac{1}{C_1} + \dfrac{1}{C_2} + \dfrac{1}{C_3}$. When the capacitors are in parallel, add the capacitances to get the equivalent capacitance. These formulas for capacitors are the reverse of the forms of the similar formulas for resistors in series and parallel.

6. E

Newton's second law is most often stated as $F = ma$, choice (A). However, it can also be written as $F = \dfrac{dp}{dt}$, where p is the momentum, choice (B). For constant mass, $\dfrac{dp}{dt} = \dfrac{mdv}{dt}$, so choice (C) is also correct. Finally, we can derive the weight $w = mg$, choice (D), from the second law. All of the choices are correct.

7. A

The average acceleration is given by the change in velocity divided by the change in time: $a = \dfrac{\Delta v}{\Delta t}$. The velocity is given by the change in *position* (not total distance traveled) over time.

Acceleration, velocity, and position are all vectors, so we are looking at net change in position. Since the location is the same at start and finish, the net change in position is zero. Hence, the average acceleration is zero.

8. B

First, find the image distance using the thin lens equation:

$$\frac{1}{s_i} + \frac{1}{s_o} = \frac{1}{f}$$

where f is the focal length, 4 cm, s_i is the unknown image distance; and s_o is the object distance, 8 cm. Substituting the numerical values and solving gives us $s_i = 8$ cm. Hence, the image is 8 cm from the lens. Following the sign conventions, the fact that the image distance is *positive* tells us that the image is on the *opposite* side of the lens from the object. The light rays will actually cross at this point, so it is a real image.

The magnification, M, is given by this equation:

$$M = -\frac{s_i}{s_o}$$

$$M = -\frac{8 \text{ cm}}{8 \text{ cm}}$$

$$M = -1$$

The fact that the magnification is −1 tells us that the image is the same size as the object but is inverted. Therefore, the image is 8 cm from the lens on the opposite side from the object, real, and inverted.

9. B

In this case, the object is the Sun, and the Sun's image is the bright spot of light on the paper. Hence, the image distance is the distance between the lens and the bright spot on the paper, which is 12 cm. The object distance is the distance to the Sun, which in this case is very close to infinity. When the object distance is infinity, its reciprocal is zero, and the thin lens equation, $\dfrac{1}{f} = \dfrac{1}{s_i} + \dfrac{1}{s_o}$, gives $f = s_i$; the focal length equals the image distance. Hence, the focal length of the lens is 12 cm.

10. E

The Lorentz force law tells us that the total electromagnetic force is the vector sum of the electric and magnetic forces acting on the charge. Both forces are proportional to the magnitude of the charge. The electric force is dependent on the electric field vector, while the magnetic force is dependent on the cross product of the velocity and magnetic field.

11. A

The ball has an initial speed of $v_o = 5$ m/s and a gravitational acceleration of $a = 10$ m/s^2. It travels a total distance $s - s_o = 10$ m. Use the kinematic equation $v^2 - v_o^2 = 2a(s - s_o)$. Substituting gives us

$$v^2 = (5 \text{ m/s})^2 + (2 \times 10 \text{ m/s}^2)(10 \text{ m} - 0 \text{ m})$$

$$v^2 = 225 \text{ m}^2/\text{s}^2$$

$$v = 15 \text{ m/s}$$

The ball hits the ground at a speed of 15 m/s.

12. D

Nearly everyone can recall that Newton's third law states, "For every action, there is an equal and opposite reaction." But what exactly does this mean? Actions and reactions come in pairs. But beware: Not all equal and opposite forces are action-reaction pairs. An action-reaction pair can involve only two objects. If object A acts on object B, the reaction required by Newton's third law is object B acting on object A with an equal and opposite force. There can be no third object involved. Hence, the reaction to Earth's gravity pulling a person down with an 800 N force is the person's gravity pulling Earth up with an 800 N force. Other forces involved in this situation may be equal and opposite to the original force, but they are not the reaction required by Newton's third law.

13. A

This question asks about the forces specifically on the weight, not those on the spring or hook, so it is crucial to consider the weight only. The Earth's gravitational force pulls on the weight downward. The spring pulls the weight upward. At the same time, the weight pulls the spring downward, but that force acts on the spring, not the weight. The hook also exerts a force on the spring but does not exert a force on the weight. Therefore, the forces on the weight are gravity acting downward and the spring acting upward.

14. E

The buoyant force is equal to the weight of the water displaced. A volume of 5 m^3 is displaced. Multiply the volume, V, by the density, $\rho = 1,000$ kg/m^3, to get the mass, m, of water displaced.

$$m = \rho V = 5,000 \text{ kg}$$

Now multiply the mass by the acceleration of gravity, $g = 10$ m/s^2, to get the weight, w, of the water displaced.

$$w = mg = (5,000 \text{ kg})(10 \text{ m/s}^2) = 50,000 \text{ N}$$

The weight of the water displaced, and therefore the buoyant force on the boat, is 50,000 N.

15. A

To find the answer, apply Hooke's law:

$$F_s = kx$$

where F_s is the force on the spring, k is the spring constant, and x is the distance the spring is, in this case, compressed. We are not worried about the direction of the force here, so you can ignore the minus sign that often appears in this equation. The force acting to compress the spring is the 80 kg driver's weight, or 800 N (using $g = 10$ m/s^2). The spring is compressed $x = 0.02$ m. Substituting the numbers and solving for k gives $k = 40,000$ N/m.

16. E

The amount of thermal expansion is given by the equation

$$\Delta \ell = \alpha \ell_0 \Delta T$$

Here $\Delta \ell$ represents the change in length as the bar is heated and is unknown. The coefficient of thermal expansion is $\alpha = 4 \times 10^{-5} (°C)^{-1}$. The bar's original length, ℓ_0, is 5 m. ΔT represents the change in temperature, which is 100°C. Putting the numbers into the equation gives us

$$\Delta \ell = (4 \times 10^{-5} (°C)^{-1})(5 \text{ m})(100°C)$$

$$\Delta \ell = 0.02 \text{ m}$$

Adding the change in length to the original length gives the final total length of 5.02 m.

17. C

There are two logically equivalent statements of the second law of thermodynamics: the efficiency of any process is always less than 1, and any process must increase the total entropy of the universe. The first statement of this law tells us that any heat engine always has an efficiency less than 1.

18. A

The magnetic force, F_B, on a particle with a charge, q, moving in a magnetic field, B, with a speed, v, is given by

$$F_B = qvB\sin\theta$$

where θ is the angle between the magnetic field and the velocity. A neutron has no charge, so it can have no magnetic force acting on it. A charged particle at rest also has no magnetic force acting on it because its speed is zero. A charged particle moving parallel to the magnetic field also experiences no force because $\sin\theta = 0$. The electron moving perpendicular to the magnetic field has none of the quantities in the above equation equal to zero, so it can have a magnetic force acting on it.

19. D

Snell's law for refraction applied to this situation gives

$$n_a \sin\theta_i = n_g \sin\theta_r$$

where n_a and n_g are the indices of refraction for air and the glass, respectively, and θ_i and θ_r are the incident and refracted angles, respectively. The index of refraction of air is very close to 1, so solving for the sine of the refracted angle gives us

$$\sin\theta_r = \frac{\sin\theta_i}{n_g}$$

20. B

When light strikes atoms of certain materials, electrons escape the atoms. If these materials are conductors, the electrons moving freely produce an electric current. This effect was explained by Albert Einstein in 1905 and is called the photoelectric effect. Solar power cells work on this principle to produce an electric current.

21. A

To answer this question, find the force, F, between two charges from Coulomb's law:

$$F = \left(\frac{1}{4}\pi\varepsilon_0\right)\frac{q_1 q_2}{r^2}$$

where q_1 and q_2 are the charges and r is the distance between them. From the question stem, $q_1 = 8 \times 10^{-4}$ C, $q_2 = 2 \times 10^{-4}$ C, and $r = 2$ m. Plugging these values into the equation for Coulomb's law gives us

$$F = (9 \times 10^9 \text{ N} \cdot \text{m}^2/\text{C}^2)\frac{(8 \times 10^{-4} \text{ C})(2 \times 10^{-4}\text{C})}{(2 \text{ m})^2}$$

$$F = 360 \text{ N}$$

So the force between the charges is 360 N.

22. B

For any wave, the wavelength times the frequency equals the speed of propagation. Hence, the equation $v = f\lambda$ applies to light waves, water waves, or any other type of waves.

23. C

Destructive interference will result in a dark band, and constructive interference will result in a bright band, so you should eliminate choices that state otherwise (choices (A) and (D)) right away. If the path length is shifted by half a wavelength, the superimposed waves will be such that the peaks of one wave line up with the valleys of the other wave (i.e., the waves are out of phase, and the waves cancel each other out. Thus the result is destructive interference, resulting in a dark band.

24. B

The capacitance, C, of a parallel plate capacitor with plates of area, A, a distance, d, apart is given by

$$C = \frac{\varepsilon_0 A}{d}$$

The form of the equation with d in the denominator tells us that if the distance doubles, the capacitance is half as much. So the capacitor with double the separation has half the capacitance.

25. E

The capacitance is $C = \frac{q}{V}$. Gauss's law tells us that $q = \varepsilon_0 EA$, and we also know $V = Ed$.

Therefore, $C = \frac{\varepsilon_0 A}{d}$ or $A = \frac{Cd}{\varepsilon_0}$. Using

$C = 0.5 \times 10^{-6}\,\text{F} = 5 \times 10^{-7}\,\text{F}$,
$d = 0.177 \times 10^{-3}\,\text{m} = 1.77 \times 10^{-4}\,\text{m}$, and
$\varepsilon_0 = 8.85 \times 10^{-12}\,\text{C}^2/\text{N} \cdot \text{m}^2$, we find that
$A = 10\,\text{m}^2$.

26. C

The brightness of the bulb will be determined by the electrical power the bulb uses. The power is the current passing through the bulb multiplied by the potential difference across the bulb. In a series circuit, the current is the same everywhere in the circuit. The potential difference across each bulb will be half the potential difference across the battery because the bulbs both have the same resistance. Thus, the bulbs will be equally bright because they have the same current passing through them and the same potential difference across them.

27. A

In this question the centripetal acceleration, a_c, is given as 9,000 m/s^2, and the radius of the circular motion is $r = 0.1\,\text{m}$. Applying the formula for centripetal acceleration gives us

$$a_c = \frac{v^2}{r}$$

$$v^2 = a_c r$$

$$v^2 = (9{,}000 \text{ m/s}^2)(0.1\,\text{m}) = 900 \text{ m}^2/\text{s}^2$$

$$v = 30 \text{ m/s}$$

Hence, the sample has a linear velocity of 30 m/s.

28. C

The energy, E, of a photon is proportional to the frequency, f, and inversely proportional to the wavelength, λ. The equation is

$$E = hf = \frac{hc}{\lambda}$$

where h is Planck's constant and c is the speed of light. Because the relationship between the energy and wavelength is inversely proportional, halving the wavelength will double the energy.

29. C

In nuclear reactions, the total mass before the reaction is slightly different from the total mass after the reaction. In reactions that release energy, such as those powering the sun, the mass is a small amount larger before the reaction; though the mass is less after the reaction, this mass is not really lost because it is converted to energy. The corresponding amount of energy is determined from the equation $E = mc^2$. Hence, the reactions powering the sun release energy because the mass before the reaction is more than after the reaction and the difference is converted into energy.

30. B

When an electron jumps to a lower energy level in an atom, it emits a photon. The photon energy equals the difference in energy between the two energy levels.

31. A

The current will be the same at all points in a series circuit. To find the current in this circuit, use Ohm's law with the equivalent resistance of the two resistors. When resistors are in series, add the resistance to find the equivalent resistance, R, so $R = R_1 + R_2$. Now Ohm's law relating the potential difference, V, current, I, and resistance, R, is

$$V = IR = I(R_1 + R_2)$$

Solving for I and noting that $I = I_1 = I_2$ because the current is the same at all points in a series circuit, we get

$$I_1 = \frac{V}{(R_1 + R_2)}$$

32. D

This question builds on the result of the previous question. The current, I_1, in resistor 1 is given in the solution to the previous question. Now apply Ohm's law to resistor 1 rather than the entire circuit:

$$V_1 = I_1 R_1$$

Substituting the expression for I_1 gives us

$$V_1 = \frac{VR_1}{(R_1 + R_2)}$$

33. B

In a series circuit, the current is the same at each point in the circuit. However, it is still true that the potential differences of the various circuit elements add up over the circuit. Note that the reverse is true for a parallel circuit; the potential differences across the various parallel circuit elements are equal, while the currents in the individual parallel circuit elements add to give the total current. In this problem, we have a series circuit rather than a parallel circuit, so

$$I = I_1 = I_2 \text{ and } V = V_1 + V_2$$

34. C

By symmetry, the downward speed is the same as the upward speed at the same level. It is possible to prove this statement by using kinematics and computing the maximum height, time of flight, and the downward speed. It is also possible to use conservation of energy arguments to reach the same conclusion. The initial kinetic energy is converted to potential energy, then converted back to kinetic energy as the bullet falls. In any case, if air resistance is neglected, the bullet, or any other similar projectile, has the same downward speed as its initial upward speed when it returns to the same level.

35. E

A falling magnet generates an electric field because the magnetic field is moving. This electric field will induce electric currents in the copper pipe, which can conduct electricity. The electric currents in the copper pipe in turn induce a magnetic field, even though the copper pipe is not magnetic. By Lenz's law, the induced magnetic fields are induced in a direction that opposes their cause. Here it is important to note that the cause of the induced magnetic field is *not* the magnetic field of the magnet. Rather it is the *change* in this magnetic field. This change is caused by the fact the magnet is moving. So by Lenz's law, the induced magnetic field is in a direction that will oppose, and therefore slow, the motion of the falling magnet. Thus, the magnet falling in the copper pipe will fall more slowly than it would otherwise.

The magnet falling in the PVC pipe will fall at the same rate as the magnet that is outside the pipes, because the PVC pipe does not conduct electricity and therefore will not induce any electric currents or magnetic fields. Hence, the magnet in the PVC pipe and the magnet outside the pipe fall at the same rate, and the magnet in the copper pipe falls more slowly.

36. C

The key here is that the question asks for the *net* work. According to the work energy theorem, the net work equals the change in kinetic energy. The weight is at rest at both the beginning and the end of the lift; thus, the change in kinetic energy is zero, and the net work of all the forces acting on the weight is zero. Note that the work done by the weight lifter is 120 J and the work done by Earth's gravity is −120 J. These values add to give a net work of 0 J.

37. B

Be very careful here. The formula for the work, W, done by a force, F, displacing an object by a displacement, s, is usually given as

$$W = Fs \cos\theta$$

where θ is the angle between the force vector and the displacement vector. Note in the figure below that this angle θ is not the same angle as the angle ϕ given in the question. In this case $\theta = 90° - \phi$, so $\cos\theta = \sin\phi$. The work done by the force is therefore

$$W = Fs \sin\phi$$

Another way of looking at this question is that the work equals the component of the force parallel

to the displacement times the displacement. Drawing the components of the force parallel and perpendicular to the displacement will show that the force needs to be multiplied by $\sin\phi$, leading to the same result as above.

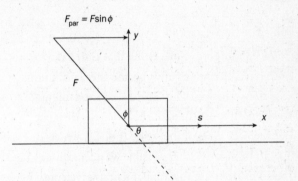

38. C

The force, F_B, on an electron with charge $-q$ and mass m moving at a speed v in a magnetic field, B, is given by

$$F_B = -qBv \sin\theta$$

where θ is the angle between the B and v vectors. In this problem, the electron is moving perpendicular to the magnetic field, so $\theta = 90°$ and $\sin\theta = 1$. This force is always perpendicular to the velocity, so it acts as a centripetal force and produces a uniform circular motion. For uniform circular motion, the required centripetal force is given by

$$F_c = \frac{mv^2}{r}$$

where r is the radius of the circular motion. Here the magnetic force supplies the centripetal force, so we can equate F_c and F_B. Ignoring the minus sign, we then get

$$eBv = \frac{mv^2}{r}$$

Solving for r gives us

$$r = \frac{mv}{eB}$$

Hence, the electron travels in a circular path of radius $\frac{mv}{eB}$.

39. B

For a standing wave in a string that is tied down at both ends, the nodes must be at either end. In this case they must be 0.5 m apart. The fundamental wavelength must therefore be twice the distance between the ends of the string (nodes), which is 1 m in this case. Other possible wavelengths can be found by dividing this fundamental wavelength by a positive integer. Hence, possible wavelengths λ are given by the formula

$$\lambda = \frac{2L}{n}$$

where L is the distance between the ends of the string and n is a positive integer. Possible values for the wavelength of a standing wave in the string are then 1 m, 0.5 m, 0.25 m, 0.125 m, and so on. The only one of these values given as a choice is 1 m.

40. E

For all waves, the speed of propagation is equal to the wavelength times the frequency. Radio waves are electromagnetic waves traveling at the speed of light, c. Hence, their frequency, f, is given by

$$f = \frac{c}{\lambda}$$

where λ is the wavelength. Using $c = 3 \times 10^8$ m/s and $\lambda = 1 \times 10^{-2}$ m gives us $f = 3 \times 10^{10}$ Hz.

41. D

Prior to the 20th century, there was much controversy about whether light was made of waves or particles. Some experiments gave evidence of light's wavelike nature, while other experiments gave evidence of its particle-like nature. The resolution to this dilemma came with the principle of wave-particle duality. According to this principle, light and particles such as electrons and protons exhibit both wavelike and particle-like properties. Subsequent experiments such as those involving the diffraction of electrons have confirmed that electrons and other particles have wavelike as well as particle-like properties. These experiments, however, do not show that aluminum atoms have wavelike properties because the diffraction pattern is from the electron beam, not the aluminum atoms.

42. C

The root mean square speed, v_{rms}, of the atoms or molecules in a gas is derived by solving

$$\frac{1}{2}\mu v_{rms}^2 = \frac{3}{2}k_B T$$

$$v_{rms} = \left(\frac{3k_B T}{\mu}\right)^{\frac{1}{2}}$$

where k_B is the Boltzmann constant, T is the temperature in Kelvins, and μ is the mass per molecule. From the equation, v_{rms} is proportional to the square root of the temperature. Note that the temperature must be in Kelvins for this equation to work.

If the temperature doubles, the root mean square velocity must increase by a factor of $\sqrt{2}$. If you cannot recall the exact form of the equation above, it is sufficient in this case to remember that the temperature, T, is proportional to the average kinetic energy of the atoms or molecules. That will tell you that the temperature is proportional to the velocity squared.

43. A

Bernoulli's equation states:

$$P + \frac{1}{2}\rho v^2 + \rho g y = \text{constant}$$

where P is the pressure, v is the speed, ρ is the density, and y is the height. From the terms of the equation, one can see that as the gas speed increases, the pressure must decrease (if the other quantities remain constant). Hence, by Bernoulli's principle, a gas that is flowing faster has a lower pressure.

44. C

An isothermal process takes place at a constant temperature; an isothermal line on a P-V diagram represents points that are all at the same temperature. In a gas, temperature increases as pressure increases, so points further from the origin than an isothermal line are at a higher temperature than the isotherm. Similarly, points closer to the origin are at a lower temperature. Hence, the temperature at point 3 is higher than the temperature at point 4; $T_3 > T_4$.

45. D

On the P-V diagram, the vertical axis represents the pressure. As the distance from the origin increases along this axis, the pressure increases. Point 3 is at a greater pressure than point 2, so $P_3 > P_2$.

46. B

This situation is a completely inelastic collision because the girl stays in the cart and they move together after the collision. In an inelastic collision, momentum is conserved, but kinetic energy is not conserved. Eliminate the choices that involve kinetic energy conservation. The vertical momentum is not conserved because the sidewalk provides an external vertical force on the girl-cart system. There are, however, no external horizontal forces on this system, so the horizontal momentum is conserved. The cart has an initial horizontal velocity, but the girl's initial horizontal velocity is zero. She hits the cart moving straight down. Hence, as mass is added to the cart, its velocity must decrease to conserve momentum. So the speed decreases to conserve momentum.

47. E

According to the second law of thermodynamics, no process can be 100 percent efficient. Energy must, however, still be conserved. The energy that cannot be used for useful work is converted into waste heat. Muscles are not 100 percent efficient, so the waste heat generated by the muscles warms up a person who is exercising.

48. A

First, look at the units. Only choices (A) and (C) have the correct units for tension, which should be force units. Eliminate the other choices. Here the tension, T, in the rope provides the centripetal force, F_c, needed for the circular motion, so the tension is equal to the centripetal force:

$$T = F_c = \frac{mv^2}{r} = \frac{mv^2}{L}$$

where m is the mass of the rock and v is the linear speed. The radius, r, of the circular motion is in this case the length, L, of the rope. To find the speed, v, of the rock, note that it travels a complete circle in a time, t. The distance traveled is just the circumference of the circle, so using the speed as the distance divided by the time gives us

$$v = \frac{2\pi L}{t}$$

Squaring the speed and substituting it into the equation for the centripetal force gives us

$$T = \frac{4\pi^2 m L^2}{t^2 L}$$

$$T = \frac{4\pi^2 m L}{t^2}$$

49. C

Pressure is defined as force divided by area, so the total force is the pressure times the area. The area of the door is 6,500 cm^2, and the pressure difference between the two sides of the door is 1,000 N/m^2. Multiplying the numbers gives the total force of 650 N.

50. B

The buoyant force equals the weight of the water displaced. In addition, it must balance the weight of the boat, so the weight of the water displaced is equal to the weight of the boat. The boat displaces a volume of 20 m^3 of water, which has a mass of 20,000 kg. To find the weight, multiply the mass by the acceleration of gravity, so the weight of the water displaced is 200,000 N. The boat must then have a weight of 200,000 N. Do not confuse weight and mass.

51. B

Einstein received the Nobel Prize for his 1905 explanation of the photoelectric effect. The light's intensity has no effect on the energy or speed of the electrons leaving the atom. A certain amount of energy is required for the electron to escape the atom; the specific amount depends on the type of atom. If a photon striking the electron has less than this amount of energy, the electron cannot escape the atom. If the photon has more than this minimum energy, then the electron can escape. The energy that the photon has in excess of the minimum needed to escape is converted into the electron's kinetic energy. So higher-energy photons will cause the electrons to escape with more kinetic energy and, therefore, greater speed. Shorter-wavelength photons have a higher frequency and more energy.

52. C

For this question, apply the kinematic equation:

$$v = v_0 + at$$

where v is the final velocity, $v_0 = 8$ m/s is the initial velocity, $a = 4$ m/s^2 is the acceleration, and $t = 5$ s is the time. Plugging these values into the equation gives us

$$v = 8 \text{ m/s} + (4 \text{ m/s}^2) \ (5 \text{ s})$$
$$v = 28 \text{ m/s}$$

So the velocity at the end of the acceleration is 28 m/s.

53. D

For this question, apply the kinematic equation for the position s:

$$s = s_0 + v_0 t + \left(\frac{1}{2}\right) at^2$$

where we take the starting position, s_0, as zero; the other values are the same as those in question 52. Plugging these values into the equation gives us

$$s = 0 + (8 \text{ m/s}) \ (5 \text{ s}) + \left(\frac{1}{2}\right) \ (4 \text{ m/s}^2) \ (5 \text{ s})^2$$
$$s = 40 \text{ m} + 50 \text{ m}$$
$$s = 90 \text{ m}$$

So the car travels a total of 90 m during the time that it is accelerating.

54. D

The energy stored in a capacitor, U_c, is given by more than one formula. The different formulas are, however, equivalent. The equations are

$$U_c = \frac{1}{2}QV = \frac{1}{2}CV^2 = \frac{1}{2}\left(\frac{Q^2}{C}\right)$$

where Q is the charge on the capacitor, C is the capacitance, and V is the potential difference across the capacitor. Using the definition of capacitance, $C = \dfrac{Q}{V}$, one can show that these formulas are all equivalent to one another. We will use the third formula to solve this problem because the charge does not change. This means that we need to worry only about the change in the capacitance as the plates are pulled farther apart. To see how the capacitance changes, consider the formula for the capacitance of a parallel plate capacitor with area A and separation d:

$$C = \frac{\varepsilon_0 A}{d}$$

If the distance, d, between the plates doubles, the capacitance will be half as much. The energy equations tell us that if the charge on the capacitor does not change, halving the capacitance will double the energy stored. Hence, the energy stored in the capacitor doubles. The extra energy comes from the work required to pulled the two charged plates apart.

55. B

Outside the sphere, the electric field is the same as if it were a point charge. Thus, the electric field lines outside this sphere will be the same as the electric field lines from an isolated positive point charge. These lines point radially outward from the point or from the center of the sphere. The direction of the electric field is the same as the direction of the force on a positive charge, which in this case is outward rather than inward.

56. A

Newton's second law states that the net force, F_{net}, equals the mass, m, times the acceleration, a:

$$F_{net} = ma$$

Here, $m = 1{,}500$ kg and $a = 5$ m/s^2. Therefore:

$$F_{net} = (1{,}500 \text{ kg}) (5 \text{ m/s}^2) = 7{,}500 \text{ N}$$

The 2,500 N frictional force, f, opposes the motion and is in the opposite direction from the accelerating force, F. Hence,

$$F_{net} = F - f$$
$$7{,}500 \text{ N} = F - 2{,}500 \text{ N}$$
$$F = 10{,}000 \text{ N}$$

The force needed to overcome friction and accelerate the car is 10,000 N.

57. E

Start with a free-body diagram as shown.

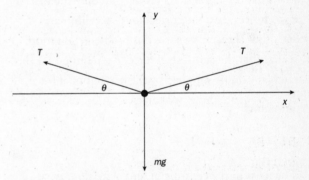

The three forces are the downward weight of the person, mg, and the tensions, T, on the rope on either side of the person. There is no acceleration in this case (unless the rope breaks), so both the x components and the y components of the forces sum to zero:

$$\Sigma F_x = 0 \qquad \text{and} \qquad \Sigma F_y = 0$$

If the person is at the center of the rope, the rope's angle, θ, from the horizontal will be the same on both sides. In this case, summing the x components of the forces tells us that the tension, T, will be the same on both sides of the rope. You may have guessed this intuitively. Summing the y components of both tensions and the person's weight, mg, and setting the sum equal to zero gives us

$$\Sigma F_y = T\sin\theta + T\sin\theta - mg = 0$$

Solving this equation for T gives us

$$T = \frac{mg}{2\sin\theta}$$

Notice that this tension can be much larger than the person's weight, especially if the angle is small.

58. A

Even though the choices look similar, close examination of the units reveals that only the first two choices have the correct units, so eliminate the other choices. To solve this problem, you must realize that the centripetal force, F_c, required for the circular orbit is supplied by the gravitational force, F_G. So equate these two forces:

$$F_c = \frac{mv^2}{r} = F_G = \frac{GMm}{r^2}$$

The mass in the centripetal force equation, m, represents the mass of the object traveling in the circle. Hence, it is the Earth's mass, m, rather than the Sun's mass, M. The radius, r, in the centripetal force equation is here the radius of Earth's orbit around the sun. Canceling the mass, m, and the radius, r, in the equation gives us

$$v^2 = \frac{GM}{r}$$

as an expression describing Earth's orbital speed around the Sun.

59. D

When light passes through a double slit, the light waves passing through each slit have different path lengths. If the path length difference at a given point equals an integer multiple of the wavelength, then the valleys and peaks of the superposed waves will line up. This causes constructive interference, and we see a bright region. If the path difference is an odd half-integer multiple of the wavelength, then the peaks of one wave will superpose on the valleys of the other wave. This causes destructive interference, and we see a dark region. The bright and dark fringes are thus caused by constructive and destructive interference. The positions for the bright and dark regions depend on the wavelength, so this double-slit experiment works best for a monochromatic light source.

60. B

Gravity is an inverse square force, so doubling the distance will give $\frac{1}{4}$ the amount of force. To show that this is correct, recall that there are two equations giving the gravitational force, F, on an object of mass, m, near the Earth's surface: $F = mg$ and $F = \frac{GMm}{R^2}$, where M is the mass of the Earth and R is the radius of the Earth. Equating these expressions and canceling m gives us

$$g = \frac{GM}{R^2}$$

Notice that g is inversely proportional to R^2. Hence, doubling R will give $\frac{1}{4}$ the acceleration of gravity.

61. D

First it is necessary to find the force on the man. The man pushing on the woman and the woman pushing on the man form an action-reaction pair. Therefore, according to Newton's third law, these are equal and opposite forces. If the man is pushing away from the woman with a force of 200 N, she is also pushing away from him with a force of 200 N in the opposite direction. Hence, the force on the man is the same as the force he exerts, 200 N. The acceleration is then this force divided by the man's mass of 100 kg, which gives $a = 2$ m/s^2.

62.   B

Initially they are both at rest, so the total momentum of the man-woman system is zero before they push apart. Their momentum is conserved if there are no external forces acting on this system, so after they push apart and break contact their total momentum is still zero. They are moving in opposite directions, so the absolute value of the man's momentum equals the absolute value of the woman's momentum. Momentum is mass times velocity. The man has twice the mass that the woman has; to have the same momentum, he must have half her speed. If her speed is 6 m/s, then his is 3 m/s in the opposite direction.

63.   C

When the child is swinging toward the musician, the sound waves are squeezed so that they have a shorter wavelength. The shorter wavelength corresponds to a higher frequency or pitch. Hence, by the Doppler effect, he will hear a higher pitch when she is moving toward him.

64.   B

The drop in frequency heard from an ambulance siren is a classic example of the Doppler effect for sound waves. Choice (C) is reversed—as celestial objects move further from us, their characteristic wavelengths increase.

65.   E

Power is the rate at which work is done. It is defined as work divided by time. An alternate formula to find power, P, is the force F times the velocity v:

$$P = Fv, \text{ or } P = \frac{W}{t} = \frac{Fd}{t} = F \times \frac{d}{t} = Fv.$$

Force and velocity must be parallel components, which they are in this question. So, in this case, there is no need to multiply by $\cos\theta$. Multiplying the numbers given in the problem gives a power output for the child of 150 W.

66.   D

The electron's velocity will have components parallel and perpendicular to the magnetic field lines. If the electron's velocity were completely parallel to the magnetic field, there would be no force on the electron, and its velocity would not change. If the electron's velocity were completely perpendicular to the magnetic field, the force would be continuously perpendicular to the velocity, resulting in circular motion around the magnetic field lines. The result when the electron's velocity has components both perpendicular and parallel to the magnetic field lines is that the electron has a helical or spiral path around the magnetic field lines.

67.   D

All conservative forces have an associated potential energy. The potential energy is the work done against the conservative force to put an object in a certain position. Because work is the force, $F = bpi$, multiplied by the distance, x, the potential energy associated with this force could have the form $bpix$. The exact value of the potential energy will, however, depend on the position chosen for the zero point of the potential energy, so this equation could differ by any constant amount.

68. A

From Coulomb's law, a uniform spherical charge distribution will have the same electric field at any point outside the sphere as a point charge of the same total charge placed at the center of the sphere. Hence, as long as $r > R$, the electric field from the spherical charge distribution and the point charge will be the same.

69. B

An inward centripetal force is required for circular motion. In this situation, the forces acting on the object are the tension in the rope and the gravitational force. At the top of the circle, they both act downward. The combination of these two forces supplies the centripetal force needed to keep the object moving in a circular path. Hence, the forces acting on the object are the tension and gravity, both downward.

70. E

A force acting on an object will cause an acceleration, $F = ma$. They are both vectors, and the direction of the force and the acceleration must be the same. An acceleration that is parallel to the velocity vector will cause an increase or decrease in the speed but will not affect the direction. An acceleration that is perpendicular to the velocity will affect the direction of the velocity but will not affect the speed. A force perpendicular to the velocity (say, for example, a sudden crosswind on an airplane) will not change the component of the velocity in the original direction. However, it will change the velocity component in the perpendicular direction, which will lead to a change in the overall velocity.

SECTION II

1. The vase falls a distance of 1.3 m. Use the kinematic equations to find the time it takes to fall and the speed when it hits the floor. After it hits the floor, you can also use kinematics to find the stopping acceleration from the known stopping distance. Once you have computed the stopping acceleration, use Newton's second law to compute the stopping force. To find the impulse required to stop the vase, use the impulse momentum relation.

(a) To find the time, t, it takes the vase to fall, use the kinematic equation:

$$s = s_0 + v_0 t + \frac{1}{2}at^2$$

where the final position, s, of the vase is 1.3 m and the initial position, s_0, is defined as zero. The initial speed, v_0, is also zero, and the downward acceleration is 9.8 m/s^2. In this case, it is easiest to define the downward direction as positive. Plugging the numbers into the equation gives us

$$1.3 \text{ m} = 0 + 0 + \left(\frac{1}{2}\right)(9.8 \text{ m/s}^2)\,t^2$$

$$t^2 = \frac{2\,(1.3 \text{ m})}{(9.8 \text{ m/s}^2)} = 0.265 \text{ s}^2$$

$$t = 0.515 \text{ s}$$

The vase takes 0.515 s to fall to the floor.

(b) Once t is known, we can find the speed, v, at which the vase hits the floor using

$$v = v_0 t + at$$
$$v = 0 + (9.8 \text{ m/s}^2)(0.515 \text{ s})$$
$$v = 5.05 \text{ m/s}$$

The vase is traveling at 5.05 m/s when it hits the floor.

(c) To stop the vase, the carpet compresses a distance $(s - s_0)$ of 0.005 m. Notice that because the vase is now stopping rather than falling downward, this is essentially a different problem, and the kinematic variables take on new values. It is a good idea to keep the downward direction positive. Here, use the kinematic equation:

$$v^2 - v_0^2 = 2a(s - s_0)$$

The initial speed, v_0, is the speed at which the vase strikes the carpet, 5.05 m/s, and the final speed, v, is zero. Inserting the numbers gives us

$$0 - (5.05 \text{ m/s})^2 = 2a\,(0.005 \text{ m})$$

$$a = -\frac{(5.05 \text{m/s})^2}{(2 \times 0.005 \text{ m})}$$

$$a = -2,550 \text{ m/s}^2$$

The negative sign here indicates that the acceleration is upward, in the opposite direction from the velocity. Thus, the vase has an upward acceleration of 2,550 m/s^2 when stopping.

(d) Find the stopping force, F, from Newton's second law, $F = ma$, where the mass, m, of the vase is 0.65 kg. Substituting the numbers gives us

$$F = (0.65 \text{ kg})(-2,550 \text{ m/s})$$
$$F = -1,660 \text{ N}$$

There is a force of 1,660 N acting upward on the vase to stop it.

(e) Find the total impulse, J, on the vase from the impulse momentum equation

$$J = F\Delta t = \Delta p$$

Here we can ignore the fact that this is a vector equation because the motion is in one dimension.

The impulse, J, equals the change in momentum, Δp. Recalling that momentum equals mass times velocity gives us

$$J = mv - mv_0$$
$$J = 0 - (0.65\ \text{kg})(5.05\ \text{m/s})$$
$$J = -3.28\ \text{kg} \cdot \text{m/s}$$

There is an upward impulse of 3.28 kg · m/s on the vase.

2. Parts (a) and (c) of this problem involve calculating the electric field of a point charge and the electric force between two point charges. Part (b) is a qualitative drawing of the electric field of two equal and opposite point charges. Part (d) requires equating the electric force to the centripetal force to compute properties of a circular orbit.

(a) The electric field, E, at a distance, r, from a point charge, q, is given by

$$E = \frac{kq}{r^2}$$

where $k = \frac{1}{4}\pi\varepsilon_0 = 9.0 \times 10^9\ \text{N·m}^2/\text{C}^2$. Substituting the values for this problem gives us

$$E = (9.0 \times 10^9\ \text{N·m}^2/\text{C}^2)\,\frac{(5 \times 10^{15}\ \text{C})}{(2 \times 10^{11}\ \text{m})^2}$$

$$E = 1.13 \times 10^3\ \text{N/C}.$$

The electric field of 1.13×10^3 N/C points outward away from the planet.

(b) The electric field between two equal and opposite charges is the dipole field shown below.

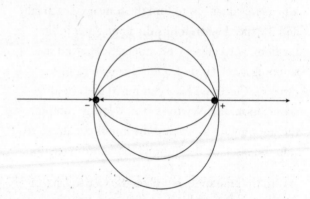

(c) The force, F, between two point charges can be found by using Coulomb's law or by multiplying the electric field of one charge by the other charge. In this case, the second approach is easier because we have already calculated the electric field in part (a). So

$$F = \frac{kq_1q_2}{r^2} = Eq_2$$
$$F = (1.13 \times 10^3\ \text{N/C})(-5 \times 10^{15}\ \text{C})$$
$$F = -5.65 \times 10^{18}\ \text{N}$$

The negative sign indicates that this force between the two charges is attractive, as would be needed if the electric force causes the planet to orbit the star.

(d) To find the orbital properties, note that the centripetal force F_c required for a circular orbit is supplied by the electrical force computed in part (c). Hence, the two should be equated:

$$F = F_c = \frac{mv^2}{r}$$

where m is the mass of the planet, r is its orbital radius, and v is its orbital speed. Substituting the numbers and solving for v gives us:

$$5.65 \times 10^{18}\ \text{N} = \frac{(6 \times 10^{24}\ \text{kg})\,v^2}{(2 \times 10^{11}\ \text{m})}$$

$$v^2 = 1.88 \times 10^5 \text{ m}^2/\text{s}^2$$

$$v = 4.33 \times 10^2 \text{ m/s}$$

To find the orbital period, T, use the relationship between distance, speed, and time and note that the distance traveled is the circumference of the circle. So

$$v = \frac{2\pi r}{T}$$

$$T = \frac{2\pi(2 \times 10^{11} \text{ m})}{(4.33 \times 10^2 \text{ m/s})}$$

$$T = 2.9 \times 10^9 \text{ s} = 94 \text{ years}$$

3. (a) The currents are calculated using Ohm's law. For example, the current for the 4.0 V case is calculated as follows:

$$I = \frac{V}{R} = \frac{(4.0\text{V})}{(50.0 \text{ ohm})} = 0.080 \text{ A}$$

Voltage Across 50-Ohm Resistor (V)	Current (A)	Magnetic Field (T)
0	0	0
4.0	0.080	1.5×10^{-5}
7.9	0.16	3.0×10^{-5}
12.3	0.25	4.8×10^{-5}
16.1	0.32	6.3×10^{-5}
20.0	0.40	7.9×10^{-5}

(b) Graph the magnetic field versus current.

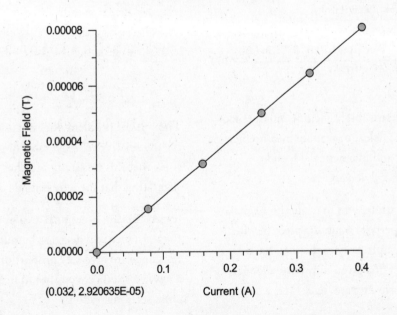

Magnetic Field versus Current

(0.032, 2.920635E-05)

(c) Slope is the rise over run of the graph, and using the first and last data points from the table, we find

$$\text{Slope} = \frac{\Delta B}{\Delta I} = \frac{(7.9 \times 10^{-5} - 0)\ \text{T}}{(0.40 - 0)\ \text{A}} = 1.98 \times 10^{-4}\ \text{T/A}$$

(d) Using $y = mx + b$, where $b = 0$,

$$B = (1.98 \times 10^{-4}\ \text{T/A})\, I$$

(e) n = Number coils/length in meters
$$= 58\ \text{loops}/0.35\ \text{m} = 166\ \text{m}^{-1}$$

(f) Comparing the theoretical equation $B = (\mu_0 n)I$ with the equation in part (d), we see that slope should be equal to $\mu_0 n$ (a constant). Thus, the permeability constant can be calculated as follows:

$$\mu_0 n = \text{slope}$$
$$\mu_0 = \text{slope}/n$$
$$\mu_0 = (1.98 \times 10^{-4}\ \text{T/A})/(166\ \text{m}^{-1})$$
$$\mu_0 = 1.19 \times 10^{-6}\ \text{T} \cdot \text{m/A}$$

(g) The theoretical value of the permeability constant may be found from the table of constants and equals $4\pi \times 10^{-7}\ \text{T} \cdot \text{m/A}$.

$$\% \text{ Error} = \frac{|\text{theoretical} - \text{experimental}|}{\text{theoretical}} \times 100$$

$$= \frac{|4\pi \times 10^{-7}\ \text{T.m/A} - 1.19 \times 10^{-6}\ \text{T.m/A}|}{4\pi \times 10^{-7}\ \text{T.m/A}} \times 100$$

$$= 5.3\%$$

Possible sources of error might include interference from stray magnetic fields; inaccurate measurements of current, voltage, and magnetic field; and variations in Slinky coil density.

4. This question requires you to relate the resistance, R, a property of a specific piece of wire, to the resistivity, ρ, a property of the material that makes up the wire. For a given resistivity, a longer piece of wire will have a greater resistance; a larger cross-sectional area will produce a smaller resistance. Part (d) is then an application of Ohm's law to find the current in the wire for a specific voltage.

(a) The resistance, R, for a piece of wire of length l and cross-sectional area A is given by

$$R = \frac{\rho l}{A}$$

In this context, ρ represents the resistivity of the material that makes up the wire. The area of a circular wire of radius r is given by

$$A = \pi r^2$$

Substituting this equation for the area into the equation for the resistance and solving for the resistivity gives us

$$R = \frac{\rho l}{\pi r^2}$$

$$\rho = \frac{R \pi r^2}{l}.$$

(b) Putting the measured values for the piece of wire into the above equation gives us:

$$\rho = \frac{R \pi r^2}{l}$$

$$\rho = \frac{(6.8 \times 10^{-4}\ \Omega)\pi(0.002\ \text{m})^2}{(0.5\ \text{m})}$$

$$\rho = 1.7 \times 10^{-8}\ \Omega \cdot \text{m}$$

Though the problem does not ask you to identify the material, checking a table of resistivities would show that this resistivity matches that of copper, suggesting that the wire is made of copper.

(c) As the cross-sectional area of the wire increases, the resistance will decrease because area and resistance are inversely proportional. Thus, to get a lower resistance, you should increase the wire's cross-sectional radius.

(d) Solving Ohm's law for the current, I, gives us

$$I = \frac{V}{R}$$

where V is the voltage or potential difference and R is the resistance. Plugging the numerical values into the equation gives us

$$I = \frac{(120 \text{ V})}{(6.8 \times 10^{-4} \text{ } \Omega)}$$

$$I = 1.76 \times 10^5 \text{ A}$$

This very high current would blow a fuse or trip the circuit breaker, thereby protecting the wiring from heating enough to start a fire. The low resistance involved is effectively a short circuit.

5. In this problem, an electron makes a transition between two energy levels in an atom. The problem asks about the properties of the photon that is absorbed or emitted.

(a) The electron is jumping to a higher energy level. From conservation of energy, this energy must come from somewhere, which tells us that the electron absorbs a photon. An electron can also gain the energy needed to jump to a higher energy level by absorbing the energy from a collision with another atom, but that is not mentioned as an option in this problem.

(b) The energy that the electron needs to absorb for this transition is the difference between the two energy levels. Hence, the energy of the absorbed photon is 18 eV − 7 eV = 11 eV.

(c) To find the frequency, f, of an 11 eV photon, use

$$E = hf$$

where the energy $E = 11$ eV and Planck's constant $h = 6.63 \times 10^{-34}$ J · s = 4.14×10^{-15} eV · s. In this case, we know the energy in eV, so the second value of Plank's constant has more convenient

units. Solving for the frequency and plugging in the numbers gives us

$$f = \frac{E}{h} = \frac{(11 \text{ eV})}{(4.14 \times 10^{-15} \text{ eV} \cdot \text{s})}$$

$$f = 2.7 \times 10^{15} \text{ Hz}$$

(d) The wavelength, λ, and frequency, f, of a photon are related by the equation:

$$c = f\lambda$$

where c is the speed of light. Solving for the wavelength and inserting the numbers gives us:

$$\lambda = \frac{c}{f}$$

$$\lambda = \frac{(3 \times 10^8 \text{ m/s})}{(2.7 \times 10^{15} \text{ Hz})}$$

$$\lambda = 1.1 \times 10^{-7} \text{ m} = 110 \text{ nm}$$

This wavelength is in the far ultraviolet region of the spectrum.

6. (a) Wavelength is defined as the length of one complete wave pattern (crest + trough). One wave is $\frac{2}{3}$ of the total length of the string.

$$\lambda = \frac{2}{3}(90 \text{ cm}) = 60 \text{ cm}$$

The amplitude of the wave is the distance from the midpoint of vibration to the top of the crest. This is half of the distance shown on the figure:

$$A = \frac{1.4 \text{ cm}}{2} = 0.7 \text{ cm}$$

(b) Use the wave equation to find speed:

$$v = f\lambda = (1200 \text{ Hz})(0.60 \text{ m}) = 720 \text{ m/s}$$

(c) The medium must be altered to change the speed of a wave. Therefore, you could increase or decrease the tension of the string to change its speed. The speed of a wave does *not* depend on wavelength or frequency of vibration.

(d) The vibration is **transverse** because the medium vibrates perpendicularly to the motion of the waves.

(e) The frequency of a wave is the same as the frequency of the vibrating source. Therefore, the sound wave has a frequency of 1,200 Hz.

(f) Sound waves are **longitudinal** because the vibration of the air molecules is in the same direction as the sound wave moves.

(g) The medium (air) must be altered to change the speed of sound. To do this, you can change the temperature or pressure of the air, or you can change the medium itself. (Note: The frequency of vibration (pitch) and the amplitude (loudness) of a sound wave have *no* effect on the speed of sound.)

7. This problem involves a convex spherical mirror. Solving it therefore requires application of the various mirror equations. Be careful not to forget the sign conventions: For a convex, or diverging, mirror, the focal length is negative. Forgetting this convention will give you the wrong answer.

(a) To find the focal length, f, recall that the focal length is half the radius of curvature, R:

$$f = \frac{R}{2}$$

Here $R = -4$ cm; because the mirror is convex, we choose the negative sign. Solving for the focal length gives us

$$f = \frac{(-4 \text{ cm})}{2}$$
$$f = -2 \text{ cm}$$

The focal length of the mirror is −2 cm.

(b) The object in this case is your face, and the object distance, s_o, is given as 10 cm. Because your face is a real object, the object distance is positive. To find the image distance, s_i, use

$$\frac{1}{s_i} + \frac{1}{s_o} = \frac{1}{f}$$

Solving for the image distance and plugging in the numerical values gives us

$$\frac{1}{s_i} = \frac{1}{f} - \frac{1}{s_o}$$
$$\frac{1}{s_i} = \frac{1}{(-2 \text{ cm})} - \frac{1}{(10 \text{ cm})} = \frac{-6}{10 \text{ cm}}$$
$$s_i = -1.7 \text{ cm}$$

The image is located 1.7 cm behind the mirror.

(c) The image magnification, M, is given by

$$M = -\left(\frac{s_i}{s_o}\right)$$

$$M = -\frac{(-1.7 \text{ cm})}{(10 \text{ cm})}$$

$$M = 0.17$$

The image of your face is 0.17 times the size of your face.

(d) The fact that the magnification in part (c) is positive indicates that the image is upright (or erect). The fact that the image distance is negative indicates that the image is behind the mirror. The light rays cannot actually cross at this point, so the image is a virtual image.

8. (a)

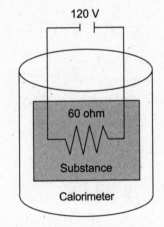

(b) Ohm's law is used to measure the current in the resistor.

$$I = \frac{V}{R} = \frac{120 \text{ V}}{60 \text{ ohm}} = 2.0 \text{ A}$$

(c) Power is the rate at which energy is transferred in a circuit. Using the power equation,

$$P = IV = (2.0 \text{ A})(120 \text{ V}) = 240 \text{ J/s} = 240 \text{ W}$$

(d) The amount of heat generated in 5 minutes (300 seconds) is given by the power equation:

Heat energy $(Q) = Pt$

Using the heat transfer formula, the substance absorbs this heat energy as follows:

$$Q = mc\Delta T$$

where m is the mass, c is the specific heat, and ΔT is the change in temperature of the substance. Equating the power and heat transfer equations, solving for c, and plugging in the numbers gives us

$$Pt = mc\Delta T$$

$$c = \frac{Pt}{m\Delta T}$$

$$c = \frac{(240 \text{ J/s})(300 \text{ s})}{0.5 \text{ kg}(85°C - 25°C)}$$

$$c = 2.4 \times 10^3 \text{ J/kg} \cdot °C$$

(e) The amount of heat energy needed to melt 0.5 kg of the substance is given by

$$Q = mL$$

where m is the mass and L is the heat of fusion of the substance. Plugging in the numbers gives us

$$Q = (0.5 \text{ kg})(1.35 \times 10^5 \text{ J/kg}) = 6.75 \times 10^4 \text{ J}$$

To get the amount of time to absorb this heat, use the power equation as follows:

$$t = \frac{Q}{P} = \frac{6.75 \times 10^4 \text{ J}}{240 \text{ J/s}}$$

$$= 280 \text{ s} = 4.7 \text{ min}$$

AP Physics C Practice Test
Answer Grid

1. Ⓐ Ⓑ Ⓒ Ⓓ Ⓔ
2. Ⓐ Ⓑ Ⓒ Ⓓ Ⓔ
3. Ⓐ Ⓑ Ⓒ Ⓓ Ⓔ
4. Ⓐ Ⓑ Ⓒ Ⓓ Ⓔ
5. Ⓐ Ⓑ Ⓒ Ⓓ Ⓔ
6. Ⓐ Ⓑ Ⓒ Ⓓ Ⓔ
7. Ⓐ Ⓑ Ⓒ Ⓓ Ⓔ
8. Ⓐ Ⓑ Ⓒ Ⓓ Ⓔ
9. Ⓐ Ⓑ Ⓒ Ⓓ Ⓔ
10. Ⓐ Ⓑ Ⓒ Ⓓ Ⓔ
11. Ⓐ Ⓑ Ⓒ Ⓓ Ⓔ
12. Ⓐ Ⓑ Ⓒ Ⓓ Ⓔ
13. Ⓐ Ⓑ Ⓒ Ⓓ Ⓔ
14. Ⓐ Ⓑ Ⓒ Ⓓ Ⓔ
15. Ⓐ Ⓑ Ⓒ Ⓓ Ⓔ
16. Ⓐ Ⓑ Ⓒ Ⓓ Ⓔ
17. Ⓐ Ⓑ Ⓒ Ⓓ Ⓔ
18. Ⓐ Ⓑ Ⓒ Ⓓ Ⓔ
19. Ⓐ Ⓑ Ⓒ Ⓓ Ⓔ
20. Ⓐ Ⓑ Ⓒ Ⓓ Ⓔ
21. Ⓐ Ⓑ Ⓒ Ⓓ Ⓔ
22. Ⓐ Ⓑ Ⓒ Ⓓ Ⓔ
23. Ⓐ Ⓑ Ⓒ Ⓓ Ⓔ
24. Ⓐ Ⓑ Ⓒ Ⓓ Ⓔ

25. Ⓐ Ⓑ Ⓒ Ⓓ Ⓔ
26. Ⓐ Ⓑ Ⓒ Ⓓ Ⓔ
27. Ⓐ Ⓑ Ⓒ Ⓓ Ⓔ
28. Ⓐ Ⓑ Ⓒ Ⓓ Ⓔ
29. Ⓐ Ⓑ Ⓒ Ⓓ Ⓔ
30. Ⓐ Ⓑ Ⓒ Ⓓ Ⓔ
31. Ⓐ Ⓑ Ⓒ Ⓓ Ⓔ
32. Ⓐ Ⓑ Ⓒ Ⓓ Ⓔ
33. Ⓐ Ⓑ Ⓒ Ⓓ Ⓔ
34. Ⓐ Ⓑ Ⓒ Ⓓ Ⓔ
35. Ⓐ Ⓑ Ⓒ Ⓓ Ⓔ
36. Ⓐ Ⓑ Ⓒ Ⓓ Ⓔ
37. Ⓐ Ⓑ Ⓒ Ⓓ Ⓔ
38. Ⓐ Ⓑ Ⓒ Ⓓ Ⓔ
39. Ⓐ Ⓑ Ⓒ Ⓓ Ⓔ
40. Ⓐ Ⓑ Ⓒ Ⓓ Ⓔ
41. Ⓐ Ⓑ Ⓒ Ⓓ Ⓔ
42. Ⓐ Ⓑ Ⓒ Ⓓ Ⓔ
43. Ⓐ Ⓑ Ⓒ Ⓓ Ⓔ
44. Ⓐ Ⓑ Ⓒ Ⓓ Ⓔ
45. Ⓐ Ⓑ Ⓒ Ⓓ Ⓔ
46. Ⓐ Ⓑ Ⓒ Ⓓ Ⓔ
47. Ⓐ Ⓑ Ⓒ Ⓓ Ⓔ
48. Ⓐ Ⓑ Ⓒ Ⓓ Ⓔ

49. Ⓐ Ⓑ Ⓒ Ⓓ Ⓔ
50. Ⓐ Ⓑ Ⓒ Ⓓ Ⓔ
51. Ⓐ Ⓑ Ⓒ Ⓓ Ⓔ
52. Ⓐ Ⓑ Ⓒ Ⓓ Ⓔ
53. Ⓐ Ⓑ Ⓒ Ⓓ Ⓔ
54. Ⓐ Ⓑ Ⓒ Ⓓ Ⓔ
55. Ⓐ Ⓑ Ⓒ Ⓓ Ⓔ
56. Ⓐ Ⓑ Ⓒ Ⓓ Ⓔ
57. Ⓐ Ⓑ Ⓒ Ⓓ Ⓔ
58. Ⓐ Ⓑ Ⓒ Ⓓ Ⓔ
59. Ⓐ Ⓑ Ⓒ Ⓓ Ⓔ
60. Ⓐ Ⓑ Ⓒ Ⓓ Ⓔ
61. Ⓐ Ⓑ Ⓒ Ⓓ Ⓔ
62. Ⓐ Ⓑ Ⓒ Ⓓ Ⓔ
63. Ⓐ Ⓑ Ⓒ Ⓓ Ⓔ
64. Ⓐ Ⓑ Ⓒ Ⓓ Ⓔ
65. Ⓐ Ⓑ Ⓒ Ⓓ Ⓔ
66. Ⓐ Ⓑ Ⓒ Ⓓ Ⓔ
67. Ⓐ Ⓑ Ⓒ Ⓓ Ⓔ
68. Ⓐ Ⓑ Ⓒ Ⓓ Ⓔ
69. Ⓐ Ⓑ Ⓒ Ⓓ Ⓔ
70. Ⓐ Ⓑ Ⓒ Ⓓ Ⓔ

AP PHYSICS C: PRACTICE TEST

Time—45 minutes 35 Questions

> **Directions:** Each of the questions or incomplete statements below is followed by five suggested answers or completions. Select the best choice in each case.
>
> To simplify calculations, use $g = 10$ m/s^2.

Use the following graph to answer question 1.

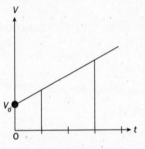

1. The graph shows the velocity as a function of time t for an object moving in a straight line. Which of the following graphs shows the corresponding acceleration a as a function of time t for the moving object in the same time interval?

(A)

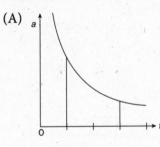

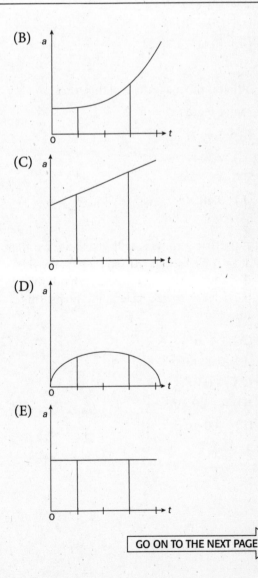

GO ON TO THE NEXT PAGE

Use the following information to answer questions 2 and 3.

A 3 kg object has a velocity at one instant of 2i m/s. Five seconds later, the object's velocity changes to (3i + 3j) m/s.

2. Find the components of the force acting on the object.

(A) (1.8i − 0.6j) N

(B) (3i + 3j) N

(C) (6i + 6j) N

(D) (9i − 9j) N

(E) (0.6i + 1.8j) N

3. What is the magnitude of this force?

(A) 4.24 N

(B) 1.90 N

(C) 2.54 N

(D) 12.73 N

(E) 8.48 N

4. A particle is moving with a velocity $v = 40$ m/s at $t = 0$. Between $t = 0$ and $t = 10$ s, the velocity decreases uniformly to zero. What is the particle's average acceleration during this time?

(A) -4 m/s^2

(B) 0.25 m/s^2

(C) 4 m/s^2

(D) -0.25 m/s^2

(E) 400 m/s^2

Use the following information to answer questions 5 and 6.

A particle moves along the x-axis with a displacement represented by the following: $x = 5t + 9t^3$.

5. What is the instantaneous velocity of this particle at some time t?

(A) $5t + 9t^2$

(B) $5t + 27t^2$

(C) $15 + 27t^2$

(D) $5 + 27t^2$

(E) $5t + 27t^3$

6. What would be the instantaneous acceleration at $t = 2$ s?

(A) 27 m/s^2

(B) 108 m/s^2

(C) 54 m/s^2

(D) 5 m/s^2

(E) 10 m/s^2

7. A person drops an object from a bridge spanning a rushing river, falling a distance h before hitting the water. Neglecting air resistance, what would be the object's speed at the moment of impact?

(A) $v = \sqrt{2gh}$

(B) $v = \sqrt{mgh}$

(C) $v = \sqrt{\dfrac{gh}{2}}$

(D) $v = \sqrt{\dfrac{mgh}{2}}$

(E) $v = \sqrt{2mgh}$

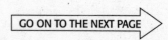
GO ON TO THE NEXT PAGE

Use the following information to answer questions 8 and 9.

The position of a ball thrown vertically up is described by the equation $y = 6t - 2.8t^2$, where y is in meters and t is in seconds.

8. What is the initial speed of the ball at $t = 0$?

   (A) −5.6 m/s

   (B) −2.8 m/s

   (C) −6 m/s

   (D) 6 m/s

   (E) 2.8 m/s

9. What is the speed of the ball at time t?

   (A) $6t - 5.6t^2$

   (B) $6t - 2.8t$

   (C) $6 - 5.6t$

   (D) $6 - 2.8t^2$

   (E) $6t^2 - 2.8t$

10. The equation of motion of a simple harmonic oscillator is $\dfrac{d^2x}{dt^2} = -25x$ where x is the displacement and t is time. The frequency of this oscillator will be

    (A) $\dfrac{5}{\pi}$

    (B) $\dfrac{5}{2\pi}$

    (C) 10π

    (D) $\dfrac{2\pi}{5}$

    (E) 5π

Use the following information to answer questions 11 and 12.

On planet X, an object weighs 5 N. On planet Y, where the acceleration due to gravity is $1.5g$, the object weighs 150 N.

11. Using the information provided, determine the object's mass.

    (A) 5.00 kg

    (B) 1.50 kg

    (C) 15.0 kg

    (D) 10.0 kg

    (E) 0.500 kg

12. What is the acceleration due to gravity on the surface of planet X?

    (A) 5.00 m/s^2

    (B) 0.500 m/s^2

    (C) 10.0 m/s^2

    (D) 15.0 m/s^2

    (E) 150 m/s^2

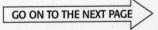
GO ON TO THE NEXT PAGE

Use the following information to answer question 13.

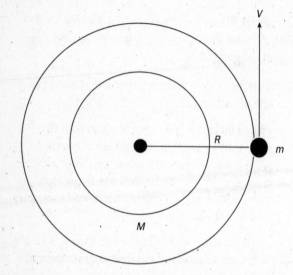

13. An object of mass m orbits a planet with mass M at a distance R, measured from the center of the planet. Assuming that the object's orbit is stable, calculate its orbital speed.

(A) $v = \sqrt{\dfrac{GMm}{R}}$

(B) $v = \sqrt{\dfrac{GM}{R}}$

(C) $v = \sqrt{\dfrac{Gm}{R}}$

(D) $v = \sqrt{\dfrac{GMm}{R^2}}$

(E) $v = \sqrt{GMR}$

Use the following information to answer question 14.

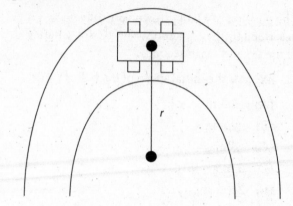

14. A car of mass m is moving down a flat road and drives around a curve with a radius r. If μ is the coefficient of static friction between the tires and the dry road, determine the maximum speed at which the car can travel such that it can make the turn and stay on the road.

(A) $\sqrt{\mu g r}$

(B) $\sqrt{\dfrac{\mu g}{r}}$

(C) $\sqrt{\dfrac{mg}{\mu r}}$

(D) $\sqrt{\dfrac{mgr}{\mu}}$

(E) $\sqrt{\dfrac{m\mu}{gr}}$

GO ON TO THE NEXT PAGE

Use the following information to answer questions 15 and 16.

A 10 kg particle moves along the x-axis. Its position varies with time according to $x = 2t + t^3$, where x is in meters and t is in seconds.

15. What is the kinetic energy of this particle at time t?

(A) $(20t + 10t^3)$ J

(B) $(20 + 30t^2)$ J

(C) $(4 + 12t^2 + 9t^4)$ J

(D) $(20 + 60t^2 + 45t^4)$ J

(E) $(2t + t^3)$ J

16. What power is required to move this particle at time t?

(A) $(120t + 180t^3)$ W

(B) $(24t + 36t^3)$ W

(C) $(20 + 120t + 180t^4)$ W

(D) $(60t)$ W

(E) $(t + 3t^2)$ W

Use the following information to answer questions 17 and 18.

A force $F = (2\mathbf{i} - \mathbf{j})$ N acts on a particle that experiences a displacement that can be represented by $s = (\mathbf{i} - 3\mathbf{j})$ m.

17. Find the magnitude of the work that was done by the force F.

(A) $\sqrt{5}$ J

(B) 2 J

(C) $\sqrt{3}$ J

(D) $2\sqrt{5}$ J

(E) 5 J

18. What is the angle between the two vectors?

(A) 90°

(B) 180°

(C) 45°

(D) 30°

(E) 60°

19. Two skaters are facing each other, initially at rest. The skaters push on each other and move away in opposite directions at constant velocities. What can be said about the total momentum of the skaters after they push away from each other?

(A) The sum of the final momentum is less than the sum of the initial momentum.

(B) The sum of the final momentum is greater than the sum of the initial momentum.

(C) The sum of the final momentum is exactly half the sum of the initial momentum.

(D) The final total momentum is zero.

(E) Each individual skater's momentum equals the sum of the total initial momentum.

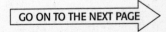

GO ON TO THE NEXT PAGE

Use the following diagram to answer question 20.

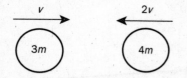

20. A disk with a mass $3m$ is moving horizontally to the right with a speed v on a frictionless surface. It collides head on with another disk with a mass $4m$ traveling to the left at a speed $2v$. Upon impact, the two disks stick together. Find the final speed of the combined masses following the collision.

(A) $-v$

(B) $\dfrac{11v}{7}$

(C) $\dfrac{-3v}{7}$

(D) $\dfrac{-5v}{12}$

(E) $\dfrac{-5v}{7}$

Use the following diagram to answer question 21.

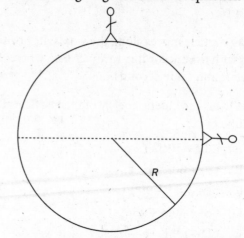

21. A person with a mass m is standing on the surface of a spherical planet of radius R and rotational speed at the equator of v. The acceleration due to gravity at the surface of this planet is g. If the person's weight at the pole is measured to be mg, what would the person's weight be at the equator?

(A) mg

(B) $\dfrac{mv^2}{R}$

(C) $m\left(\dfrac{g - v^2}{R}\right)$

(D) $m\left(\dfrac{v^2}{R - g}\right)$

(E) $m(gR - v^2)$

GO ON TO THE NEXT PAGE

Use the following information to answer question 22.

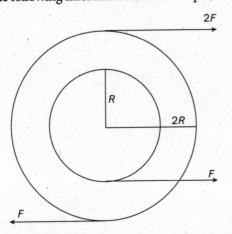

22. A solid cylinder is pivoted about a frictionless axle as shown. Three forces are exerted tangentially to the cylinder. A force of $2F$ and another force of F are applied to a region of the cylinder where the radius is $2R$. A single force of magnitude F is applied to a part of the cylinder where the radius is R. Determine the magnitude of net torque on this system and the direction of rotation about the axle.

   (A) $1FR$ clockwise

   (B) $5FR$ counterclockwise

   (C) $7FR$ clockwise

   (D) $7FR$ counterclockwise

   (E) $5FR$ clockwise

Use the following information to answer questions 23 and 24.

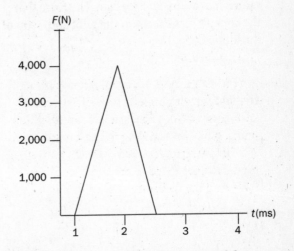

The graph above represents an estimated force-time curve for a tennis racket hitting a tennis ball. Time is plotted in milliseconds on the horizontal axis, and force in newtons is plotted on the vertical axis.

23. What impulse was delivered to the tennis ball by the tennis racket?

   (A) $3 \text{ kg} \cdot \text{m/s}$

   (B) $3{,}000 \text{ kg} \cdot \text{m/s}$

   (C) $6{,}000 \text{ kg} \cdot \text{m/s}$

   (D) $6 \text{ kg} \cdot \text{m/s}$

   (E) $4.5 \text{ kg} \cdot \text{m/s}$

24. What is the average force exerted on the tennis ball?

   (A) $1{,}000 \text{ N}$

   (B) $2{,}000 \text{ N}$

   (C) $2{,}500 \text{ N}$

   (D) $5{,}000 \text{ N}$

   (E) $4{,}000 \text{ N}$

GO ON TO THE NEXT PAGE

25. A pilot of mass m flies a jet plane on a "loop-the-loop" course by flying in a vertical circle of radius R at a constant speed v. The force that the cockpit seat exerts on the pilot at the top of the loop is

(A) mg.

(B) $mg\left(\dfrac{v^2}{R+1}\right)$.

(C) $\dfrac{mv^2}{R(g+1)}$.

(D) $m\left(\dfrac{v^2}{R-g}\right)$.

(E) $\dfrac{mv^2}{R}$.

Use the following diagram to answer question 26.

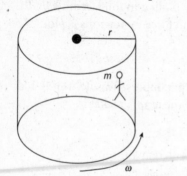

26. A popular amusement park ride consists of a cylinder that spins so rapidly that the floor can drop away, leaving the riders pressed against the wall. If the cylinder above has a radius of r and rotates with an angular speed of ω, what is the minimum coefficient of friction μ between the rider and the wall required to keep the rider from slipping?

(A) $\mu = mr\omega^2$

(B) $\mu = \dfrac{g}{r\omega^2}$

(C) $\mu = mg$

(D) $\mu = mrg$

(E) $\mu = \dfrac{mg}{r\omega^2}$

Use the following diagram to answer question 27.

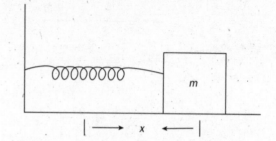

27. The mass m is attached to a spring with a spring constant k. The mass is initially displaced from equilibrium by a distance A and released to oscillate back and forth on a frictionless surface. Determine the speed v of the mass at any given displacement x.

(A) $v = mkxA$

(B) $v = \sqrt{\left(\dfrac{m}{k}\right)(A^2 - x^2)}$

(C) $v = \dfrac{mx}{kA}$

(D) $v = \sqrt{\left(\dfrac{k}{m}\right)(A^2 - x^2)}$

(E) $v = \dfrac{mkA}{x}$

28. A satellite orbits a planet at a distance R as measured from the center of the planet. The satellite maintains a stable altitude h at a constant velocity v. If the satellite's mass is m and the planet's mass is M, determine the amount of work being done by the pull of gravity on the satellite.

(A) $2\pi mv^2$

(B) $\dfrac{2\pi GmM}{R}$

(C) $\dfrac{GmMh}{R^2}$

(D) $\dfrac{2\pi mv^2 R}{h}$

(E) 0

GO ON TO THE NEXT PAGE

Use the following information to answer questions 29 and 30.

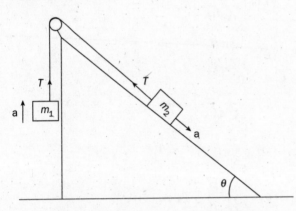

Two masses are attached by a light string that passes over a frictionless, massless pulley without sliding. Mass 2 sits on a frictionless inclined plane that makes an angle θ with the horizontal. Mass 2 is heavier than mass 1, and when released, the masses accelerate in the directions indicated by the arrows. The tension in the string produced by the pull of the masses is labeled T.

29. What is the acceleration of the two masses?

(A) $\dfrac{m_1 g \sin\theta - m_2 g}{m_1 + m_2}$

(B) $\dfrac{m_2 g \sin\theta - m_1 g}{m_1 + m_2}$

(C) $\dfrac{m_1 g - m_2 g \sin\theta}{m_1 + m_2}$

(D) $\dfrac{m_1 + m_2}{m_2 g \sin\theta - m_1 g}$

(E) $\dfrac{m_1 + m_2}{m_1 g - m_2 g \sin\theta}$

30. What is the tension in the string?

(A) $\dfrac{m_1 m_2 (1 + g \sin\theta)}{m_1 + m_2}$

(B) $\dfrac{(m_1 + m_2)(1 + g \sin\theta)}{m_1 m_2}$

(C) $\dfrac{m_1 m_2 g (1 + \sin\theta)}{m_1 + m_2}$

(D) $\dfrac{g (m_1 + m_2)(1 + \sin\theta)}{m_1 m_2}$

(E) $\dfrac{(m_1 + m_2)(1 + \sin\theta)}{g m_1 m_2}$

Use the following information to answer questions 31 and 32.

An oscillating particle experiences a displacement that is given by $x = (4 \text{ m}) \cos\left(2t + \dfrac{\pi}{3}\right)$, where x is in meters and t is in seconds.

31. At $t = 0$, determine the velocity of the particle.

(A) 3.46 m/s

(B) 6.92 m/s

(C) −3.46 m/s

(D) −6.92 m/s

(E) −8 m/s

32. At $t = 0$, determine the acceleration of the particle.

(A) −8 m/s^2

(B) 8 m/s^2

(C) −4 m/s^2

(D) 4 m/s^2

(E) −16 m/s^2

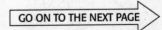
GO ON TO THE NEXT PAGE

33. A meteorite with a mass m_1 has a speed of v_1 when it collides head-on with the Earth (mass m_2). What would the recoil speed of the Earth be at the moment of impact?

(A) $v_2 = \dfrac{m_1(m_1 + m_2)}{v_1}$

(B) $v_2 = \dfrac{(m_1 + m_2)v_1}{m_1}$

(C) $v_2 = \dfrac{m_1 + m_2}{m_1 v_1}$

(D) $v_2 = \dfrac{m_2 v_1}{m_1 + m_2}$

(E) $v_2 = \dfrac{m_1 v_1}{m_1 + m_2}$

Use the following information to answer questions 34 and 35.

A mass m hangs from a cable tied to two other cables connected to a support. The upper cables make angles θ_1 and θ_2 with the horizontal.

34. What is the sum of the horizontal components of the tensions?

(A) $T_2\cos\theta_2 - T_1\sin\theta_1 - mg = 0$
(B) $T_1\sin\theta_1 + T_2\sin\theta_2 - mg = 0$
(C) $T_1\cos\theta_1 - T_2\sin\theta_2 = 0$
(D) $T_2\cos\theta_2 - T_1\cos\theta_1 = 0$
(E) $T_1\cos\theta_1 + T_2\sin\theta_2 + mg = 0$

35. Determine the magnitude of T_1 in terms of T_2.

(A) $\left(\dfrac{\cos\theta_2}{\cos\theta_1}\right)T_2 = T_1$

(B) $\left(\dfrac{\sin\theta_2}{\sin\theta_1}\right)T_2 = T_1$

(C) $\left(\dfrac{\cos\theta_1}{\cos\theta_2}\right)T_2 = T_1$

(D) $\left(\dfrac{\sin\theta_1}{\sin\theta_2}\right)T_2 = T_1$

(E) $\left(\dfrac{\sin\theta_1}{\cos\theta_2}\right)T_2 = T_1$

IF YOU FINISH BEFORE TIME IS UP, YOU MAY CHECK YOUR WORK ON THIS SECTION ONLY. DO NOT TURN TO ANY OTHER SECTION IN THE TEST.    STOP

SECTION I: ELECTRICITY AND MAGNETISM
Time—45 Minutes 35 Questions

Directions: Each of the questions or incomplete statements below is followed by five suggested answers or completions. Select the best choice in each case.

To simplify calculations, use $g = 10$ m/s^2.

Use the following information to answer questions 36, 37, and 38.

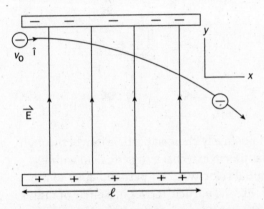

An electron enters the region of a uniform electric field as illustrated with an initial velocity $v_o = 2.5 \times 10^6$ m/s and $E = 300$ N/C. The width of the plates is $l = 0.1$ m.

36. Find the acceleration of the electron while in the electric field.

 (A) 7.2×10^{13} ĵ m/s^2

 (B) 4.3×10^{13} ĵ m/s^2

 (C) 3.2×10^{13} ĵ m/s^2

 (D) $- 1.3 \times 10^{13}$ ĵ m/s^2

 (E) $- 5.3 \times 10^{13}$ ĵ m/s^2

37. Find the time it takes the electron to travel through the region of the electric field.

 (A) 1×10^{-8} s

 (B) 2×10^{-8} s

 (C) 3×10^{-8} s

 (D) 4×10^{-8} s

 (E) 5×10^{-8} s

38. What is the vertical displacement y of the electron while it is in the electric field?

 (A) 0.0230 m

 (B) 0.0445 m

 (C) −0.0750 m

 (D) −0.0424 m

 (E) −0.0340 m

GO ON TO THE NEXT PAGE
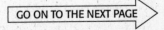

39. Which of the following combinations of 3 Ω resistors would dissipate 36 W of power when connected to a 12 V battery?

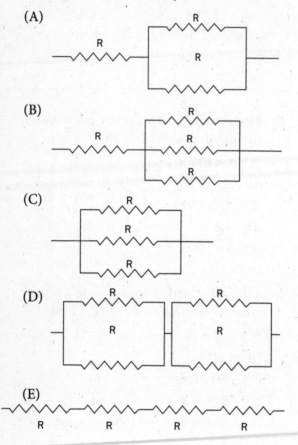

(A)

(B)

(C)

(D)

(E)

Use the following diagram to answer question 40.

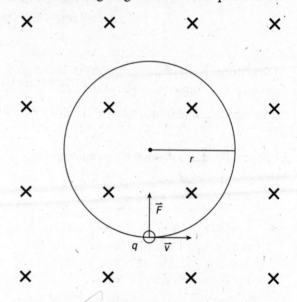

40. A positively charged particle moves through a uniform external magnetic field with its initial velocity vector perpendicular to the field. The particle moves in a circle of radius r at a constant velocity v. If the charge on the particle is q and its mass is m, what would be the magnitude of the particle's velocity in the magnetic field?

(A) $\dfrac{qBr}{m}$

(B) $\dfrac{mr}{qB}$

(C) $\dfrac{qB}{m}$

(D) $\dfrac{m}{qBr}$

(E) $\dfrac{qm}{Br}$

GO ON TO THE NEXT PAGE

Use the following diagram to answer question 41.

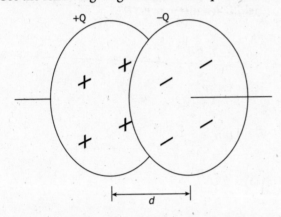

41. A capacitor with capacitance C consists of two circular plates. All else being the same, what is the capacitance of a capacitor that has plates with twice the diameter?

(A) C

(B) $2C$

(C) $4C$

(D) $0.5C$

(E) $0.25C$

42. What combination of 3 μF capacitors would give an equivalent capacitance of 9 μF between the points a and b?

(A)

(B)

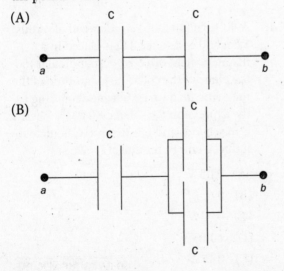

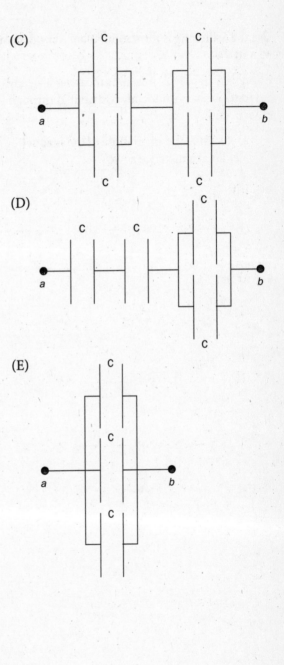

GO ON TO THE NEXT PAGE

Use the following information to answer questions 43 and 44.

A long, straight wire of radius R carries a steady current I_0 that is uniformly distributed through the cross section of the wire.

43. Determine the magnitude of the magnetic field B at a distance $r > R$.

(A) $\dfrac{2\pi r}{\mu_0 I_0}$

(B) $\dfrac{\mu_0 I_0 r}{2\pi}$

(C) $\dfrac{2\pi \mu_0 I_0}{r}$

(D) $\dfrac{\mu_0 I_0}{2\pi r}$

(E) $\dfrac{I_0}{2\pi r \mu_0}$

44. Determine the magnitude of the magnetic field B at a distance $r < R$.

(A) $\dfrac{2\pi r}{\mu_0 I_0 R^2}$

(B) $\dfrac{\mu_0 I_0 r}{2\pi R^2}$

(C) $2\pi \mu_0 I_0 \left(\dfrac{R^2}{r} \right)$

(D) $\dfrac{\mu_0 I_0 R^2}{2\pi r}$

(E) $\dfrac{I_0 R^2}{2\pi r \mu_0}$

45. A wire of resistance R dissipates power P when it is connected in a circuit with a battery whose voltage is V. If this battery is replaced with a 3 V battery, the power dissipated by the resistor will be

(A) P.

(B) $3P$.

(C) $\sqrt{3P}$.

(D) $6P$.

(E) $9P$.

46. A 40-turn (40 loop) circular coil of radius 5 cm and resistance 2 Ω is placed in a magnetic field directed perpendicular to the plane of the coil. The magnitude of the magnetic field varies in time according to the expression $B = -0.02t - 0.03t^2$, where t is in seconds and B is in tesla. Calculate the induced emf in the coil at $t = 2.5$ s.

(A) 0.08 V

(B) 0.05 V

(C) 0.03 V

(D) −0.03 V

(E) 0.17 V

GO ON TO THE NEXT PAGE

Use the following diagram to answer question 47.

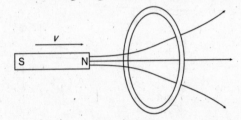

47. The bar magnet shown above is moved rapidly toward a stationary loop of wire with a velocity of magnitude v. The plane of the wire loop is perpendicular to the path of the magnet. As the magnet gets closer, the magnetic flux Φ_m through the loop increases with time. This flux causes

(A) an induced current with a clockwise rotation in the wire loop.

(B) an induced current with a counterclockwise rotation in the wire loop.

(C) an induced current perpendicular to the plane of the wire loop.

(D) an induced current parallel to the plane of the bar magnet.

(E) no induced currents in the wire loop.

48. A negative particle with charge q and mass m is near the surface of a planet with gravitational field g. Determine the magnitude and direction of the *electric field* that will keep the negative particle at rest.

(A) 0

(B) mg Up

(C) mg Down

(D) $\dfrac{mg}{q}$ Up

(E) $\dfrac{mg}{q}$ Down

Use the following information to answer questions 49 and 50.

A single loop circuit contains two external resistors and two sources of emf. The internal resistances of the batteries are negligible.

49. Find the current in the circuit.

(A) 0.25 A

(B) 0.5 A

(C) 0.6 A

(D) 1 A

(E) 8 A

50. What is the power lost in resistors 1 and 2, respectively?

(A) 8 W and 24 W

(B) 1 W and 3 W

(C) 0.5 W and 1.5 W

(D) 0.25 W and 0.083 W

(E) 0.125 W and 0.417 W

GO ON TO THE NEXT PAGE

Use the following diagram to answer question 51.

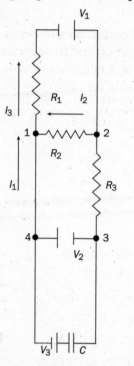

Use the following diagram to answer question 52.

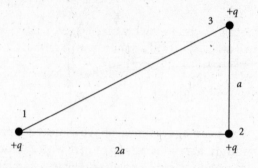

51. The multiloop circuit above contains three resistors, three batteries, and one capacitor. Which of the following equations shows the change in potential around the loop indicated by the numbers 1 through 4?

(A) $V_2 - R_3I_1 + R_2I_2 = 0$

(B) $V_2 - R_2I_1 + R_3I_2 = 0$

(C) $V_2 - R_3I_2 + R_2I_3 = 0$

(D) $V_2 - R_2I_3 + R_3I_2 = 0$

(E) $V_2 - R_2I_1 + R_2I_2 = 0$

52. Three point charges, each with a charge $+q$, are arranged at the corners of a triangle as shown in the figure above. Determine the magnitude of the electrostatic force only between the charges at positions 1 and 3.

(A) $\dfrac{kq^2}{3a^2}$

(B) $\dfrac{kq^2}{a^2}$

(C) $\dfrac{kq^2}{4a^2}$

(D) $\dfrac{kq^2}{9a^2}$

(E) $\dfrac{kq^2}{5a^2}$

GO ON TO THE NEXT PAGE

Use the following information to answer questions 53 and 54.

Three point charges, each with charge $+q$, are located at the center of a sphere with a radius r.

53. Determine the net electric flux through the Gaussian sphere.

(A) $\dfrac{3q}{\varepsilon_0}$

(B) $\dfrac{q}{\varepsilon_0}$

(C) $\sqrt{\dfrac{3q}{\varepsilon_0}}$

(D) $\dfrac{2q}{\varepsilon_0}$

(E) $\sqrt{\dfrac{2q}{\varepsilon_0}}$

54. Calculate the electric field intensity at a point on the surface of the Gaussian sphere that surrounds the three point charges.

(A) $\dfrac{q}{4\pi\varepsilon_0 r^2}$

(B) $\dfrac{3q}{2\pi\varepsilon_0 r^2}$

(C) $\dfrac{3q}{4\pi\varepsilon_0 r^2}$

(D) $\dfrac{9q}{2\pi\varepsilon_0 r^2}$

(E) $\dfrac{9q}{4\pi\varepsilon_0 r^2}$

Use the following information to answer questions 55 and 56.

A proton $(+e)$ enters a uniform magnetic field B directed at an angle θ from the x-axis as shown above. The proton enters the field with a velocity v in the direction of the positive x-axis direction.

55. In what direction would the proton $(+e)$ be accelerated?

(A) $+y$
(B) $+x$
(C) $+z$
(D) $-y$
(E) $-z$

56. If we replaced the proton $(+e)$ with an electron $(-e)$, in what direction would the electron be accelerated?

(A) $+z$
(B) $-z$
(C) $+x$
(D) $-x$
(E) $+y$

GO ON TO THE NEXT PAGE

Use the following diagram to answer question 57.

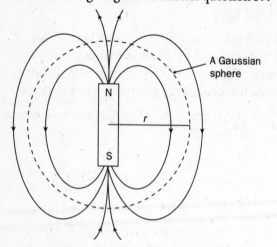

A Gaussian sphere

Use the following diagram to answer question 58.

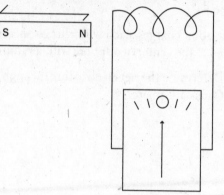

Galvanometer

57. A spherical Gaussian surface of radius r surrounds a bar magnet with a magnetic field strength B. True statements about the magnetic flux through this Gaussian surface include which of the following?

I. The net magnetic flux through the Gaussian surface is zero.

II. The net magnetic flux equals the net electric flux through the Gaussian surface.

III. The Gaussian surface appears to surround a magnetic monopole from a distance.

(A) I only
(B) II only
(C) III only
(D) I and II only
(E) II and III only

58. A bar magnet is moved very rapidly back and forth through a wire coil that is connected to a galvanometer. As the magnet moves, the needle on the galvanometer will

(A) experience a deflection towards the right.
(B) experience a deflection towards the left.
(C) oscillate between zero and deflection right.
(D) oscillate between zero and deflection left.
(E) oscillate between defection left and deflection right.

GO ON TO THE NEXT PAGE

Use the following diagram to answer question 59.

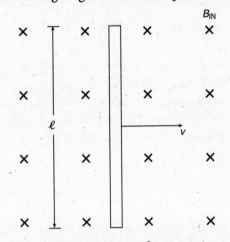

Use the following diagram to answer question 60.

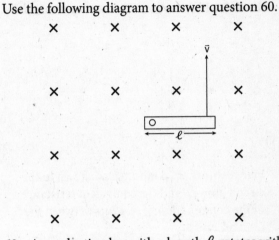

59. A conducting bar of length ℓ is moved at a velocity v perpendicular to a magnetic field directed into the page. Assuming that the bar is accelerated to the right, electrons in the bar would

(A) move to the upper end of the bar.

(B) move to the lower end of the bar.

(C) move to the middle of the bar.

(D) oscillate between both ends of the bar.

(E) remain motionless.

60. A conducting bar with a length ℓ rotates with a constant angular velocity ω in a uniform magnetic field B directed into the page. The bar rotates about a fixed pivot point located at one end of the bar. The magnetic field is perpendicular to the rotational plane of the bar. What emf would be induced between the ends of the bars?

(A) $\frac{1}{4}B\omega^2\ell^2$

(B) $\frac{1}{4}B\omega\ell^2$

(C) $\frac{1}{2}B\omega\ell^2$

(D) $\frac{1}{3}B^2\omega\ell$

(E) $\frac{1}{4}B^2\omega\ell^2$

Use the following diagram to answer question 61.

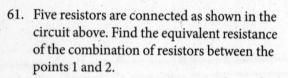

61. Five resistors are connected as shown in the circuit above. Find the equivalent resistance of the combination of resistors between the points 1 and 2.

(A) $1\ \Omega$

(B) $2\ \Omega$

(C) $6\ \Omega$

(D) $10\ \Omega$

(E) $14\ \Omega$

Use the following diagram to answer question 62.

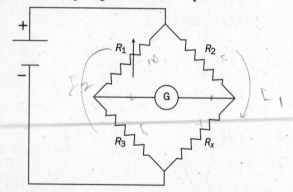

62. Four resistors are arranged as shown in the circuit above. R_1 is a variable resistor and adjusted until the galvanometer in the circuit reads zero. If $R_1 = 10\ \Omega$, $R_2 = 5\ \Omega$, and $R_3 = 6\ \Omega$, what would R_x equal?

(A) $12\ \Omega$

(B) $8.33\ \Omega$

(C) $3\ \Omega$

(D) $1.1\ \Omega$

(E) $0.33\ \Omega$

Use the following diagram to answer question 63.

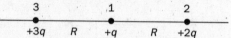

63. A test charge $+q$ is moved directly between two charges as shown. Charges $+3q$ and $+2q$ are located on either side of $+q$ at a distance of R. Determine the magnitude and direction of the net electrostatic force on $+q$.

(A) $\dfrac{kq}{R}$; right

(B) $\dfrac{kq}{R^2}$; right

(C) $\dfrac{kq}{R^2}$; left

(D) $\dfrac{kq^2}{R^2}$; right

(E) $\dfrac{kq^2}{R^2}$; left

64. An object with a total charge of 10 C has zero acceleration in a uniform vertical electric field of 500 N/C. What is the mass of this object?

(A) 500 kg

(B) $\dfrac{1}{5}$ kg

(C) 5 kg

(D) 0.51 kg

(E) 10 kg

Use the following diagram to answer question 65.

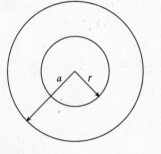

65. An insulating sphere of radius a encloses a uniform charge density ρ with a total charge Q. Determine the electric field E at a point inside the sphere where $r < a$.

(A) $\dfrac{kr}{Qa^3}$

(B) $\dfrac{kQ}{ra^3}$

(C) $\dfrac{kQa}{r^3}$

(D) $\dfrac{kQr}{a^3}$

(E) $\dfrac{Qr}{ka^3}$

Use the following information to answer questions 66 and 67.

$V = 5x^2y - 2xy^2 + 2z^2$ volts is the electric potential over a certain region of space.

66. Determine the electric field over this region.

(A) $(10xy - 2y^2)\,\mathbf{i} + (5x^2 - 4xy)\,\mathbf{j} + (2z^2)\,\mathbf{k}$

(B) $(7xy - 2y^2)\,\mathbf{i} + (4x^2 - 4xy)\,\mathbf{j} + (4z)\,\mathbf{k}$

(C) $(2y^2 - 10xy)\,\mathbf{i} + (4xy - 5x^2)\,\mathbf{j} + (-4z)\,\mathbf{k}$

(D) $(10xy - 2y^2)\,\mathbf{i} + (5x^2 - 2xy)\,\mathbf{j} + (4z)\,\mathbf{k}$

(E) $(10xy - 2y^2)\,\mathbf{i} + (10x - 4y)\,\mathbf{j} + (4z)\,\mathbf{k}$

67. What is the magnitude of the field at a point that has coordinates (in meters) (1, 0, 0)?

(A) 5 N/C

(B) 3 N/C

(C) 1 N/C

(D) 2 N/C

(E) 7 N/C

GO ON TO THE NEXT PAGE

68. A dielectric is inserted between the plates of a charged capacitor that is not connected to anything. True statements about this capacitor include which of the following?

I. The capacitance of the capacitor increases.

II. The charge on the capacitor decreases.

III. The voltage drop across the capacitor decreases.

(A) I only

(B) II only

(C) III only

(D) I and III only

(E) II and III only

69. A narrow beam of electrons produces a current of 1.6×10^{-5} A. There are 10^{10} electrons in each meter of the beam. Which of the following would be the best estimate of the average speed of the electrons in the beam?

(A) 10^{-15} m/s

(B) 10^{-10} m/s

(C) 10^{-4} m/s

(D) 10^{4} m/s

(E) 10^{10} m/s

Use the following diagram to answer question 70.

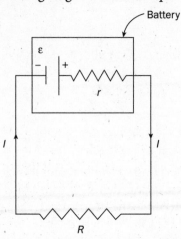

70. The battery in the circuit above is an emf source of magnitude ε with an internal resistance of r. The battery is connected to a resistor R, which results in a current I. What is the internal resistance of the battery?

(A) $\dfrac{I}{\varepsilon - IR}$

(B) $\dfrac{IR - \varepsilon}{I}$

(C) $\dfrac{\varepsilon - IR}{I}$

(D) $\dfrac{\varepsilon}{IR}$

(E) $\dfrac{IR}{\varepsilon}$

Time—45 Minutes 4 Questions

> **Directions:** Answer all four questions. The suggested time is about 11 minutes for answering each of the questions. Show all your work. Calculators are permitted, except for those with typewriter (QWERTY) keyboards.

1. An 80 kg college student climbs to the top of a 15 m flagpole on a dare. Unfortunately, the student becomes frightened at the top of the flagpole and refuses to come back down. One of his friends—a physics major—comes up with a plan to rescue him. She places a 25 kg wood platform at the base of the pole; a large spring with a spring constant of 3,500 N/m is attached to the bottom. She also determines that if he slides down the pole, there should be a constant frictional force of 375 N between his body and the pole. When everything is in place, she convinces him to slide down the pole towards the spring-loaded platform.

    (a) Assuming that the frictional force acts during the entire motion as the student slides down the pole, determine the student's speed just before he hits the wood platform.

    (b) Determine the maximum distance that the spring will be compressed. Assume the student continues to hold onto the pole while being slowed down by the platform.

2. A planet is observed to travel around a star in a nearly circular orbit with a constant speed v at a distance r from the star. The planet has an orbital period of T. During close flybys of the planet and the star, scientists are able to measure each object's density. From this information, they are able to determine the mass of the planet to be m_p and the mass of the star to be m_s.

    (a) Based on this information, develop an expression for the orbital speed of the planet around the star.

    (b) Use these astronomical measurements to derive Kepler's third law from Newton's laws of motion.

    (c) If the planet's orbital period is measured to be 11.86 years and its average distance from the star is 7.78×10^{11} meters, what is the mass of the star?

GO ON TO THE NEXT PAGE

3. A rigid rod of mass M and length ℓ rotates in a vertical plane about a frictionless pivot point in the center of the rod. A mass m_1 is placed on one end of the rod, while a mass m_2 is placed on the other end of the rod. The rod rotates with an angular velocity ω. Note: The moment of inertia for a rod that rotates about its center is given by $I = \frac{1}{12}M\ell^2$.

(a) Determine the angular momentum (L) of the system.

(b) When the rod makes an angle of θ to the horizontal, determine the angular acceleration α of the system as a function of the masses, ℓ, θ, and gravity (g).

4. The position over time of an object of mass 2,000 g is described by the equation $x(t) = 6t^2 - 2t^3$, where x is measured in meters and t in seconds.

(a) At what time is the object at its maximum positive position?

(b) Calculate the displacement $\mathbf{d}$ (a vector) of the particle at $t = 4$ sec and the total length L (a scalar) travelled by the particle in the first 4 seconds.

(c) At what time does the velocity $v = -18$ m/s?

(d) At what time is the force exerted on the object -96 N?

SECTION II: ELECTRICITY AND MAGNETISM
Time—45 Minutes 3 Questions

Directions: Answer all three questions. The suggested time is about 15 minutes for answering each of the questions. Show all your work. Calculators are permitted, except for those with typewriter (QWERTY) keyboards.

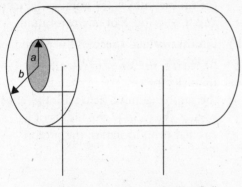

inner wire   outer conducting shell

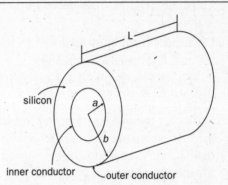

1. (a) A long, straight coaxial cable consists of a solid inner wire with radius a and an outer conducting shell with radius b as shown above. The inner wire carries a current I in one direction, and the outer shell carries the same current in the opposite direction. Assume that a vacuum exists in the gap between the inner conductor and the outer conductor.

    i.   If the inner wire's current I is distributed uniformly over the circular cross section of the wire, use Ampere's law to develop an expression for the magnetic field for $r \le a$, where r is the radial distance from the center of the wire.

    ii. Now develop an expression for the magnetic field for $a \le r \le b$.

    iii. Now develop an expression for the magnetic field for $r \ge b$.

    iv. Sketch a graph of magnetic field versus radial distance from the center of the wire (no specific values for the magnetic field are required).

(b) Now consider a length of the cable as shown in the figure above. The gap between the inner and outer conductor is now completely filled with silicon. The radius of the inner conductor is measured to be $a = 0.25$ cm, the radius of the outer conductor is $b = 2.00$ cm, and the length of the tube is 20 cm. If the resistivity of silicon is $640\ \Omega \cdot$ m, determine the resistance of the silicon as measured between the inner and outer conductors.

GO ON TO THE NEXT PAGE ▷

2. A battery has an emf of 20 V and an internal resistance of 0.08 Ω. The battery's terminals are connected to a load resistance of 5 Ω.

(a) Find the current in the circuit.

(b) Determine the terminal voltage of the battery.

(c) Calculate the power dissipated in the load resistor.

(d) Determine the power dissipated by the internal resistance of the battery.

(e) Find the power that is generated by the battery.

(f) Show that the maximum power lost in the load resistance R occurs when $R = r$; that is, when the load resistance matches the internal resistance of the battery.

3. A proton is moving in a circular orbit of radius 12 cm in a uniform magnetic field of magnitude 0.45 T directed perpendicular to the velocity of the proton.

(a) For this proton of mass m in a magnetic field B, moving at a velocity v with an orbital radius r, develop expressions for the momentum p, the angular velocity ω, and the period T of one revolution.

(b) Find the orbital speed of the proton and its kinetic energy in eV.

(c) If an electron moves perpendicular to the same magnetic field with this speed, what is the ratio of the radius of its circular orbit to that of the proton?

IF YOU FINISH BEFORE TIME IS CALLED, YOU MAY CHECK YOUR WORK ON THIS SECTION ONLY. DO NOT TURN TO ANY OTHER SECTION IN THE TEST.

AP Physics C Practice Test: **Answer Key**

SECTION I: MECHANICS

1. E	21. C
2. E	22. E
3. B	23. A
4. A	24. B
5. D	25. D
6. B	26. B
7. A	27. D
8. D	28. E
9. C	29. B
10. B	30. C
11. D	31. D
12. B	32. A
13. B	33. E
14. A	34. D
15. D	35. A
16. A	
17. E	
18. C	
19. D	
20. E	

SECTION II: ELECTRICITY AND MAGNETISM

36. E	56. B
37. D	57. A
38. D	58. E
39. B	59. B
40. A	60. C
41. C	61. B
42. E	62. C
43. D	63. D
44. B	64. A
45. E	65. D
46. B	66. C
47. B	67. A
48. E	68. D
49. B	69. D
50. C	70. C
51. A	
52. E	
53. A	
54. C	
55. C	

ANSWERS AND EXPLANATIONS

SECTION I: MECHANICS

1. E

The object's velocity plotted as a function of time indicates that the object's velocity is increasing in a linear fashion. This is evidenced by the straight line plotted on the velocity versus time graph. Because the object's velocity is increasing with time in a linear way, the slope of this line will yield the object's acceleration; the slope of the velocity versus time line is a constant. We can assume that the object's acceleration is constant since we can measure the same slope at any point on the line. Therefore, we would select choice (E) as the answer because this shows an acceleration versus time graph with a horizontal line plotted on it. This horizontal line means that the object's acceleration is constant.

2. E

The 3 kg object has a velocity at one instant of $2\mathbf{i}$ m/s. Then 5 seconds later, the object's velocity changes to $(3\mathbf{i} + 3\mathbf{j})$ m/s. This means that the object experienced acceleration. The acceleration vector for an object in motion is equal to the change in the object's velocity vector divided by the change in time:

$$\mathbf{a} = \frac{\Delta v}{\Delta t}$$

Substituting our velocity values over the time interval in question, we get

$$\mathbf{a} = \frac{\Delta v}{\Delta t} = \frac{((3\mathbf{i} + 3\mathbf{j})\text{m/s} - 2\mathbf{i}\text{ m/s})}{5\text{ s}}$$
$$= (0.2\mathbf{i} + 0.6\mathbf{j})\text{ m/s}^2$$

According to Newton's laws of motion, an object experiences acceleration when it is acted upon by an external force. The vector representing the net force on the object is equal to the mass of the object multiplied by the acceleration vector.

$$\mathbf{F} = m\mathbf{a}$$

Substituting for mass and the acceleration vector, we get

$$\mathbf{F} = m\mathbf{a} = (3\text{ kg})(0.2\mathbf{i} + 0.6\mathbf{j})\text{ m/s}^2 = (0.6\mathbf{i} + 1.8\mathbf{j})\text{ N}.$$

3. B

As determined in question 2, the force vector is represented by $\mathbf{F} = (0.6\mathbf{i} - 1.8\mathbf{j})$ N.

Remember that the magnitude of any vector of the form $\mathbf{A} = A_x\mathbf{i} + A_y\mathbf{j}$ can be found by

$$A = \sqrt{(A_x^2 + A_y^2)}$$

Therefore, the magnitude of this vector can then be determined by

$$F = \sqrt{(0.6)^2 + (-1.8)^2} = 1.9\text{ N}$$

Although you don't have a calculator, 1.9 N is the only answer that is in the ballpark.

4. A

The average acceleration is equal to the change in velocity divided by the change in time.

$$a = \frac{\Delta v}{\Delta t}$$

At $t = 0$, the velocity of the particle is 40 m/s, and at $t = 10$ s, the velocity of the particle is 0 m/s. Substituting, we get

$$a = \frac{\Delta v}{\Delta t} = \frac{(0\text{ m/s} - 40\text{ m/s})}{(10\text{ s} - 0\text{ s})} = -4\text{ m/s}^2$$

5. D

The instantaneous speed is defined as the object's speed at a given time t. It can be determined by taking the first derivative with respect to time of the function that represents the particle's displacement:

$$v(t) = \frac{dx}{dt}$$

where x is the particle's displacement.

Taking the derivative of $x = 5t + 9t^3$ with respect to time yields

$$v(t) = \frac{dx}{dt} = \frac{d}{dt}(5t + 9t^3) = 5 + 27t^2$$

6. B

The instantaneous acceleration at some time t can be determined by taking the first derivative with respect to time of the function that represents the particle's instantaneous speed.

$$a(t) = \frac{dv}{dt}$$

For this situation, $v(t) = 5 + 27t$. Substituting this into the following yields

$$a(t) = \frac{dv}{dt} = \frac{d}{dt}(5 + 27t^2) = 54t$$

We are asked to evaluate this at a specific time. Therefore, at $t = 2s$, we would get

$$a(2) = 54(2) = 108 \text{ m/s}^2$$

7. A

To determine the object's speed at the time of impact, we need to consider the object's energy during the fall. As the object falls, the potential energy it had standing on the bridge is changed into kinetic energy. At the moment of impact, all of the object's energy is kinetic energy. Therefore, the potential energy before the jump will equal the kinetic energy at impact.

$$U_g = K$$

Substituting the expressions for potential energy and kinetic energy, we get

$$\frac{1}{2}mv^2 = mgh$$

Now we are going to solve this equation for v. We notice that we can cancel m on both sides of the equation.

$$\frac{1}{2}v^2 = gh$$

This means that the speed of a falling object (neglecting air resistance) doesn't depend upon the object's mass.

Simplifying and solving for velocity, we get $v = \sqrt{2gh}$.

8. D

Take the first derivative with respect to time for the displacement function $y = 6t - 2.8t^2$ to obtain the velocity function for the ball.

$$v(t) = \frac{dy}{dt} = \frac{d}{dt}(6t - 2.8t^2) = 6 - 5.6t$$

Now determine the velocity at $t = 0$ to determine the ball's initial velocity.

$$v(0) = 6 - 5.6(0) = 6 \text{ m/s}$$

9. C

Take the first derivative with respect to time for the displacement function $y = 6t - 2.8t^2$ to obtain the velocity function for the ball.

$$v(t) = \frac{dy}{dt} = \frac{d}{dt}(6t - 2.8t^2) = 6 - 5.6t$$

10. B

The displacement of a simple harmonic oscillator can be written as

$$x = A\cos(\omega t + \delta)$$

The acceleration of a simple harmonic oscillator can be determined by taking the second derivative of this displacement function with respect to time.

$$\frac{d^2x}{dt^2} = \frac{d^2}{dt^2}(A\cos(\omega t + \delta))$$

$$= -\omega^2 A\cos(\omega t + \delta)$$

Since $x = A \cos (\omega t + \delta)$, we can simplify to

$$\frac{d^2 x}{dt^2} = -\omega^2 A \cos (\omega t + \delta) = -\omega^2 x$$

In our question, we are given that the equation of a simple harmonic oscillator is

$$\frac{d^2 x}{dt^2} = -25x$$

We can see that this looks very much like the equation for the acceleration of a simple harmonic oscillator: $\frac{d^2 x}{dt^2} = -\omega^2 x$. So, by inspection, we can assume that angular frequency ω is

$$\omega^2 = 25 \text{ or } \omega = 5$$

The equation for the frequency of a simple harmonic oscillator is

$$f = \frac{\omega}{2\pi}$$

Substituting our value for angular frequency, $\omega = 5$, we find that the frequency of this harmonic oscillator is

$$f = \frac{\omega}{2\pi} = \frac{5}{2\pi}$$

11. D

Since we know the weight of the object and the acceleration due to gravity on planet Y, we can use this information to determine the mass of the object.

The weight of the object is determined by

$$w_y = mg_y$$

Solving for m, we get

$$m = \frac{w_y}{g_y}$$

Substituting $w_y = 150$ N and $g_y = 1.5g$ (let $g = 10$ m/s²) yields

$$m = \frac{w_y}{g_y} = \frac{150 \text{ N}}{1.5(10 \text{ m/s}^2)} = 10 \text{ kg}$$

12. B

In the previous problem, we determined the mass to be 10 kg. We can use this to determine the acceleration due to gravity on planet X.

$$g_x = \frac{w_x}{m} = \frac{5 \text{ N}}{10 \text{ kg}} = 0.5 \text{ m/s}^2$$

13. B

To determine the object's orbital speed, we need to consider the forces acting on the object and determine which of these is providing the centripetal acceleration to maintain the circular path. The only force acting on the object is gravitation. Since this force is responsible for the object's circular motion around the planet, it follows that the centripetal force exerted on the orbiting object will be given by the gravitational force acting on the object.

$$F_C = F_G$$

Substituting in the equations for centripetal force and gravitational force, we get

$$\frac{mv^2}{R} = \frac{GMm}{R^2}$$

We will now solve this equation for speed v. It is worth noting that we can cancel the orbiting object's mass m from both sides of the equation. This indicates that the orbital speed of the object is independent from the object's mass.

$$\frac{v^2}{R} = \frac{GM}{R^2}$$

Simplifying and solving for v, we get

$$v = \sqrt{\frac{GM}{R}}$$

14. A

To determine the speed of the car as it makes the turn, we need to consider the forces working on the car. For the car to make the turn safely and stay on the road, the force of friction between the tires and the road must provide the centripetal force necessary to pull the car into the circular path as it travels around the curve.

$$F_c = f_s$$

Substituting the relationships for centripetal force and static friction

$$\frac{mv^2}{r} = \mu F_N$$

F_N is the normal force and can be expressed as the weight of the car on the level road. Therefore,

$$\frac{mv^2}{R} = \mu mg$$

We can cancel m on both sides. When we simplify and solve for v, we get

$$v = \sqrt{\mu g r}$$

15. D

The kinetic energy is defined as follows:

$$K = \frac{1}{2}mv^2$$

Since v is the first derivative with respect to time of the displacement, it follows that

$$K = \frac{1}{2}mv^2 = \frac{1}{2}m\left(\frac{dx}{dt}\right)^2$$

In the problem, our displacement is defined as $x = 2t + t^3$. The first derivative of this yields

$$v = \frac{dx}{dt} = \frac{d}{dt}(2t + t^3) = 2 + 3t^2$$

Since the mass of the particle is 10 kg and $v = 2 + 3t^2$, we can substitute these values into our equation for kinetic energy.

$$K = \frac{1}{2}mv^2 = \frac{1}{2}(10 \text{ kg})(2 + 3t^2)^2$$
$$= \frac{1}{2}(10 \text{ kg})(4 + 12t^2 + 9t^4)$$
$$= 20 + 60t^2 + 45t^4$$

Therefore, the kinetic energy of this particle at time t is

$$K = (20 + 60t^2 + 45t^4) \text{ J}$$

16. A

Power can be defined as the first derivative with respect to time of the work done in moving the particle.

$$P = \frac{dW}{dt}$$

Since work is being done to move the particle, and the particle gains kinetic energy due to this motion, the following must be true:

$$W = K = 20 + 60t^2 + 45t^4$$

Taking the first time derivative of this expression will give us the power at time t.

$$P = \frac{dW}{dt} = \frac{d}{dt}(20 + 60t^2 + 45t^4)$$
$$= (120t + 180t^3) \text{ W}$$

17. E

To solve this problem, you need to use the definition of the vector dot product.

$$\mathbf{A} \cdot \mathbf{B} = A_x B_x + A_y B_y$$

The magnitude of the work done by $\mathbf{F}$ can be determined by calculating the dot product of the vectors $\mathbf{F}$ and $\mathbf{s}$.

$$W = \mathbf{F} \cdot \mathbf{s} = F_x s_x + F_y s_y$$

Since $\mathbf{F} = (2\mathbf{i} - \mathbf{j})$ N and $\mathbf{s} = (\mathbf{i} - 3\mathbf{j})$ m, it follows that the dot product of $\mathbf{F}$ and $\mathbf{s}$ will be

$$W = \mathbf{F} \cdot \mathbf{s} = (2)(1) + (-1)(-3) = 5 \text{ J}$$

The magnitude of the work done by the force in this system is 5 J.

18. C

The dot product of two vectors can also be expressed as

$$\mathbf{A} \cdot \mathbf{B} = AB \cos \theta$$

where θ is the angle between $\mathbf{A}$ and $\mathbf{B}$.

Therefore, the work expressed as a dot product of $\mathbf{F}$ and $\mathbf{s}$ can be further defined as

$$W = \mathbf{F} \cdot \mathbf{s} = Fs \cos \theta$$

where θ is the angle between $\mathbf{F}$ and $\mathbf{s}$.

Since $\mathbf{F} = 2\mathbf{i} - \mathbf{j}$, then the magnitude of $\mathbf{F}$ will be

$$F = \sqrt{(2)^2 + (-1)^2} = \sqrt{5}$$

Also, since $\mathbf{s} = \mathbf{i} - 3\mathbf{j}$, then the magnitude of $\mathbf{s}$ will be

$$s = \sqrt{(1)^2 + (-3)^2} = \sqrt{10}$$

Solving for θ and substituting the values for $F = \sqrt{5}$ and $s = \sqrt{10}$, we get

$$\theta = \cos^{-1}\left(\frac{W}{Fs}\right) = \cos^{-1}\left(\frac{5 \text{ J}}{(\sqrt{5} \times \sqrt{10})}\right)$$
$$= 45°$$

19. D

Initially, the skaters aren't moving, so the total sum of the momentum is zero because each skater's momentum is zero. When the skaters push away from each other, they each gain the same amount of momentum but directed in opposite directions. Therefore, when you sum the final momentum of the system, the individual momentums of the skaters

cancel each other out, giving us a sum of zero for the final momentum of the system.

20. E

This problem deals with the conservation of momentum. The sum of the momentums before the collision will equal the momentum of the composite body after the collision.

$$\Sigma p_i = \Sigma p_f$$
$$m_1 v_1 + m_2 v_2 = (m_1 + m_2)v_f$$

where v_f is the final velocity of the combined masses.

Substituting our values for the masses and their initial velocities,

$$(3m)v + (4m)(-2v) = (3m + 4m)v_f$$

Note that v_2 is expressed as $-2v$ since it is traveling toward the left.

$$3mv - 8mv = 7mv_f$$
$$-5mv = 7mv_f$$

This yields an expression for the final velocity of the combined mass of the two disks:

$$v_f = \frac{-5v}{7}$$

21. C

At the pole, the person weighs herself and determines that her weight is simply

$$w = mg$$

However, at the equator, the person is experiencing the rotational velocity of the planet. Hence, part of the gravitational force that pulls on the person will actually go into providing the centripetal force that keeps the person going around the Earth at the equator.

This can be represented as follows:

$$w_{equator} = w - F_c = mg - \frac{mv^2}{R}$$

This means the person actually weighs slightly less at the equator than she does at the poles of the planet.

Therefore, the weight of the person at the equator is

$$w_{\text{equator}} = m\left(\frac{g - v^2}{R}\right)$$

22. E

The net torque on this cylinder can be determined by considering the sum of all the individual torques produced by all of the forces at work on this solid cylinder.

$$\tau_{\text{net}} = \tau_1 + \tau_2 + \tau_3$$

The forces at work on the outermost diameter of the cylinder at $2R$ are $2F$ and F.

The torques produced by these forces are

$$\tau_1 = -(2F)(2R) = -4FR \text{ and } \tau_2 = -F(2R) = -2FR$$

Please note that these torques are negative because they will produce a clockwise rotation of the cylinder. The force at work on the innermost diameter of the cylinder at R is F. This force produces a torque:

$$\tau_3 = FR$$

Note that this torque is positive because it will produce a counterclockwise rotation. Therefore, the net torque on this system will be

$$\tau_{\text{net}} = \tau_1 + \tau_2 + \tau_3 = -4FR - 2FR + FR = -5FR$$

Thus, the magnitude is $5FR$.

Since the sign on this torque is negative, the rotation will be clockwise.

23. A

Impulse is defined as the area under the force-time graph.

For the triangular graph given, the area is $\frac{1}{2}$ (base) × (height).

The base of the this triangle is the duration of the impulse. This can be read off the graph as $\Delta t = 1.5$ ms. The height can also be read as 4,000 N.

The time is in milliseconds, so we'll need to convert it to seconds.

$$\Delta t = 1.5 \text{ ms} = 1.5 \times 10^{-3} \text{ s}$$

Therefore, the impulse can be determined as follows:

$$J = \frac{1}{2} \text{ (base) (height)}$$

$$= \frac{1}{2} (1.5 \times 10^{-3} \text{ s}) (4,000 \text{ N}) = 3 \text{ kg} \cdot \text{m/s}$$

24. B

The average force is defined to be the constant force that would supply the same impulse over the same period of time as the nonconstant pulse given in a force-time graph.

$$F_{\text{avg}} = \frac{J}{\Delta t} = \frac{3 \text{ kg} \cdot \text{m/s}}{1.5 \times 10^{-3} \text{ s}} = 2,000 \text{ N}$$

25. D

As the pilot flies the plane through the "loop-the-loop" maneuver, two forces are at work on the pilot's body: the normal force exerted by the seat and the pilot's weight. Both of these forces are along the radial direction, so they both contribute to the overall centripetal force required to pull the pilot around in the circular path dictated by the flight pattern. At the top of the loop, both the normal force exerted by the seat on the pilot and the pilot's weight are pointed downward, toward the center of the circular flight path, so they add to provide the centripetal acceleration. Hence, the sum of the two forces is just equal to the product of the mass and the centripetal acceleration.

$$F_{\text{N}} + mg = \frac{mv^2}{R}$$

Therefore, the force that the cockpit seat exerts on the pilot at the top of the loop is

$$F_N = \frac{mv^2}{R} - mg = m\left(\frac{v^2}{R-g}\right)$$

26. B

To set up this problem, we need to remember that the force of friction is no more than the coefficient of friction multiplied by the normal force. In this case, the normal force is going to be the force that keeps you moving in circular motion. Also, the force of friction that keeps you on the wall is going to have to be equal to the weight of the rider.

With this in mind,

$$F_f = \mu F_N$$
$$mg = \mu mr\omega^2$$

Solving for μ gives us this:

$$\mu = \frac{g}{r\omega^2}$$

27. D

To determine the velocity of the mass, we need to consider the sum of the kinetic and potential energies of the oscillating mass and spring. This sum will equal the initial energy as represented by the work done to give the mass its initial displacement to A. At this initial displacement, the mass is not moving, so all the energy is potential energy, given by

$$E_{total} = \frac{1}{2}kA^2$$

For any other displacement, x, this total energy is split between kinetic and potential energies according to

$$E_{total} = \frac{1}{2}kA^2 = \frac{1}{2}mv^2 + \frac{1}{2}kx^2$$

Solving this for v yields

$$v = \sqrt{\left(\frac{k}{m}\right)(A^2 - x^2)}$$

28. E

Your first instinct might be to apply the definition of work to this problem, thinking that the force of gravity multiplied by the circumference of the satellite's orbit would yield the amount of work being done. However, this would *not* be correct.

The force of gravity is directed toward the planet. The satellite's displacement vector would be drawn at a tangent to the satellite's orbital path. This means that the displacement vector would be perpendicular to the force vector.

By the definition of work,

$$W = \mathbf{F} \cdot \mathbf{s} = Fs\cos\theta$$

Since $\theta = 90°$,

$$W = Fs\cos 90° = Fs\,(0) = 0 \text{ J}$$

The gravitational force between the planet and the satellite doesn't do any work on the satellite. That is why the satellite maintains a constant altitude and a constant speed in orbit. If the satellite were to lose altitude, then gravity would be doing work on the satellite.

29. B

The acceleration of the masses in the system can be determined by examining the forces that are acting upon each of the masses. Mass 1 has the force of its weight m_1g pulling downward at the same time that the force of string tension T is pulling it upward. This mass is also accelerating upward by the resultant force m_1a. These forces on mass 1 are related to each other as follows:

(1) $T - m_1g = m_1a$

The mass sliding on the inclined plane, mass 2, has a component of its weight $m_2g \sin \theta$ pulling it down the incline while the string tension T is pulling in the opposite direction. This mass is accelerating down the inclined plane with an acceleration also given by a. It follows that these forces on mass 2 are related to each other by the following.

(2) $m_2g \sin \theta - T = m_2a$

We can solve (1) and (2) for T and get the following:

$T = m_1a + m_1g = m_2g \sin \theta - m_2a$

Let's rearrange this equation by placing all of the expressions that involve a on the left and those involving g on the right.

$m_1a + m_2a = m_2g \sin \theta - m_1g$

$a(m_1 + m_2) = g(m_2 \sin \theta - m_1)$

Therefore, the acceleration of the masses is

$$a = \frac{m_2g \sin \theta - m_1g}{m_1 + m_2}$$

30. C

We can use the results from the previous problem to help us solve this one. By considering the forces acting upon mass 1, we determined that these forces are related to one another by

$T - m_1g = m_1a$

This means that the string tension T is

$T = m_1g + m_1a = m_1(g + a)$

Substituting the relationship we found for the acceleration of the two masses

$$a = \frac{m_2g \sin \theta - m_1g}{m_1 + m_2}$$

into this equation gives

$$T = m_1(g + a) = m_1 \left(\frac{g + (m_2g \sin \theta - m_1g)}{m_1 + m_2} \right)$$

Simplifying this yields

$$T = \frac{m_1m_2g \sin \theta - m_1^2g + m_1^2g + m_1m_2g}{m_1 + m_2}$$

$$= \frac{m_1m_2g(1 + \sin \theta)}{m_1 + m_2}$$

Therefore, the string tension is

$$T = \frac{m_1m_2g(1 + \sin \theta)}{m_1 + m_2}$$

31. D

We take the first time derivative of the displacement function to get the velocity.

$$v(t) = \frac{dx}{dt} = \frac{d}{dt} 4 \cos \left(2t + \frac{\pi}{3} \right) = -8 \sin \left(2t + \frac{\pi}{3} \right)$$

At $t = 0$, this works out to be

$$v(0) = -8 \sin \left(2(0) + \frac{\pi}{3} \right) = -6.92 \text{ m/s}$$

32. A

We take the first time derivative of the velocity function to get the acceleration.

$$a(t) = \frac{dv}{dt} = \frac{d}{dt} \left(-8 \sin \left(2t + \frac{\pi}{3} \right) \right)$$

$$= -16 \cos \left(2t + \frac{\pi}{3} \right)$$

Now we calculate the acceleration at $t = 0$.

$$a(0) = -16 \cos \left(2(0) + \frac{\pi}{3} \right) = -8 \text{ m/s}^2$$

33. E

The momentum of the meteorite before impact is equal to $m_1 v_1$, while the Earth's momentum is zero during this time. At the moment of impact, the meteorite's momentum is transferred to the Earth/meteorite conglomerate such that the final momentum will be $(m_1 + m_2) v_2$, where v_2 is the recoil speed of the Earth.

Since momentum is conserved, we can set up the problem as follows:

Initial Momentum = Final Momentum

or

$$m_1 v_1 = (m_1 + m_2) v_2$$

Solving for v_2 we get the recoil speed of the Earth following the impact.

$$v_2 = \frac{m_1 v_1}{m_1 + m_2}$$

34. D

Let's begin by determining each of the component tensions for each of the three cable tensions in the system.

F	x component	y component
T_1	$-T_1 \cos \theta_1$	$T_1 \sin \theta_1$
T_2	$T_2 \cos \theta_2$	$T_2 \sin \theta_2$
T_3	0	$-mg$

Notice the sign on each component. A negative sign on a vertical component indicates that the component is directed downward, while a positive sign indicates that the component operates upward. Similarly, a negative sign on a horizontal component indicates that it is directed to the left, while a positive sign indicates the component is directed to the right.

Because the mass is hanging from the cables motionless, the net tension must be zero. We can then assume that because there is no motion in the horizontal direction or the vertical direction, the sum of the forces in these directions will add to zero:

$$\Sigma F_x = 0 \text{ and } \Sigma F_y = 0$$

Therefore, the sum of the horizontal tension components can be written as follows:

$$\Sigma F_x = T_2 \cos \theta_2 - T_1 \cos \theta_1 = 0$$

35. A

We can use the result obtained for the sum of the horizontal tension components in the problem above to find an expression for T_1 in terms of T_2.

$$T_2 \cos \theta_2 - T_1 \cos \theta_1 = 0$$
$$T_1 \cos \theta_1 = T_2 \cos \theta_2$$

The tension T_1 is

$$T_1 = \left(\frac{\cos \theta_2}{\cos \theta_1} \right) T_2$$

SECTION II: ELECTRICITY AND MAGNETISM

36. E

The force exerted on the electron is due to the electric field that is present between the two plates. The magnitude of this field is given by

$$F = qE$$

Since the charge is an electron, we can substitute e (the symbol of an electron) for the charge q.

$$F = eE$$

The problem asks us to determine the acceleration of the electron while in the field. We can calculate the acceleration of the electron from

$$F = ma = eE$$

where m is the mass of the electron.

Solving for the acceleration, we get

$$a = \frac{eE}{m}$$

The electric field between the plates is directed along the positive y-axis. This means that the electron will experience the acceleration in the direction of the negative y-axis.

$$\mathbf{a} = -\frac{eE}{m}\mathbf{j}$$

Substituting the values for $e = 1.6 \times 10^{-19}$ C (provided on the AP Physics exam equation tables), $E = 300$ N/C, and $m = 9.11 \times 10^{-31}$ kg:

$$\mathbf{a} = -\frac{eE}{m}\mathbf{j} = -\frac{(1.6 \times 10^{-19}\text{ C})(300\text{ N/C})}{(9.11 \times 10^{-31}\text{ kg})\,\mathbf{j}}$$

$$= -5.3 \times 10^{13}\,\mathbf{j}\text{ m/s}^2$$

37. D

Because the length of the plates is 0.1 m, this is the distance that the electron travels while in the plates'

electric field. The initial velocity of the electron is 2.5×10^6 m/s. This velocity is directed along the x-axis. The acceleration that the electron experiences once in the field is directed along the vertical. Hence, this initial velocity is the component of the velocity that carries the electron through the field. With this in mind, the distance the electron travels in time t is given by

$$x = v_0 t$$

Solving for t and substituting l for x, we can calculate the time it takes the electron to travel through the plates.

$$t = \frac{l}{v_0} = \frac{0.1\text{ m}}{2.5 \times 10^6\text{ m/s}} = 4 \times 10^{-8}\text{ s}$$

38. D

Because the electron is being accelerated by the electric field, the displacement y can be determined from the following:

$$y = \frac{1}{2}at^2$$

Substituting $a = 5.3 \times 10^{13}$ m/s^2 and $t = 4 \times 10^{-8}$ s, we determine the vertical displacement:

$$y = \frac{1}{2}\,at^2 = \frac{1}{2}\,(-5.3 \times 10^{13}\text{ m/s}^2)(4 \times 10^{-8}\text{ s})^2$$

$$= -0.0424\text{ m}$$

39. B

To solve this problem, it is worth remembering the rules for determining the equivalent resistance of a combination of resistors in a circuit.

The equivalent resistance of a combination of resistors in series can be determined by

$$R_{eq} = R_1 + R_2 + \cdots$$

The equivalent resistance of a combination of resistors in parallel can be determined by

$$\frac{1}{R_{eq}} = \frac{1}{R_1} + \frac{1}{R_2} + \cdots$$

For this problem, the correct choice is (B). This configuration of resistors has an equivalent resistance of 4 Ω. The three 3 Ω resistors in parallel have an equivalent resistance that can be determined as follows:

$$\frac{1}{R_p} = \frac{1}{R} + \frac{1}{R} + \frac{1}{R} = \frac{1}{3} + \frac{1}{3} + \frac{1}{3} = \frac{3}{3} = 1$$

$$R_p = 1 \; \Omega$$

We now add this to the resistor in series to get the total equivalent resistance of the circuit.

$$R_{eq} = R_p + R = 3 \; \Omega + 1 \; \Omega = 4 \; \Omega$$

The other equation to keep in mind is $P = IV$. Since you can calculate the power dissipated from the equivalent circuit, and the total voltage has to be 12 V, you just need to figure out which circuit has a current of 3 A. This only occurs in the circuit in choice (B).

40. A

The centripetal force is whatever radially directed force keeps the particle moving in a circle. In this problem, this is provided by the force that the magnetic field exerts on the particle. This force can be set equal to the product of the mass and the centripetal acceleration.

$$\frac{mv^2}{r} = qvB\sin\theta$$

where m is the mass of the particle, v is the velocity of the particle, r is the radius of the circle, q is the particle's charge, B is the magnitude of the magnetic field, and θ is the angle between the velocity and the magnetic field.

Since the velocity and magnetic field are perpendicular to one another, the angle between v and B is 90 degrees. With $\theta = 90°$, the expression above reduces to

$$\frac{mv^2}{r} = qvB\sin\theta = qvB(\sin)\,(90°) = qvB\,(1) = qvB$$

Canceling v on both sides gives us

$$\frac{mv}{r} = qB$$

Solving for v gives us

$$v = \frac{qBr}{m}$$

41. C

The capacitance of a parallel-plate capacitor is proportional to the area of its plates and inversely proportional to the plate separation. The expression for the capacitance of a parallel-plate capacitor is

$$C = \frac{\varepsilon_0 A}{d}$$

where d is the distance between the plates, A is the area of the capacitor's plate, and ε_0 is the vacuum permittivity constant and is equal to $8.85 \times 10^{-12} \; C^2/N \cdot m^2$. Capacitance is directly proportional to the area of the plates. If the plate radius is doubled, then the area (πr^2) must quadruple. Thus, the capacitance quadruples.

42. E

In order to solve this problem, it is worth remembering the rules for determining the equivalent capacitance of a combination of capacitors in a circuit.

The equivalent capacitance of a combination of capacitors in series can be determined by

$$\frac{1}{C_{eq}} = \frac{1}{C_1} + \frac{1}{C_2} + \cdots$$

The equivalent capacitance of a combination of capacitors in parallel can be determined by

$$C_{eq} = C_1 + C_2 + \cdots$$

The correct combination of capacitors is given in choice (E). This configuration of three 3 μF capacitors in parallel has an equivalent capacitance of

$$C_{eq} = C + C + C = 3 \ \mu F + 3 \ \mu F + 3 \ \mu F = 9 \ \mu F$$

43. D

Ampere's law tells us that the line integral of the dot product of the magnetic field vector **B** and the displacement vector d**s** around any closed path will equal the product of the permeability of free space μ_o and the current I_o.

$$\int \mathbf{B} \cdot d\mathbf{s} = \mu_o I_o$$

For this problem, we are integrating around the wire at a distance r from the center of the wire such that $r \geq R$. This means that the path we are following is simply the circumference of the circle at radius r. Applying this to Ampere's law, we get

$$\int \mathbf{B} \cdot d\mathbf{s} = B \int ds = B \ (2\pi r) = \mu_o \ I_o$$

Dividing both sides of the equation by the circumference gives us the magnitude of the magnetic field B at $r > R$.

$$B = \frac{\mu_o I_o}{2\pi r}$$

44. B

We will employ Ampere's law again to solve this problem. In this situation, the integral path will be around a circle of radius r that is inside the radius R of the wire. The magnetic field in this region is produced by the fraction of the current I_o that moves through this part of the wire.

Since we have been told to assume that the current is uniform through a cross section of the wire, the fraction of the current in the radius $r < R$ can be determined by considering the ratios of the areas in question, πr^2 and πR^2. This ratio will equal the ratios of the currents through the areas I and I_o.

$$\frac{I}{I_o} = \frac{\pi r^2}{\pi R^2}$$

Solving for I, we find that

$$I = \left(\frac{r^2}{R^2}\right) I_o$$

Applying Ampere's law around the path $2\pi r$ with a current I

$$\int \mathbf{B} \cdot d\mathbf{s} = B \int ds = B \ (2\pi r)$$
$$= \mu_o \ I_o = \mu_o \left(\frac{r^2}{R^2}\right) I_o$$

Dividing both sides by the circumference and solving for B, we find that the magnitude of the magnetic field can be obtained from

$$B = \frac{\mu_o I_o r}{2\pi R^2}$$

45. E

The power dissipated in a resistor is dependent upon the amount of resistance, the current, and the voltage drop across the resistor. This power relation can be expressed in the following ways:

$$P = IV = I^2 R = \frac{V^2}{R}$$

For this problem, we'll use the following power expression:

$$P = \frac{V^2}{R}$$

For the simple circuit given (one resistor and one battery), this gives the power dissipated by a resistor R in a circuit with a battery with voltage V (all of the emf provided by the battery must be depleted in the sole resistor). Now, we will replace this battery with a 3 V battery.

$$P_{new} = \frac{(3V)^2}{R} = \frac{9V^2}{R}$$

Since $P = \dfrac{V^2}{R}$, it follows that

$$P_{\text{new}} = \frac{9V^2}{R} = 9\left(\frac{V^2}{R}\right) = 9P$$

Therefore, the power dissipated by the resistor in the 3 V circuit would increase by a factor of 9.

46. B

Faraday's law of induction states that the induced emf in a circuit is directly proportional to the rate of change of magnetic flux over time through the circuit. This statement can be expressed as

$$\varepsilon = -\frac{d\Phi_{\text{m}}}{dt}$$

where ε is the emf in volts and Φ_{m} is the magnetic flux in webers (Wb).

The magnetic flux can be determined by $\Phi_{\text{m}} = BA$, where B is the magnetic field in webers per meter squared $\left(\dfrac{\text{Wb}}{\text{m}^2}\right)$ and A is the area of the coil in m^2.

Since we are dealing with a coil made of a number of loops of the same area, where the magnetic flux moves through all of the loops, the induced emf is

$$\varepsilon = -\frac{Nd\Phi_{\text{m}}}{dt}$$

where N is the number of loops.

Substituting in $\Phi_{\text{m}} = BA$ in the equation above gives us

$$\varepsilon = -\frac{Nd\Phi_{\text{m}}}{dt} = N\frac{d}{dt}(BA)$$

Since the A of the coil isn't time dependent, we can factor that out as follows:

$$\varepsilon = -NA\frac{d}{dt}(B)$$

In the problem, we are given that $B(t) = 0.02t + 0.03t^2$. We can substitute this expression into the

equation above and take the time derivative of the magnetic field to obtain

$$\varepsilon = -NA\frac{d}{dt}(B) = -NA\frac{d}{dt}(-0.02t - 0.03t^2)$$

$$= -NA(-0.02 - 0.06t)$$

Now we can insert the values for coil turns $N = 40$ and area $A = \pi(0.05 \text{ m})^2$ in order to determine the emf at time $t = 2.5$ s.

$$\varepsilon = NA(0.02 + 0.06t)$$

$$= (40)\pi(0.05 \text{ m})^2(0.02 + 0.06(2.5 \text{ s})) = 0.05 \text{ V}$$

47. B

The changing magnetic flux through a wire loop will cause an induced emf and an induced current in the wire loop. The direction of the current can be deduced from Lenz's law, which states that the polarity of the induced emf will produce a current in the wire loop that will create a magnetic flux to oppose the change in magnetic flux through the loop.

In this situation, as the bar magnet rapidly approaches the wire loop, the magnetic flux through the loop increases with time. A current is produced in the wire loop that produces a magnetic flux in the opposite direction from that produced by the bar magnet. This means that the induced current is produced in the wire loop in the counterclockwise direction.

48. E

The electric force ($F_{\text{e}} = qE$) must balance the weight ($F_{\text{g}} = mg$).

$$F_{\text{e}} = F_{\text{g}}$$
$$qE = mg$$
$$E = \frac{mg}{q}$$

The electric force on the charge must be up to balance the downward weight. However, since the charge is negative, the electric field must be *down* because negative charges experience electric forces in a direction opposite to the electric field (recall that the electric field is defined for positive test charges).

49. B

To determine the current in the circuit, we'll need to examine the total voltage drops as we travel around the circuit following the current. Moving around the circuit in a clockwise direction starting from the upper left corner of the circuit, we'll see a potential increase of $+\varepsilon_1$, a drop in potential of $-IR_1$ across the first resistor, a potential drop across the second emf source of $-\varepsilon_2$, and another drop in potential across the second resistor of $-IR_2$. The total drop in potential will be

$$\Sigma\Delta V = +\varepsilon_1 - IR_1 - \varepsilon_2 - IR_2 = 0$$

Solving this equation for the current I, we get the following expression:

$$I = \frac{\varepsilon_1 - \varepsilon_2}{R_1 + R_2}$$

Substituting our emf and resistance values, we find that the current in the circuit will be

$$I = \frac{\varepsilon_1 - \varepsilon_2}{R_1 + R_2} = \frac{10\text{ V} - 6\text{ V}}{6\ \Omega + 2\ \Omega} = \frac{4}{8} = 0.5\text{ A}$$

50. C

In the previous problem, we determined that the current in the circuit was 0.5 A. We can use this value to determine the power dissipated by each resistor by using

$$P = I^2 R$$

Therefore, the power dissipated by each resistor can be calculated as follows:

$$P_1 = I^2 R_1 = (0.5\text{ A})^2 (2\ \Omega) = 0.5\text{ W}$$

and

$$P_2 = I^2 R_2 = (0.5\text{ A})^2 (6\ \Omega) = 1.5\text{ W}$$

51. A

A review of Kirchoff's rules is handy in the analysis of this circuit loop:

1. The sum of all of the currents entering any junction in an electric circuit must equal the sum of all the currents leaving the junction.

2. The sum of all changes in potential around any closed circuit loop will be zero.

If we move around the loop from 1 to 4 in a clockwise direction and apply the second rule, we find an increase in potential across V_2, an increase in potential across R_2 because we are crossing against the current I_2, and a decrease in potential as we move across R_3 because we are crossing in the same direction as current I_1.

Therefore, the potential change around the loop from 1 to 4 can be expressed as

$$V_2 - R_3 I_1 + R_2 I_2 = 0$$

52. E

We'll use Coulomb's law for electrostatic force between two point charges to solve this problem:

$$F_{13} = \frac{kq_1 q_3}{r^2}$$

where q_1 is the charge at position 1, q_3 is the charge at position 3, and r is the distance between the two charges.

The magnitude of the distance between the charges 1 and 3 can be determined by applying the Pythagorean theorem.

$$r = \sqrt{(2a)^2 + a^2} = \sqrt{5a^2}$$

Inserting all of our known values, we can determine the Coulomb force between the charges 1 and 3.

$$F_{13} = \frac{kq_1q_3}{r^2} = \frac{k(+q)(+q)}{(\sqrt{5a^2})^2} = \frac{kq^2}{5a^2}$$

53. A

Gauss's law states that the net electric flux through any closed Gaussian surface is equal to the net charge contained inside the surface divided by the permittivity of free space ε_o. This can be expressed as the following equation:

$$\Phi_c = \int \mathbf{E} \cdot d\mathbf{A} = \frac{q_{in}}{\varepsilon_o}$$

where $\mathbf{E}$ is the electric field, $\mathbf{A}$ is the area of the Gaussian surface (in this case it will be a sphere), and q_{in} is the net charge inside the Gaussian surface. Applying this to our problem, we find that

$$\Phi_c = \int \mathbf{E} \cdot d\mathbf{A} = \frac{q_{in}}{\varepsilon_o} = \frac{3q}{\varepsilon_o}$$

Therefore, the net electric flux through the Gaussian surface due to the three point charges of $+q$ each will be

$$\Phi_c = \frac{3q}{\varepsilon_o}$$

54. C

We will apply Gauss's law for electric flux to determine the magnitude of the electric field at the surface of the Gaussian sphere.

$$\Phi_c = \int \mathbf{E} \cdot d\mathbf{A} = \frac{q_{in}}{\varepsilon_o}$$

Since we are integrating over a sphere, then $A = 4\pi r^2$. It follows that

$$\Phi_c = \int \mathbf{E} \cdot d\mathbf{A} = E \int dA = E(4\pi r^2) = \frac{3q}{\varepsilon_o}$$

Solving for E, we find that the electric field at the surface of the Gaussian sphere of radius r that surrounds 3 point charges of $+q$ each will be

$$E = \frac{3q}{4\pi\varepsilon_o r^2}$$

55. C

The magnetic force exerted on a charged particle moving through a magnetic field with a velocity v is given by

$$\mathbf{F} = q\mathbf{v} \times \mathbf{B}$$

The proton will be accelerated by the magnitude of the magnetic force acting upon it. Since $\mathbf{v}$ and $\mathbf{B}$ are both in the xy plane, and there is an angle of θ between them, the magnetic force will be perpendicular to the plane containing $\mathbf{v}$ and $\mathbf{B}$. By applying the right-hand rule to the cross product of $\mathbf{v}$ and $\mathbf{B}$, we can see that the magnetic force $\mathbf{F}$ will be in the positive z direction. Therefore, the proton will experience an acceleration in the $+z$ direction.

56. B

In the previous problem, we determined that the direction for the acceleration of the proton in question was in the $+z$ direction. Now we are replacing the proton with an electron. Since the charge on the electron $(-e)$ is opposite to the charge on a proton $(+e)$, the acceleration that the electron experiences must be in the opposite direction to that of the acceleration experienced by the proton. Therefore, since the electron must experience an acceleration in the direction of the negative z-axis $(-z)$.

57. A

Gauss's law states that the net magnetic flux through any closed surface is always zero.

$$\Phi_c = \int \mathbf{B} \cdot d\mathbf{A} = 0$$

Central to the validity of this statement is the fact that no isolated magnetic poles or monopoles have been detected in the universe (perhaps they don't even exist!). All magnetic phenomena are observed to have a dipole nature. This means that the magnetic field lines coming from the north-oriented part of the magnet will always fall back onto the south-oriented part of the magnet.

If we surround a bar magnet with a Gaussian sphere, then the number of lines coming from the north end of the magnet and exiting the Gaussian surface will equal the number of lines entering the Gaussian surface and coming back onto the south-oriented part of the magnet. Therefore, the net magnetic flux through the Gaussian surface will be zero.

58. E

According to Faraday's law of induction, the emf induced in the circuit is directly proportional to the rate of change of the magnetic flux over time through the circuit. As the magnet moves rapidly to the right, the needle on the galvanometer will be deflected from zero. As the magnet moves rapidly to the left, the galvanometer's needle will be deflected in the opposite direction. Therefore, as the magnet moves rapidly back and forth through the coil, the needle on the galvanometer will be deflected full-scale in the both directions.

59. B

The electrons in the conducting bar will experience a force along the conductor that can be determined from $\mathbf{F} = q\mathbf{v} \times \mathbf{B}$. Because the magnetic field B is directed into the page and the velocity vector is perpendicular to the magnetic field, when we calculate the cross product of $\mathbf{v}$ and $\mathbf{B}$, we find that the electrons in the conducting bar will experience downward force (by the right-hand rule). This force will cause the electrons to move to the lower end of the conducting bar as the bar starts moving through the magnetic field.

60. C

If we consider emf induced in small segments of the bar as it rotates at velocity v, then we can determine the total emf induced between both ends of the bar.

$$\varepsilon = \int Bv\,dr$$

where dr is the length of a small segment of the bar.

The velocity v can be written as the product of the angular velocity and a length r along the bar.

$$v = \omega r$$

Evaluating the integral yields:

$$\varepsilon = \int_0^\ell Bv\,dr = \int_0^\ell B(\omega r)\,dr$$

$$= B\omega \int_0^\ell r\,dr$$

$$= \frac{1}{2} B\omega r^2 \int_0^\ell$$

$$= \frac{1}{2} B\omega \,(\ell^2 - 0^2)$$

Therefore, the emf induced between the two ends of the bar as it rotates in the magnetic field B with a velocity v is

$$\varepsilon = \frac{1}{2} B\omega \,\ell^2$$

61. B

We can assume that the current entering junction 1 will be the same current that is exiting the circuit via junction 2. With this in mind, we can see that current through each 2 Ω resistor is the same. This means that the potential drop across each 2 Ω resistor is the same. Thus, the potential at the points between each pair of 2 Ω resistors (where the ends of the 6 Ω resistor connect to the circuit) are the same. Therefore, these two points can be connected without affecting the circuit. We could actually remove the 6 Ω resistor and not affect the circuit.

The circuit can then be reduced to a situation where we have two 2 Ω in parallel that are in series with another set of 2 Ω resistors also in parallel.

The equivalent resistances of two 2 Ω resistors in parallel is

$$\frac{1}{R} = \frac{1}{2}\,\Omega + \frac{1}{2}\,\Omega$$

which gives us $R = 1\ \Omega$.

Since we have two sets of two 2 Ω resistors in parallel with each other, the equivalent resistance between points 1 and 2 will be

$$R_{eq} = 1\ \Omega + 1\ \Omega = 2\ \Omega$$

62. C

Resistances can be measured using this device, known as a Wheatstone bridge. A resistor of unknown resistance is placed at position R_x. R_1 has an adjustable resistance, which is changed until the current across the bridge is zero and the galvanometer reads zero.

If the current passing through R_1 is I_1 and the current passing through R_2 is I_2, then we can say the following concerning the potential across each resistor in the circuit:

$$I_1 R_1 = I_2 R_2 \quad \text{and} \quad I_1 R_3 = I_2 R_x$$

Therefore, the following ratios apply to this circuit:

$$\frac{I_1}{I_2} = \frac{R_2}{R_1} = \frac{R_x}{R_3}$$

Solving for R_x, we get

$$R_x = \frac{R_2 R_3}{R_1}$$

Substituting our values for R_1, R_2, and R_3, we can determine that

$$R_x = \frac{R_2 R_3}{R_1} = \frac{(5\ \Omega)(6\ \Omega)}{(10\ \Omega)} = 3\ \Omega$$

63. D

The Coulomb force can determine the electrostatic force between two charges:

$$F_{12} = \frac{kq_1 q_2}{r^2}$$

where q_1 and q_2 are the charges that are separated by a distance r.

The test charge $+q$ experiences a force as a result of the $+3q$ charge and the $+2q$ charge. These forces are determined as follows:

$$F_{13} = \frac{k(3q)(q)}{R^2} = \frac{3kq^2}{R^2}$$

This repulsive force is exerted toward the right.

$$F_{12} = \frac{k(2q)(q)}{R^2} = \frac{2kq^2}{R^2}$$

This repulsive force is exerted toward the left.

The net electrostatic force acting upon the test charge $+q$ will be

$$F_{net} = F_{13} - F_{12} = \frac{3kq^2}{R^2} - \frac{2kq^2}{R^2} = \frac{kq^2}{R^2}$$

Because the net force is positive, this force will be directed toward the right.

64. A

The object in question has two forces working upon it: its weight due to the force of gravity pulling it downward and the electrical force pushing it upwards. Since the object isn't moving, the forces are in balance and must equal each other.

$$F_g = F_E$$

Substituting the proper expressions, we find that

$$mg = qE$$

The mass of the object can be found by solving for m:

$$m = \frac{qE}{g}$$

Substituting in our values for q, E, and g, we can calculate the mass of the object:

$$m = \frac{qE}{g} = \frac{(10 \text{ C})(500 \text{ N/C})}{10 \text{ m/s}^2} = 500 \text{ kg}$$

65. D

We will apply Gauss's law to this situation.

$$\int \mathbf{E} \cdot d\mathbf{A} = \frac{q_{in}}{\varepsilon_o}$$

Since we are integrating over a spherical surface with a radius of r, the area will be $4\pi r^2$. Applying this to Gauss's law, we get

$$E(4\pi r^2) = \frac{q_{in}}{\varepsilon_o}$$

Solving for E gives us

$$E = \frac{q_{in}}{\varepsilon_o}(4\pi r^2)$$

The charge contained inside the sphere with radius a has a total charge of q. However, we are considering only a portion of that charge, which is contained within a sphere with a radius $r < a$. The charge contained within this region is given by

$$q_{in} = \rho V = \rho \left(\frac{4}{3}\pi r^3\right)$$

Inserting this into our electric field equation gives us

$$E = \frac{q_{in}}{\varepsilon_o}(4\pi r^2) = \rho \frac{\left(\frac{4}{3}\pi r^3\right)}{\varepsilon_o(4\pi r^2)}$$

This expression reduces to

$$E = \rho \frac{r}{3\varepsilon_o}$$

The charge density ρ is for the entire sphere of charge Q with a radius a. The charge density can be expressed as

$$\rho = \frac{Q}{\frac{4}{3}\pi a^3}$$

We now can make this substitution into our electric field equation.

$$E = \rho \frac{r}{3\varepsilon_o} = \frac{\left(\dfrac{Q}{\left(\frac{4}{3}\pi a^3\right)}\right)r}{3\varepsilon_o} = \frac{Qr}{4\pi\varepsilon_o a^3}$$

Since $k = \dfrac{1}{4\pi\varepsilon_o}$, our electric field expression can be further simplified as

$$E = \frac{kQr}{a^3}$$

66. C

The electric field can be defined as the negative gradient of the electric potential function.

$$\mathbf{E} = -\nabla V$$

We can determine $\mathbf{E}$ by taking the negative derivative of V with respect to each component.

$$E_x = -\frac{dV}{dx} = -\frac{d}{dx}(5x^2y - 2xy^2 + 2z^2)$$
$$= -10xy + 2y^2$$
$$E_y = -\frac{dV}{dy} = -\frac{d}{dy}(5x^2y - 2xy^2 + 2z^2)$$
$$= -5x^2 + 4xy$$
$$E_z = -\frac{dV}{dz} = -\frac{d}{dz}(5x^2y - 2xy^2 + 2z^2)$$
$$= -4z$$

Therefore, the electric field for this potential will be

$$\mathbf{E} = (2y^2 - 10xy)\,\mathbf{i} + (4xy - 5x^2)\,\mathbf{j} + (-4z)\,\mathbf{k}$$

67. A

We are now asked to evaluate the magnitude of our electric field $\mathbf{E}$ at a point $(1, 0, 0)$. We'll begin by calculating each component of $\mathbf{E}$ at this point.

$$E_x = 2y^2 - 10xy = 2(0)^2 - 10(1)(0) = 0 \text{ N/C}$$

$E_y = 4xy - 5x^2 = 4(1)(0) - 5(1)^2 = -5$ N/C

$E_z = 4z = 4(0) = 0$ N/C

The magnitude of E is given by

$$E = \sqrt{(E_x^2 + E_y^2 + E_z^2)}$$

Substituting in our magnitudes for each component, we find that

$$E = \sqrt{(E_x^2 + E_y^2 + E_z^2)}$$
$$= \sqrt{(0)^2 + (-5)^2 + (0)^2} = 5 \text{ N/C}$$

68. D

A dielectric is a nonconducting material (rubber, wood, glass, etc.) that can be inserted between the plates of a capacitor. Before the dielectric is added, the capacitance of the capacitor can be found by the following equation:

$$C_o = \varepsilon_o \frac{A}{d}$$

When the dielectric is inserted between the plates, the voltage drops between the plates by a factor of κ:

$$V = \frac{V_o}{\kappa}$$

where V_o is the initial voltage of the system.

The charge on the capacitor doesn't change with the addition of the dielectric. Therefore, the capacitance changes to make this change in voltage occur.

$$C = \frac{Q_o}{V} = \frac{Q_o}{\left(\dfrac{V_o}{\kappa}\right)} = \frac{\kappa Q_o}{V_o} = \kappa C_o$$

This means that the capacitance of a dielectric increases by a factor of κ.

$$C = \frac{\kappa \varepsilon_o A}{d}$$

In short, a dielectric will cause the voltage of a capacitor to decrease. This decrease in voltage is made possible by an increase in the capacitance of the capacitor.

69. D

The current in a conductor I is defined as the amount of charge that flows through a conductor in a certain amount of time.

$$I = \frac{\Delta Q}{\Delta t}$$

The current I can also be represented by a combination of the number n of charges q that travel with a velocity v_d through a cross sectional area A.

$$I = nqv_d A$$

Solving for velocity v_d (also called the "drift velocity"), we get

$$v_d = \frac{I}{nqA}$$

We are given the product $nA = 10^{10}$ electrons/m of beam.

Substituting in our values for I, n, and q (the charge on an electron),

$$v_d = \frac{I}{nAq} = \frac{(1.6 \times 10^{-5}\text{A})}{(10^{10})(1.6 \times 10^{-19}\text{C})} = 10^4 \text{ m/s}$$

Therefore, the best estimate for the speed of these electrons is 10^4 m/s.

70. C

The terminal voltage of the battery is given by $V = \varepsilon - Ir$, where V is the terminal voltage, ε is the battery's emf, and Ir is the voltage drop across the battery's internal resistance. Since V is also the voltage drop across the resistor R, we see that

$$IR = \varepsilon - Ir$$

Solving for r, we obtain

$$r = \frac{\varepsilon - IR}{I}$$

SECTION II: MECHANICS

1. (a) At the top of the flagpole, the student has potential energy (U_s) that can be determined by the following:

$$U_s = m_s gh$$

where m_s is the mass of the student, g the acceleration due to gravity, and h the height of the flagpole.

As the student slides down the flagpole, this energy is converted into kinetic energy. However, friction robs some of this energy as it performs work on the sliding student and slows his descent. Therefore, the final kinetic energy will actually be less than the beginning potential energy.

$$U_s - W_f = K_s$$

where W_f is the work done by the frictional force (F_f) and K_s is the final kinetic energy of the student. When we substitute in the various formulas for potential energy, work, and kinetic energy, we get the following:

$$m_s gh - F_f h = \frac{1}{2} m_s v^2$$

Solving the above for v, we get an expression for the final velocity of the student sliding down the flagpole just before he hits the platform.

$$v = \sqrt{\frac{2\,(m_s gh - F_f h)}{m_s}}$$

Substituting in the all of the known values gives us

$$v = \sqrt{\frac{2(80 \text{ kg})(9.8 \text{ m/s}^2)(15 \text{ m}) - (375 \text{ N})(15 \text{ m})}{80 \text{ kg}}}$$

$$= 12.38 \text{ m/s}$$

(b) The collision of the student and platform is completely inelastic. At the moment of impact, the new velocity v_T can be determined by conservation of momentum:

$$m_s v_s = m_T v_T = (m_s + m_p) v_T$$

where v_s is the student's velocity at impact, calculated in (a). Solve for v_T:

$$v_T = \left(\frac{m_s}{m_s + m_p}\right) v_s = \left(\frac{80}{105}\right)(12.38) = 9.43 \text{ m/s}$$

We can now use conservation of energy to determine the displacement y of the spring. The energy absorbed by the spring as it reaches its maximum compression is $\frac{1}{2} ky^2$. The student/platform has a kinetic energy of $\frac{1}{2} m_T v_T^2$ and potential energy of $m_T gy$. However, since the student is still holding onto the pole, energy in the amount of $F_f y$ is lost, where $F_f = 375$ N.

By combining,

$$\frac{1}{2} ky^2 = \frac{1}{2} m_T v_T^2 + m_T gy - F_f y$$

Substituting the numbers, we get

$$1750 y^2 = 4668.56 + 1029 y - 375 y$$

or

$$y^2 - 0.37 y - 2.67 = 0$$

Use the quadratic equation ($y > 0$ only) to find $y = 1.83$ m.

2. (a) The shape of the planet's orbit is observed to be nearly circular. This means that the gravitational force between the planet and star is providing the centripetal force that keeps the planet going in a circle around the star.

$$F_{\text{gravity}} = F_{\text{centripetal}}$$

Substituting in the expressions for Newton's universal law of gravitation and centripetal force, we get

$$\frac{Gm_s m_p}{r^2} = \frac{m_p v^2}{r}$$

Notice that we can cancel the planet's mass m_p from both sides of the equation. This means that the orbital speed of a planet around a star is independent from the planet's mass.

$$\frac{Gm_s}{r^2} = \frac{v^2}{r}$$

Simplifying and solving for v, we get the following expression for orbital speed:

$$v = \sqrt{\frac{Gm_s}{r}}$$

We see from this expression that the orbital speed of a planet depends only upon the mass of the star and the planet's distance from the star.

(b) Again, we will start by comparing Newton's universal law of gravitation to the centripetal force.

$$\frac{Gm_s m_p}{r^2} = \frac{m_p v^2}{r}$$

Again, we will cancel the planet's mass m_p from both sides of the equation. We can also make a substitution here by realizing that the orbital speed of the planet v can be expressed as $\frac{2\pi r}{T}$. This gives us

$$\frac{Gm_s}{r^2} = \frac{\left(\frac{2\pi r}{T}\right)^2}{r}$$

Simplifying and solving for T^2, we can derive Kepler's third law.

$$T^2 = \left(\frac{4\pi^2}{Gm_s}\right) r^3$$

(c) We will start with the expression for Kepler's third law we derived in the previous problem.

$$T^2 = \left(\frac{4\pi^2}{Gm_s}\right) r^3$$

We are looking for the mass of the star, so we must solve this expression for m_s.

$$m_s = \frac{4\pi^2 r^3}{GT^2}$$

Substituting all of the known values, we obtain the following result for the mass of the star. Note: You must convert the period in years to seconds before performing the calculations. A handy rule of thumb that provides adequate accuracy is that $1 \text{ yr} \approx \pi \times 10^7$ s.

$$m_s = \frac{4\pi^2 (7.78 \times 10^{11} \text{m})^3}{\left(6.67 \times 10^{-11} \frac{\text{N} \cdot \text{m}^2}{\text{kg}^2}\right)(11.86\pi \times 10^7 \text{s})^2}$$

$$= 2.00 \times 10^{30} \text{ kg}$$

3. (a) Angular momentum is equal to the moment of inertia of the system multiplied by the angular velocity. To determine the moment of inertia of the system, you need to add up the moments of inertia of all of the components in the system. The system consists of the rod, mass 1, and mass 2.

The motion of the masses leads to a moment of inertia for each mass, given by

$$I = m\left(\frac{\ell}{2}\right)^2, \text{ where } \frac{\ell}{2} \text{ is the distance from the mass}$$

to the center of rotation.

It follows that the moment of inertia for the whole system is

$$I = \frac{1}{12}M\ell^2 + m_1\left(\frac{\ell}{2}\right)^2 + m_2\left(\frac{\ell}{2}\right)^2$$

$$= \left(\frac{\ell^2}{4}\right)\left(\frac{M}{3} + m_1 + m_2\right)$$

When the angular velocity is ω, the angular momentum will be

$$L = I\omega = \left(\frac{\ell^2}{4}\right)\left(\frac{M}{3} + m_1 + m_2\right)\omega$$

(b) To find the angular acceleration α of this system, we must first determine the net torque τ_{net} of the system. We do not know which mass is creating a positive torque or negative torque (it depends on which one is on which side of the pivot point at a given time). However, we *do* know that they provide torques in opposite directions from each other. Therefore, we will assume that one is positive and the other is negative.

Torque produced by the force m_1g about the pivot point is

$$\tau_1 = m_1 g\left(\frac{1}{2}\right)\ell\cos\theta$$

The torque produced by the force m_2g about the pivot point is in the opposite direction:

$$\tau_2 = -m_2 g\left(\frac{1}{2}\right)\ell\cos\theta$$

Therefore, the net torque of the system is

$$\tau_{net} = \tau_1 + \tau_2 = m_1 g\left(\frac{1}{2}\right)\ell\cos\theta - m_2 g\left(\frac{1}{2}\right)\ell\cos\theta$$

$$= \frac{1}{2}(m_1 - m_2)g\ell\cos\theta$$

Using $\tau_{net} = I\alpha$, we can get the following expression for angular acceleration:

$$\alpha = \frac{\tau_{net}}{I} = \frac{2(m_1 - m_2)g\cos\theta}{\ell\left(\dfrac{M}{3} + m_1 + m_2\right)}$$

4. (a) To find the maximum or minimum position, take the first derivative of the position and set it equal to zero:

$$x'(t) = 12t - 6t^2 = 0$$

For $t > 0$, $12 - 6t = 0 \Rightarrow t = 2$ s.

To prove it is a maximum rather than a minimum, you must show that the second derivative is negative:

$x''(t) = 12 - 12t$, which is negative for $t > 1$.

(b) $d = x(4) = 6(4^2) - 2(4^3)$
$= 6(16) - 8(16) = -32$ m

For $0 < t \leq 2$, the object travels
$x(2) = 6(4) - 2(8) = 24 - 16 = 8$ m.
For $2 < t \leq 4$, the object travels
$x(2) - x(4) = 16 - (-32) = 48$ m.
Therefore, $L = 8 + 48 = 56$ m.

(c) $v = \dfrac{dx}{dt} = 12t - 6t^2 = -18$ m/s.

Therefore, $6t^2 - 12t - 18 = 0$, or $t^2 - 2t - 3 = 0$.

Factoring: $(t - 3)(t + 1) = 0$. For $t > 0$, the solution is $t = 3$ s.

(d) $F = ma = m\dfrac{dv}{dt} = 2$ kg $(12 - 12t) = 24 - 24t$.

Setting the expression for F equal to the given value and solving for t,

$$24 - 24t = -96 \text{ N}$$

$$1 - t = -4$$

$$t = 5 \text{ s}$$

SECTION II: ELECTRICITY AND MAGNETISM

1. (a)

i. We can use Ampere's law because of the high degree of symmetry. At a distance r, we know that the magnetic field is tangential to a circle of radius r about the wire and that B is constant everywhere on the circle. Ampere's law is given as follows:

$$\oint \vec{B} \bullet \vec{dl} = \mu_0 I_c$$

Since the field is constant, the B pulls out in front of the integral, and the closed integral becomes the circumference of the circle:

$$B\oint \vec{dl} = B(2\pi r) = \mu_0 I_c$$

The current, I_c, enclosed by the circle is determined by the ratio of the amount of area enclosed by the circle and the cross-sectional area of the inner conductor:

$$I_c = \left(\frac{\pi r^2}{\pi a^2}\right) I = \frac{r^2}{a^2} I$$

If we plug this into Ampere's law,

$$B(2\pi r) = \mu_o \frac{r^2}{a^2} I$$

$$B = \frac{\mu_o}{2\pi} \frac{I}{a^2} r$$

ii. When $a \le r \le b$, the total current enclosed is I. Therefore,

$$B(2\pi r) = \mu_o I$$

$$B = \frac{\mu_o}{2\pi} \frac{I}{r}$$

iii. When $r \ge b$, the total current enclosed is zero because the currents flow in opposite directions; therefore, the magnetic field is *zero*. This magnetic shielding helps explain why coaxial cables are used in so many applications.

iv. The graph of B versus r looks like this:

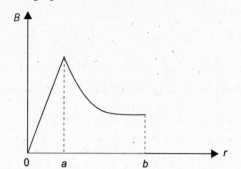

(b) The resistance of a uniform conductor can be determined by

$$R = \frac{\rho L}{A}$$

where ρ is the resistivity, L is the length of the conductor, and A is the cross-sectional area of the conductor.

In order to determine the resistance of the silicon, we need to envision it as a hollow tube with a surface area given by $A = 2\pi r L$.

We want to integrate over the region bounded by radius a at the bottom and radius b at the top. To do this, we need to represent the resistance of our hollow cylinder as

$$dR = \frac{\rho dr}{2\pi r L} = \left(\frac{\rho}{2\pi r L}\right) dr$$

We can now set up our integral:

$$R = \int_a^b dR = \int_a^b \left(\frac{\rho}{2\pi r L}\right) dr$$

$$= \left(\frac{\rho}{2\pi L}\right) \int_a^b \frac{1}{r} dr = \left(\frac{\rho}{2\pi L}\right) \ln\left(\frac{b}{a}\right)$$

Substituting all of our values for $a = 0.25$ cm, $b = 2.00$ cm, $\rho = 640 \, \Omega \cdot$ m, and $L = 20$ cm:

$$R = \left(\frac{\rho}{2\pi L}\right) ln\left(\frac{b}{a}\right)$$

$$= \left(\frac{(640 \, \Omega \cdot \text{m})}{2\pi(0.20 \, \text{m})}\right) ln\left(\frac{0.02 \, \text{m}}{2.5 \times 10^{-3} \, \text{m}}\right)$$

Therefore, the resistance of the silicon in the coaxial cable will be:

$$R = 1,059 \, \Omega$$

2. (a) To determine the current I in the circuit, let's first consider that the sum of all the potential drops around the circuit must equal zero. This gives us

$$\varepsilon - Ir - IR = 0$$

where ε is the emf of the battery, $-Ir$ is the potential drop across the internal resistance of the battery, and $-IR$ is the voltage drop across the resistor.

Solving this for I gives us

$$\varepsilon - Ir - IR = 0$$
$$\varepsilon - I(r + R) = 0$$
$$\varepsilon = I(r + R)$$
$$I = \frac{\varepsilon}{r + R}$$

Therefore, the current in the circuit is

$$I = \frac{\varepsilon}{r + R} = \frac{20 \, \text{V}}{0.08 \, \Omega + 5 \, \Omega}$$

$$= \frac{20 \, \text{V}}{5.08 \, \Omega} = 3.94 \, \text{A}$$

(b) The terminal voltage of the battery in the circuit can be determined in the following ways.

The difference between the emf of the battery and the potential drop of its internal resistance will give you the terminal voltage of the battery.

$$V = \varepsilon - Ir = 20 \, \text{V} - (3.94 \, \text{A})(0.08 \, \Omega) = 19.68 \, \text{V}$$

The voltage drop across the load resistor will give you a similar result.

$$V = IR = (3.94 \, \text{A})(5 \, \Omega) = 19.7 \, \text{V}$$

(c) The power dissipated in the load resistor can be determined as the product of the square of the current in the circuit multiplied by the magnitude of the load resistor.

$$P_R = I^2 R = (3.94 \, \text{A})^2 (5 \, \Omega) = 77.62 \, \text{W}$$

(d) The power dissipated by the internal resistance of the battery can be calculated as follows:

$$P_r = I^2 r = (3.94 \, \text{A})^2 (0.08 \, \Omega) = 1.24 \, \text{W}$$

(e) The power delivered by the battery is the sum of the powers that are dissipated in both the load resistor and the internal resistance of the battery.

$$P_B = P_R + P_r = 77.62 \, \text{W} + 1.24 \, \text{W} = 78.86 \, \text{W}$$

We can also determine this value by considering that the power delivered by the battery could be expressed as

$$P_B = I\varepsilon = (3.94 \, \text{A})(20 \, \text{V}) = 78.8 \, \text{W}$$

(f) The maximum power lost in the load resistance R occurs when $R = r$. This situation occurs when the load resistance matches the internal resistance of the battery. To prove this, we must first develop an expression for power in the circuit as a function of both the load resistance R and the internal resistance r:

$$P = I^2 R = \left(\frac{\varepsilon}{R + r} \right)^2 R = \frac{\varepsilon^2 R}{(R + r)^2}$$

The point at which the maximum power is lost in the load resistance can be determined by taking the first derivative with respect to R of the power function above and setting it equal to zero.

$$P'(R) = \frac{dP}{dR} = 0$$

The first derivative for the power function with respect to R is

$$P'(R) = \frac{(R + r)^2 (\varepsilon^2) - \varepsilon^2 R \, (2(R + r)(1))}{((R + r)^2)^2} = 0$$

For the above fraction to equal zero, it follows that

$$(R+r)^2(\varepsilon^2) - \varepsilon^2 R\,(2(R+r)(1)) = 0$$

In the next steps, we will simplify this and solve for R.

$$(R+r)^2(\varepsilon^2) = \varepsilon^2 R\,(2(R+r)(1))$$

Canceling an $\varepsilon^2(R+r)$ on both sides yields

$$R+r = 2R$$
$$r = 2R - R = R$$

This means that the maximum power lost occurs when the load resistance matches the internal resistance of the battery. The maximum power will have the following value:

$$P(r) = \frac{\varepsilon^2(r)}{(r+r)^2} = \frac{\varepsilon^2(r)}{(2r)^2} = \frac{\varepsilon^2}{4r}$$

3. (a) If the proton is moving in a circular path with a radius r at a velocity v, then the proton is experiencing a centripetal acceleration with the following magnitude:

$$a_c = \frac{v^2}{r}$$

The source of the acceleration that is exerted on the proton is the magnetic field force

$$F_M = qvB$$

where q is the charge on the proton and B is the magnetic field strength.

This force is then equal in magnitude to the mass of the proton times the centripetal acceleration, as required by Newton's second law:

$$\frac{mv^2}{r} = qvB$$

Solving for v, we get

$$v = \frac{qBr}{m}$$

so $p = mv = qBr$.

The angular velocity is $\omega = \dfrac{v}{r} = \dfrac{qB}{m}$.

The period is $T = \dfrac{2\pi}{\omega} = \dfrac{2\pi m}{qB}$.

(b) We can use the following expression to determine the orbital speed of the proton:

$$v = \frac{qBr}{m}$$

Substituting our values for $q = 1.6 \times 10^{-19}$ C, $B = 0.45$ T, $r = 0.12$ m, and $m = 1.67 \times 10^{-27}$ kg, we get the following value for velocity:

$$v = \frac{qBr}{m} = \frac{(1.6 \times 10^{-19} \text{ C})(0.45 \text{ T})(0.12 \text{ m})}{(1.67 \times 10^{-27} \text{ kg})}$$

$$= 5.17 \times 10^6 \text{ m/s}$$

The kinetic energy is expressed as

$$K = \frac{1}{2}mv^2$$

$$= (0.5)(1.67 \times 10^{-27} \text{ kg})(5.17 \times 10^6 \text{ m/s})^2$$

$$= 2.23 \times 10^{-14} \text{ J} = 1.39 \times 10^5 \text{ eV}$$

(c) We can solve for r in the expression for velocity that we developed above to determine the orbital radius of the electron.

$$v = \frac{qBr}{m} \rightarrow r = \frac{vm}{qB}$$

Substituting our values for $q = 1.6 \times 10^{-19}$ C, $B = 0.45$ T, $m = 9.11 \times 10^{-31}$ kg, and $v = 5.17 \times 10^6$ m/s, we get the following value for the orbital radius:

$$r = \frac{vm}{qB} = \frac{(5.17 \times 10^6 \text{ m/s})(9.11 \times 10^{-31} \text{ kg})}{(1.6 \times 10^{-19} \text{ C})(0.45 \text{ T})}$$

$$= 6.54 \times 10^{-5} \text{ m}$$

The ratio is therefore $\dfrac{6.54 \times 10^{-5} \text{ m}}{1.2 \times 10^{-1} \text{ m}} = 5.45 \times 10^{-4}$.

GLOSSARY

A

absolute pressure measurement of pressure that includes the contribution of atmospheric pressure; *always* a positive quantity

absolute zero 0 K or –273.15°C; the temperature at which the kinetic energy of molecules is at a minimum

absorption process by which an incoming photon elevates an electron from a lower orbit to a higher orbit if its energy is exactly right

acceleration describes the change in velocity within a time interval in units of meters per second per second (m/s^2)

action-reaction pairs sets in which forces are always found, reflecting the fact that for every action there is a simultaneous equal and opposite reaction

adiabatic process a process in which there is no heat transfer between the gas and the outside world

alpha decay type of nuclear reaction occurring in heavy nuclei where a heavy nucleus spontaneously emits an alpha particle and transmutes into another element

ammeter device for measuring current that is connected in series with an element of a circuit

ampere A, the unit for current; equivalent to coulombs per second

Ampere's law addresses the contour integral of the magnetic field in the direction of the path of integration and relates this result to the enclosed current; the magnetic analogue of Gauss's law

Amperian loop Ampere's law requires you to integrate the magnetic circulation around a closed path called an Amperian loop

amplitude measurable feature of a wave corresponding to the maximum value that the medium is disturbed from equilibrium (i.e., its condition when there is no wave present)

angular acceleration rate at which changes in angular velocity occur

angular velocity measure of changes in angular displacement as they occur over time

antinode in a standing wave pattern, points along the medium that undergo maximum displacement; there is always one antinode halfway between two consecutive nodes

apparent weight the difference between the actual weight of a body and the buoyant force; is always less than the actual weight

Archimedes' principle buoyant force is equal to the weight of the fluid displaced by a body

atomic energy levels the energies associated with stable orbits because of quantized angular momentum

atomic number Z, the number of protons in a given nucleus; the quantity that defines and distinguishes each element

atomic spectra also called spectral lines, the characteristic light frequencies observed when tubes filled with various gases, such as hydrogen, are excited and the light emitted is passed through a prism; instead of a continuous band of colors (as one would see with "white" light), only very distinct lines of specific colors (or frequencies) are observed when this light undergoes dispersion

average acceleration acceleration measured over a substantive time interval

average velocity the vector form of speed and the displacement per time interval; a measurement taken over the entire interval

Avogadro's number the number of atoms, molecules, or other particles that make up one mole, which has been found to be 6.02×10^{23} (mol)$^{-1}$; this number of *things*/mol is known as Avogadro's number (N_0, also often denoted by the symbol N_A)

B

Bernoulli's equation $P_1 + \rho g y_1 + \frac{1}{2}\rho v_1^2 = P_2 + \rho g y_2 + \frac{1}{2}\rho v_2^2 = const.$; a form of the conservation of mechanical energy for the fluid flowing through a pipe

binding energy the energy required to free an electron from a state; to free an electron from state n, this is simply $-E_n$

Biot-Savart law a formula that determines the magnetic field near a current; the magnetic analogue of Coulomb's law

buoyant force net upward force on an object equal to the weight of the fluid displaced

C

calorie quantity of heat equal to 4.186 J

capacitance the ratio of the potential difference between a capacitor's plates to the charge on either plate of the capacitor; if this potential is known to be V volts, we can define the capacitance as $C = \dfrac{Q}{V}$, measured in coulombs per volt, or farads (F); Q is the charge on the positively charged conductor, by convention, and V is the potential difference between the conductors

capacitor a device that stores energy in an electric field

Carnot cycle special set of processes depicting the most efficient heat engine cycle

Cartesian coordinate system defined by three mutually perpendicular axes; a point in space is described by noting how far you must move parallel to each of these axes (also called the rectangular coordinate system)

center of mass point at which an object's weight is concentrated or centered; point used to describe the object's responses to external forces

centripetal acceleration acceleration toward the center of a circular path

centripetal force the force that maintains circular motion; equal to mass × centripetal acceleration

charge matter's intrinsic affinity for electromagnetic interaction; a property that comes in positive and negative types

charged an object is charged if it has an excess number of charge carriers on it

coefficient of area expansion γ, used to calculate changes in area in two-dimensional expansion, $\Delta A = \gamma A_0 \Delta T$; to a good approximation, $\gamma \cong 2\alpha$

coefficient of linear expansion α; the constant of proportionality for expansion in one dimension, $\Delta \ell = \alpha_0 \ell_0 \Delta T$

coefficient of volume expansion β, used to calculate changes in volume in two-dimensional expansion $\Delta V = \beta V_0 \Delta T$; to a good approximation, $\beta \cong 3\alpha$

components vectors that can be combined to find a resultant vector

Compton shift (also called the Compton effect) the increase in wavelength that occurs when X-ray photons interact with electrons in a material

Compton wavelength for an electron, this shift is defined by the formula $\dfrac{h}{m_e c}$; a Compton shift can be caused by a photon's collision with any particle; the only difference would be that the mass term would need to be the mass of the particle in question.

concave lens a lens that is thinner in the middle than it is on the edges

conductor an object or substance that allows charges to move through its volume (e.g., many metals and salt solutions)

conservative force a force that creates no net loss of energy when used to do work (e.g., gravity)

constructive interference interaction between two physical waves in the same medium that are at the same place at the same time; the crest of one wave passes through the crest of another wave, resulting in a wave that is larger than either individual wave

contact force a force that results from actual contact

continuity equation a statement of the conservation of mass for an ideal fluid flowing in a pipe of varying cross-sectional area, which says that if the fluid meets a constriction in a given pipe, it must speed up

converging lens a lens that focuses parallel rays of light onto one point, also called a positive lens

convex lens a lens that is thicker in the middle than it is on the edges

Coulomb's law a mathematical description of the way a point charge's electrostatic forces interact and affect one another

crest the high point of a wave

critical angle the incident angle at which total internal reflection first occurs, given by using Snell's law and setting θ_2 equal to 90 degrees: $\theta_{\text{critical}} = \sin^{-1} \dfrac{n_2}{n_1}$; total internal reflection occurs when $\theta_1 > \theta_{\text{critical}}$

cross product a way of multiplying vectors that gives a vector perpendicular to both vectors being multiplied; often used in angular momentum and electromagnetism, it results in a vector product and often describes two force fields and how they affect one another

curved mirror a mirror designed so that incoming parallel light is reflected to a single point or is reflected such that it appears to come from a single point

cutoff frequency f_c, a light frequency below which no photoelectrons are observed, no matter how intense the light source is; the frequency above which electrons can be freed from the material (i.e., the frequency of radiation having an energy just equal to the work function)

cylindrical coordinates a three-dimensional extension of polar coordinates in which the third axis is simply z (r, θ, z)

D

de Broglie wavelength the wavelength associated with a moving particle as given by $\lambda = \dfrac{h}{p}$

decay process by which an excited electron falls back to its normal, lowest energy (i.e., its ground state), emitting an appropriate-frequency photon

density mass per unit volume

destructive interference interaction between two physical waves in the same medium that are at the same place at the same time; the crest of one wave passes through the trough of another wave, resulting in a wave that is smaller than one of the other waves

dielectrics insulating materials that react to an applied electric field; this reaction is called "polarization" because the charges separate slightly, creating dipoles all throughout the material

diffuse reflection reflection in which parallel incident rays do not result in parallel reflected rays, such as that which happens when light shines on any surface that is not flat and shiny

displacement a vector form of distance; an object's change in position

displacement vector the resultant vector between two points: $\Delta r = r_1 - r_2$

diverging lens a lens that defocuses parallel rays of light such that they appear to come from one point; also called a negative lens

dot product a way of multiplying vectors that describes the projection of one vector onto another and therefore results in a scalar product; often used in physics problems associated with work and power

dynamics the study of forces and their effects on the motion of objects

E

eddy currents tiny loops of current created in conductors in response to changing magnetic fields

effective resistance the result of combining resistors; also called reduced circuit resistance

electric circuit a system through which electric current flows, causing changes in pieces of the system and (quite often) in its surroundings

electric current the rate of flow of a positive charge across a surface per unit time

electric dipole composed of two charges (q_1 and q_2) that are equal in magnitude and opposite in sign, separated by a distance, d; the line connecting the charges is often called the axis of the dipole

electric field the electric force per unit charge acting on a point charge

electric flux Φ_E, the "flow" of electric field across a surface; $\Phi_E = EA\cos\theta$, where E is the magnitude of the electric field, A is the area of the surface, and Φ is the angle between the field vector and the area vector, an arrow perpendicular to the surface whose length represents the size of the surface; flux is also the dot product of the field and the area: $\Phi_E = \mathbf{E} \bullet \mathbf{A}$

electric potential the work $(F \times d\cos\theta)$ required to bring a small positive test charge from infinity to a particular point

electromotive force emf, energy per unit charge that is supplied by a source of electric current

energy level diagram a series of lines depicting atomic energy levels, running from the ground state $(n = 1)$ through all the excited states $(n = 2, 3, \ldots, \infty)$

equations of state equations describing relationships in gases between various quantities of interest (e.g., pressure, volume, temperature, quantity of gas, etc.)

equilibrium in a medium the condition of the medium when there is no wave present

excitation an electron's change of state as a result of absorption (i.e., its movement from a lower orbit to a higher orbit)

F

face one surface of a lens

Faraday's law relates the change in magnetic flux through a surface bounded by a path to an induced potential called the electromotive force, or emf, which takes volts as its units

field force force that creates effects without visible contact

first law of thermodynamics a statement of the conservation of energy for a system that accounts not only for mechanical energy but also for thermal energy and internal energy; the first law deals with macroscopic quantities

fluid anything that is capable of flowing; both liquids and gases are included under this definition

focal length f, the distance from the reflecting surface of a spherical mirror to the point where parallel rays converge

focal point the point to which incoming parallel light is reflected in a curved mirror, also called its focus

force push or pull exerted on an object that causes acceleration

free-body diagram a diagram that represents all forces acting on a single object with arrows; the length of the arrows represents magnitude, and these arrows point in the direction of the vectors, drawn radially from the center of the object

frequency the number of wave cycles occurring per unit of time; the inverse of period; in its simplest form, frequency represents the number of wave crests that pass a point in 1 second

friction the opposing force between two surfaces that are sliding or rolling parallel to one another that opposes the applied force on an object

fundamental wave the first standing wave in a standing wave pattern, also called the first harmonic; the fundamental is the standing wave with the longest wavelength

G

gauge pressure measurement of pressure equal to absolute pressure minus atmospheric pressure; this quantity is often negative

Gauss's law the electric flux through a closed surface is proportional to the total charge enclosed by that surface; $Q_{enc} = \varepsilon_0 \int E \cdot dA$

Gaussian surfaces imaginary closed surfaces that are the subject of Gauss's law

geometric optics the branch of optics that studies light as a stream of particles or a ray

gravitational force attractive force between any two objects that depends on both the masses of the objects and the distance between the objects; the equation describing Newton's law of gravity is $F_G = \dfrac{-Gm_1m_2}{r^2}$

ground state an electron's normal, lowest energy state

H

heat energy that is transferred into or out of something, caused by differences in temperature

heat engine a device that absorbs heat from the environment and transforms some portion of this heat into mechanical work

heat transfer occurs when energy is transferred into or out of something, caused by differences in temperature

Hooke's law strain is proportional to stress on a spring below the elastic limit of the spring

hydraulic lever mechanism in which a fluid is enclosed between two pistons having different cross-sectional areas; when an outside force is applied to the piston having the smaller area, the resulting pressure increase is transmitted throughout the fluid, including (and most importantly) to the larger piston

I

ideal fluid a fluid in which the flow meets four conditions: it is nonviscous, it is incompressible, its flow is steady, and the flow is irrotational

ideal gas law $PV = nRT$, where P is the absolute pressure of the gas, V is the volume of the gas, n is the amount of gas in moles, and T is the temperature of the gas

ideal gas a gas with a relatively low pressure or density

image the convergence or apparent convergence of light

impulse force multiplied by the time over which the force is applied

incompressible unable to be compressed; fluids that have a constant density

inductance property of an electric circuit by which changing magnetic flux induces an electromotive force in the circuit

inertia the tendency of an object to resist a change in motion

index of refraction n, the ratio of the speed of light in vacuum to the speed of light in a given material, given by $n = \dfrac{c}{v}$, where c equals the speed of light in vacuum and v equals the speed of light in the material

inductor an electrical device that stores energy in a magnetic field

instantaneous acceleration acceleration measured at a particular instant in time

instantaneous velocity vector form of speed measured at a particular instant in time

insulator an object or substance that prevents charges from moving through its volume, such as glass or plastic

interfere when two physical waves in the same medium pass through one another, they are said to interfere; interference can be either constructive or destructive

internal energy energy that is tied up in a substance in the various types of motion and bonds of its atoms and molecules

internal reflection reflection that occurs when incident light encounters a medium of lower optical density; the ray bends away from the normal, and the exit angle is greater than the incident angle

ionized converted to an ion; when an electron is lifted to the highest ($n = \infty$) energy level or state, it has been set free from the atom altogether, and the atom is said to be ionized

irrotational term describing a fluid flow that is smooth without any swirling or turbulence; this does not mean that the flow goes only in straight lines but simply that there are no vortices along the flow

isobaric process a process in which pressure is constant; represented by a horizontal line on a P-V diagram, and the work can be calculated directly from its definition, $W = -P\Delta V$

isothermal process a process in which the temperature is held constant; such a process is represented by a "$\frac{1}{V}$" type of curve on a P-V diagram

isotopes atoms of the same element that have different numbers of neutrons

isovolumetric process a process in which the volume remains constant (sometimes also called an **isochoric process**); represented as vertical lines on a P-V diagram; because there is no change in volume, no work is done during this type of process

J

joule J, the unit in which work and energy are measured; one joule is the product of one newton acting over one meter

K

kinematics analysis of motion without paying attention to the forces causing the motions

kinetic friction friction that occurs when there is relative motion between two surfaces (e.g., a book sliding across a table)

L

law of conservation of angular momentum in the absence of external torques, the total angular momentum of a system remains constant

law of reflection the angle of incidence equals the angle of reflection

law of refraction when light travels from one medium to another, it bends toward a line drawn normal to the interface; the exact relationship between the angle of incidence and the angle of refraction is Snell's law, $n_1 \sin\theta_1 = n_2 \sin\theta_2$, where subscript 1 refers to the first medium and subscript 2 refers to the second medium

lens an object that can focus parallel rays of light onto one point or can defocus parallel rays of light such that they appear to come from one point

Lenz's law induced current must oppose the change in flux (conservation of energy)

longitudinal a wave that creates a disturbance that is parallel to the direction it moves

Lorentz equation an equation describing the magnitude of the force exerted on a charge when a charge, q, moves through a region of space with a magnetic field, B

Lorentz force the force exerted on a charge when a charge, q, moves through a region of space with a magnetic field, B; the magnitude of the force is given by the Lorentz equation

M

magnet an object that has a magnetic field around it

magnetic constant the permeability of free space, symbolized by μ_0

magnetic field B, the field that produces a force on a moving charge at right angles to the motion of the charge

magnetic monopole a magnet with only one end, none of which exist

magnification M, the ratio of the image height to the object height; $M = \dfrac{h_i}{h_o}$ (where h is height), which can be shown to be equivalent to $M = -\dfrac{s_i}{s_o}$, where s refers to distance along the optical axis from the optical element

magnitude the absolute value of the length of a vector

manometer an instrument for measuring pressure, frequently presented as a hollow piece of tubing bent into some shape (typically one or more "U" shapes) and filled with one or more fluids separated by distinct interfaces

mass the measurement of inertia

mass number A, the total number of protons and neutrons in a nucleus, which is equal to $Z + N$ (i.e., atomic number + neutron number)

mass-energy equivalence the total energy of a particle of mass m and momentum p is given by $E^2 = p^2c^2 + (mc^2)^2$, where c is the speed of light in vacuum and has a value of approximately of 3×10^8 m/s; this formula implies that even a stationary particle ($p = 0$) has some nonzero energy, which is proportional to its mass

mechanical equivalent of heat a constant factor relating the calorie to the joule, which is 1 calorie = 4.186 J

medium something that can transmit a wave

mirror an object designed to reflect light in a particular way

mirror equation a simple equation that relates focal length to object and image distances: $\frac{1}{f} = \frac{1}{s_o} + \frac{1}{s_i}$; this equation is called the thin lens equation when applied to lenses

moles the SI unit of amount of a substance; the abbreviation is mol

momentum mass × velocity

N

net external force the sum of all vector forces acting on an object

net work the total work done by all the forces acting on the system (i.e., the work done by the net force)

neutron number N, the number of neutrons in a nucleus

Newton's first law an object at rest will remain at rest, and an object in motion will continue in motion with a constant velocity unless the object experiences a net external force

Newton's second law the magnitude of the acceleration of an object is directly proportional to the resultant force acting on it and inversely proportional to its mass; the direction of the resultant acceleration is in the direction of the resultant force

Newton's third law if two objects interact, the magnitude of the force exerted on object A by object B is equal to the magnitude of the force simultaneously exerted on object B by object A; these forces are opposite in direction

nodes in a standing wave pattern, nodes are the points along the medium that appear to be standing still (i.e., points that do not deviate from equilibrium); there is always exactly one node halfway between two antinodes

nonconservative force a force that results in a net loss of energy when used to do work (e.g., friction that converts some energy to heat that can no longer be used to do useful work)

nonviscous term describing a fluid that flows freely with no friction

normal force the force that is exerted by any object in contact with another object

nucleon the generic term given to particles that can be in a nucleus; both neutrons and protons are considered nucleons

O

observed frequency measurable characteristic of a wave, described by the number of wave crests an observer encounters divided by the time required to encounter them

ohm Ω, the unit of electrical resistance; one ohm is the resistance created when a 1-volt potential difference is applied to a sample and a 1-amp current is produced

Ohm's law the definition of electrical resistance; a ratio between the current in a circuit and the emf applied to a closed circuit

ohmmeter device for measuring resistance, which can be constructed from a battery, a voltmeter, and an ammeter

ohmic resistor an object for which resistance is constant

opaque an opaque object is made of a medium that does not transmit any light

optics the study of how light travels through various media

P

P-V diagrams graphic representations of changes in the state of a fixed sample (constant n) of an ideal gas, in which the changes are tracked by plotting the sequence of (P, V) ordered pairs for each step in the process as the gas undergoes the change from one state to another

parabolic mirror a mirror with a surface that is a portion of a parabola rotated about its symmetry axis

parallel plate capacitor consists of two conducting plates, each of area A, separated by a distance d; if the separation is small (d is significantly less than $\sqrt{A}$), then the capacitor is considered ideal

Pascal's principle when pressure is applied to an *enclosed* static fluid, it will be instantaneously transmitted undiminished throughout the fluid and to the walls of the enclosing container

period the time it takes an object or system to complete one cycle (e.g., a spring or any other simple harmonic oscillator)

permeability of free space $4\pi \times 10^{-7} \dfrac{\text{T} \cdot \text{m}}{\text{A}}$ (tesla-meter per amp)

photoelectric effect when light is shone onto certain metallic targets, photoelectrons are ejected and can be used to create an electric current in a circuit; this effect provides evidence for the existence of photons and points to important problems in the classical Maxwell wave theory of light

photoelectron an electron that is ejected from a substance (often an alkali metal) after absorbing electromagnetic radiation

photon a discrete bundle of electromagnetic radiation

physical optics the branch of optics that treats light as a wave

physical wave a wave transmitted through a massive medium, such as sound waves, water waves, and waves on a string

plane mirror a mirror intended to cause specular reflection over its entire surface

Planck's constant $h = 6.63 \times 10^{-34}$ J $\cdot$ s; an experimentally determined value

point charges imaginary objects that carry charge (and possibly mass) but have no physical size

polar coordinate system two-dimensional (r, θ) coordinate system; r is the distance from the origin to the point (radial distance), and θ is the angular coordinate (the polar angle), which is the measure of the angle from the x-axis counterclockwise to the point

poles opposite ends of a magnetic object; one pole is "north," the other is "south," and the magnetic field points from north to south

position vector the vector that points from the origin of the axes to a point or particle: $\mathbf{r} = x\,\hat{\mathbf{i}} + y\,\hat{\mathbf{j}}$

potential difference the work done per unit charge as the charge moves from one point to another in an electric field; when working with electric currents and circuits, it is common to refer to the potential difference as the "applied" or "impressed" voltage

power measure of the rate at which work is done; work (or change in energy) divided by time interval

pressure force per unit area

process something that is done to cause a sample of gas to change from one state to another

projectile motion a term that describes the movement of an object that has been launched with any initial velocity and allowed to follow its trajectory without any outside forces working on it

Q

quantize to limit to a discrete set of values using quantum mechanical rules

quantum a packet of energy identical to other packets in the same emission; a unit of energy

R

radius of curvature R, the radius of the sphere from which a spherical mirror is taken

resultant a vector that represents the sum of multiple vectors in multiple dimensions

reflection one of two things that can happen when a ray of light encounters the boundary between two different materials; reflection can be thought of as light bouncing off a surface

refraction one of two things that can happen when a ray of light encounters the boundary between two different materials; refraction bends the ray as it enters the new material

resistance the ratio of applied voltage to current produced

resistivity an intrinsic property of all materials, which is measured in ohms per meter (Ω/m); a proportionality constant relating cross-sectional area and length for a given electric conductor to its resistance at a given temperature

resistor resistive element in a circuit

root-mean-square (rms) velocity velocity of gas particles defined as the square *root* of the *mean* of the *squares* of the velocities:

$$v_{rms} = \sqrt{\frac{3k_B T}{\mu}} = \sqrt{\frac{3RT}{M}}$$

S

scalar a quantity that can completely be specified by a magnitude

simple harmonic motion the oscillating motion that occurs when the force restoring the system to its equilibrium position is directly proportional to its displacement from the equilibrium position

Snell's law the law of refraction

specific gravity the ratio of a substance's density to the density of water (assumed to be 1,000 kg/m^3)

spectral lines see **atomic spectra**

specular reflection reflection in which parallel incident rays result in parallel reflected rays, such as that which happens with typical bathroom mirrors

speed of light in vacuum c, approximately 3×10^8 m/s

spherical coordinates coordinate system that defines a point in space by giving a radial distance from the origin, a longitude angle, and a latitude angle; there is not one standard notation

spherical mirror a mirror with a surface that is a portion of a sphere

standing wave a wave that appears to be stationary

static friction friction that occurs when there is no relative motion between two surfaces (e.g., a book sitting on an incline)

steady occurring at a constant rate, as with the flow of a fluid

stopping potential ΔV_s, the size of the potential difference required to stop all the emitted electrons from crossing the gap in a circuit, thus dropping the current to zero

T

test charge a small positive charge used to establish the nature of an electric field; it is so small that its presence does not affect the rest of the world

thermal equilibrium two objects are said to be in thermal equilibrium if no changes in their temperatures occur when they are brought into contact

thermodynamic cycle any set of sequential processes that eventually lead a gas back to its initial state so the sequence may be repeated

thermodynamic efficiency a measure of how much useful energy (in the form of work) is derived from a thermodynamic cycle divided by how much energy is supplied in the form of heat

thermodynamic state the condition of a sample of gas in terms of the quantities that are important in a thermodynamic description; the state variables are, in turn, related to one another through the relevant equation of state

thermometer object that can be used to test whether two independent, unconnected objects are in thermal equilibrium (i.e., are at the same temperature)

thin lens a lens that can be treated as a single surface

thin lens equation a simple equation that relates the focal length to the object and image distances: $\frac{1}{f} = \frac{1}{s_o} + \frac{1}{s_i}$; this equation is called the mirror equation when applied to mirrors

torque the rotational analogue to force; product of force and the perpendicular distance between the axis of rotation and the extended line of the force vector

total internal reflection reflection of the total amount of incident light at the boundary between two media

transients short-term behaviors in circuits, those that appear before the circuit settling into a steady state

transitions changes in an electron's state resulting from excitation and decay

translucent a translucent object is made of a medium that allows light to pass through but tends to scatter some part of the light in all directions

transparent a transparent object is made of a medium that lets light pass through without causing a narrow beam to spread

transverse wave a wave that creates a disturbance perpendicular to the direction the wave moves

traveling wave a wave that travels or moves through a medium; a wave that can be seen to move

trough the low point of a wave

U

unified atomic mass unit a unit of mass, given by the abbreviation u, that simplifies the arithmetic in reaction calculations; it is approximately, but not exactly, the mass of a proton, and the conversion is $1\,u = 1.66 \times 10^{-27}$ kg

universal gravitational constant G, a number that had to be measured experimentally; its value is $G = 6.67 \times 10^{-11}$ N $\cdot$ m^2/kg^2

V

vector a quantity that is completely specified by a magnitude and a direction; the direction of a vector can be given in several formats, including positive and negative values; trigonometric functions; north, south, forward, up, and down coordinates; or any term useful for the physical situation

voltmeter device for measuring voltage that is connected across the ends of an element in a circuit

W

watt W, unit in which power is measured

wave a disturbance that moves through a medium, transferring energy and momentum but not mass

wave-particle duality the fact that light behaves as both wave and particle, depending on the situation

wave fronts a set of parallel lines representing plane waves

wavelength usually the length between two points on the wave that have the same y-value and are on the same slope as one another at the same time; it is generally easiest to talk about wavelength as the distance between two successive maxima (crests) or two successive minima (troughs)

wave speed a constant that is the product of wavelength and frequency for a given medium

weight the force of gravity on an object

work the force applied multiplied by the displacement over which the force is applied

work energy theorem theorem stating that the net work, W_{net}, done on an object equals the change in kinetic energy, ΔK, of the object: $W_{net} = \Delta K = K - K_0$

work function ϕ, the minimum amount of energy needed to pop an electron free of a material's surface

working fluid a fluid used to absorb and transfer heat energy

X

X-ray radiation that is generated through the interaction of charged particles with an atom's electrons; a generic name given to E-M radiation having wavelengths ranging between approximately 1 pm (= 10^{-3} nm) and approximately 10 nm

Z

zero-th law of thermodynamics if two objects are both in thermal equilibrium with some third object, the two objects will then be in thermal equilibrium with one another (i.e., the two objects are at the same temperature)